EARTHLY THINGS

Earthly Things

IMMANENCE, NEW MATERIALISMS, AND PLANETARY THINKING

Karen Bray, Heather Eaton, and Whitney Bauman,
Editors

FORDHAM UNIVERSITY PRESS   NEW YORK 2023

Fordham University Press has no responsibility for the persistence or
accuracy of URLs for external or third-party Internet websites referred
to in this publication and does not guarantee that any content on such
websites is, or will remain, accurate or appropriate.

Fordham University Press also publishes its books in a variety of electronic
formats. Some content that appears in print may not be available in
electronic books.

Visit us online at www.fordhampress.com.

Library of Congress Cataloging-in-Publication Data available online
at https://catalog.loc.gov.

Printed in the United States of America
25 24 23   5 4 3 2 1
First edition

Contents

EARTHLY THINGS

Introduction

Karen Bray, Heather Eaton, and Whitney Bauman

Over the past fifty years there has been a resurgence in both new and old forms of immanent ways of thinking about metaphysical and ontological realities. To some extent, these modes of thinking about the link between mind/body, culture/nature, and spirit/matter have coincided with the emergence of an ecological crisis (climate change, species extinction) and an awareness of our own embeddedness within the rest of the natural world (evolution and ecology—not to mention a much longer cosmic expansion). These immanent frames for understanding how our ideas materialize in the world and how our entanglement with other bodies in an evolving planetary community shape our ideas, have great potential for rethinking human-technology-animal-Earth relationships. They can assist in addressing the problems created by what Hunter Lovins calls, "global climate weirding" and other forms of ecological degradation.[1] Older (not better or worse) forms of thought from animisms, shamanisms, and other religious traditions are joined by these more recent (again, not better or worse) forms of thinking with immanence—sometimes referred to as a new turn to ontology—such as the universe story, process thought, emergence theory, the new materialisms (NMs), object-oriented ontologies (OOO's), and even queer theory; yet, these older forms of thinking (still active today) and the new philosophies of immanence rarely meet in conversations surrounding religion and ecology/nature.

This volume enters this lacuna in search of dialogue, collaboration, and challenge across modes of thinking, disciplines, methodologies, and embodiments. It is an attempt to map out some of the connections and differences between these immanent frameworks, in an effort to provide some eco-intellectual commons for thinking within and about the planetary community.

While there is a growing body of literature on the New Materialisms (NMs is our common umbrella category for these newer theories of immanence) and Religion and Ecology / Religion and Nature, little has been done to bring these discourses together. Few sources explore their interconnections, possible historical connections, and points of productive tension. Joerg Rieger's and Edward Waggoner's edited volume *Religious Experience and the New Materialism*, and Catherine Keller and Mary-Jane Rubenstein's edited volume *Entangled Worlds*, begin to explore the relevance of NMs for religious studies and theology.[2] Graham Harvey's work on *Neo-Animisms* is also important for bringing Indigenous knowledge into dialogue with these NMs.[3] Timothy Morton's recent book on Buddhism, and Karen Barad's *Meeting the Universe Halfway*, begin to describe how older thought-streams influence their own forms of immanent thinking.[4] Emerging scholar, Mathew Arthur traces the links between feminist science studies, NMs, affect theory, and Indigenous-futurism and animism, arguing against a colonial separation between theology/philosophy and animistic thinking.[5] Likewise, scholars such as William Connolly and Donna Haraway admit the influence that process theology/philosophy has had on their own thinking.[6] However, none of these reflections attempts to think about these in a systematic way that explores the deep connections between, for example, certain understandings of emergence theory and Daoism,[7] or certain NMs and Process Thought. Nor do they begin to flesh out how such reflection could push the boundaries of thinking about humans as planetary creatures. Key to (and largely unique to) how this volume pushes these boundaries is a methodology in which dialogue across essays was encouraged. Many of the chapters herein began as paper presentations during a multi-year seminar of the American Academy of Religion. Authors infused their chapters, sometimes explicitly and sometimes implicitly, with the thinking of the other contributors. In other words, our dialogue happens, not merely from the juxtaposition of differing attempts at thinking immanence, but also through intentional intertextuality.

This volume brings scholars working primarily in religion and ecology/ nature and "religious studies" into dialogue with NM scholarship in an effort to map out our combined resources for thinking immanence. The authors explore some of the deep historical and conceptual connections between newer forms of thinking immanence, often positioned within Western frameworks, and some of the older religious and philosophical traditions, from West, East, and the Global South. Finally, we offer some critical and constructive suggestions for future thinking about humans as planetary creatures and about the necessity of crossing disciplinary, methodological, and geographic boundaries for this work. The essays in this volume cover three main, intersecting themes:

Immanent Religiosities, New Materialisms and other theories of Immanence, and Planetary Thinking.

Immanent Religiosities

In conceptualizing *religiosity*, we recognize that the very term "religion" is caught up in Western colonialism.[8] We affirm that there ought to be a variety of names to describe what has, in the Western academy, been called religion since at least the nineteenth century. Lifeways, cultures, philosophies, healing practices and meaning-making practices are all part of what we mean by religiosity, even if it still derives from the colonial religion. Indeed, the dialogue performed in this volume hopes to challenge disciplinary boundaries, which resulted in part from coloniality. Religiosity hopefully conveys not just the comparative approach to "world religions," but the aesthetics, values, practices, and meaning-making efforts that people employ in their daily lives.

Having said that, a variety of forms and concepts of immanence can be found within historical expressions of religiosity. Though often not the dominant strand of a given tradition or meaning-making practice, especially for traditions deemed worthy of the category world religion, themes of immanence are always present in the more revolutionary and mystical components of what we think of as a "tradition." For example, the dialectic between immanence and transcendence, particularly in monotheisms that stress transcendence, is a well-worn topic of concern, conversation, and debate. Even though many traditions have emphasized either immanence or transcendence, such an emphasis is never monolithic or innocent. Both foci have been exploited for political, cultural, colonial, and oppressive purposes in various eras and contexts. The concept of transcendence has been roundly critiqued from within the field of religious studies and from deconstructive and postmodern theories. It has been argued by feminist, liberation, post-colonial, queer, and other critical interpretations, and analysis of the history of various religious ideas that *transcendence* has been used to justify a whole host of oppressions, including slavery, racism, genocide, sexism, heterosexism, and anthropocentrism.

One form of such transcendence is found in the hierarchical structure that many know by the name *patriarchy*. This is a set of value assumptions that are attached to bodies and argue first that the highest value goes to (a male) God or some sort of immaterial ultimate reality. From here, things that are associated with the immaterial (spirit, reason, soul) are valued over things associated with the material (bodies). White, male, heterosexual bodies are valued over women's bodies, and bodies that are Black and Brown. Humans are valued over all other animals, and the rest of the natural world. And key to this volume, the emphasis on transcendence has also been used to fortify the colonial categories of religion and philosophy, which have served to hierarchize modes of religiosity and, perhaps ironically, of thinking and doing immanence, placing those more closely aligned with European men at the top. This type of value hierarchy hinges upon the idea that what is "really real" and most valuable is also transcendent from the material world. Once one begins to critique any piece of the patriarchal ladder, it throws the whole thing into question. The critique of transcendence is an old one that can be found in many traditions. In the modern Western world (and its attending monotheisms), Nietzsche's declaration that "God is Dead" is one such critique that set the wheels in motion for thinking without transcendence. This type of thinking, without a metaphysics of transcendence, led to the "post" (modern / colonial / structuralist) emphasis on the co-construction of knowledge and the importance of contextual thinking. And yet, such frames also frequently remained within the realm of Western academic discourse, and occasionally took flight from the material world and minoritized spiritual practices.

At the same time, there are significant critiques of thinking about the world without transcendence. Some have argued that this collapse into immanence allows for a reductive materialistic approach to the world: everything is just a matter of matter. The mind can be reduced to neurons, emotions can be reduced to chemicals, and all sorts of behaviors can be reduced to genetics. Others, like the critique of pantheism launched by panentheism, argue that without a little transcendence, novelty would be impossible. If there is no point that transcends the everyday flux of material, energy, and information then, these critics argue, we are left with complete relativity. Anything goes, including alternative facts and fake news. This is all that is left without some transcendent point from which to judge actions and knowledge claims.

Some religious responses to these critiques, although not all apologetic, have demonstrated that there are traditions of *immanent thinking* or even *immanent religiosities* throughout the histories of most cultures. Such traditions resist the binary between universalism or relativity, preferring contextuality and multiperspectivalism. Scholars from the field of religion and ecology have, for instance, been articulating immanent religiosities over the past fifty years in non-reductive and wholistic ways that animate the rest of the natural world. Such work has been bolstered by a growing awareness that a focus on immanence is useful, even necessary, in response to ecological issues. This belief is merging with insights from the deep and extensive interconnections brought forth within new materialisms, whole-systems, and planetary thinking. This work is also being incorporated into multiple efforts to articulate and develop Earth democracies, ecological civilizations, and other initiatives seeking common political and ecological ground. Immanence, while ancient and contemporary, is becoming a significant emphasis across multiple conceptual frameworks.

New Materialisms and Other Theories of Immanence

Whereas some of the essays in this volume draw mostly from immanent religiosity within extant, living religious rituals and traditions, other essays draw, primarily, from new, critical theories of immanence. New materialisms, Object-Oriented Ontology (OOO), emergence theories, neo-animisms, affect theory, neo-Marxisms, and critical romanticisms are all used in these essays. Although these different efforts have been collectively referred to as part of a return to ontology, we reference them here collectively as *new materialisms*. We do so in order to acknowledge the wider understanding of new materialisms that are emerging in theoretical circles. Yet we do not limit them to just those theoretical issues. Rather, we hope to think beyond canonized fields of thought and across varieties of attempts to weave the material and ideal, the dust and the spirit, the body and the mind, back together again after the Cartesian humpty dumpty fell off the wall. We also embrace, but remain critical of, the word "new" in New Materialisms. New here should not imply that there have not been immanent, pantheist, and animist ways of thinking throughout human history. In this sense they are not new, but, they are new for those who have been born into, and shaped by, Western, modern, and colonial ontologies that separate humans from the rest of the material world; which is seen as mostly inert, inanimate, and dead. They are also new in that they might incorporate and combine understandings from what we call modern Western science with older ways of thinking immanence. Finally, they are new in the ways in which they challenge an overly simplified division between historical materialism and poststructuralism, arguing that to immerse oneself in the story of matter is to encounter contingency and unknowability. At its core, our deployment of NMs help us judiciously understand how ideas materialize in the world around us and how the world around and within us materially impacts our ideas. Some of these newer theories find explicit grounding in the older immanent religiosities that came before them, and others do not.

There is no single definition of New Materialisms. It is not our aim to combine or unify distinct interpretations used in this volume. We appreciate a proliferation of meanings and interpretations. However, the approach taken by all the essays here does assume that there is something like a single plane of existence from which bodies and ideas, matter and spirit, brain and mind all emerge.[9] If we are unified in anything, it is a commitment to anti-dualism. At the same time, such anti-dualism does not result in a closed monism or wholism, rather it might better be captured by thinking of relational, open, entangled, and evolving systems. How thoughts matter, and how matter thinks, one might say, is a question that involves a type of open-ended dialectic. Ideas, cultures, values, hopes, and dreams collectively emerge out of matter, and "return" to shape the many bodies that make up said matter. Of course, bodies are shaped differently according to the norms and values of the worlds in which they find themselves, and at any given time, the many worlds in which bodies are constructed, and deconstructed, make up what we are calling the planetary.[10]

Planetary Thinking

Finally, many of the essays in this volume draw on immanent thinking wherever it can be found, in an effort to think ethically, aesthetically, and politically about what planetary communities might look like. These essays draw from newer, older, and ongoing traditions of immanence to imagine what it might mean to live first and foremost as planetary creatures. Such creatures start from multiple grounds and contexts, and thus the imagined futures and solutions to problems will also be open and plural. These essays purposefully do not seek a common outcome or single future; rather, we imagine a planet made up of many possible worlds.

The idea of the planetary and planetarity draws from concepts Gayatri Chakravorty Spivak articulated in *Death of a Discipline*.[11] She juxtaposes the planetary with the global. To paraphrase: if the global is a universal view from nowhere that is imposed over the face of the planet, the planetary is an embodied, contextual view "from below" that links up with other bodies to create the open evolving planetary system. Such planetary thinking does not gloss over the critical and historical differences articulated, and begins to create hybrid, alternative futures for thinking with these differences. The planetary includes the critical presence of the other plants, animals, and entities that make up the open planetary system.

Planetary thinking and planetarity also draw from evolutionary biology and astronomy, and the current knowledge that the planet (materiality and biosphere) is differentiated, yet profoundly and mysteriously interconnected. Planetary thinking, while diverse, needs to address these evolutionary and biological processes, in whatever context. Another aspect of planetary thinking is that of proposing some "common ground" for a planetary future.[12] This common ground can stimulate ethical and political projects. Such common grounds do not mean universal foundations or any sort of ultimate *telos* toward which everything is moving. Crucially, the "undercommons" created by any common grounds proposed must always be critically analyzed and rethought to include as many, previously ignored, voices as possible.[13] These voices will inevitably shift the ground in ways that will require new common grounds to be co-constructed and articulated. This process continues ad infinitum.

As Karen Barad, a New Materialist, Queer Theorist, and Particle Physicist, argues in *Meeting the Universe Halfway*, there is co-construction, agency, and indeterminacy "all the way down" in the universe. What we think of as universal laws are more like habits of becoming, even at the biological, chemical, and physical levels. We might say that common grounds from our earthly, planetary perspective, include forces and concepts such as: gravity, humans are mammals, mammals are animals, and we humans and other animals need oxygen, water, and sunlight to live. However, our Sun was not always here and will not always be here—nor our Earth, nor humans, nor mammals. These things are all "for a time" and in a context, even if that context is billions of years. The point is that whatever we think of as thought, as human, as culture, emerges out of, and returns to, affects this very same process of planetary becoming.

The goal of this volume is to begin thinking ethically, aesthetically, and politically with the planetary imaginary. Whatever else globalization and climate change might mean, it includes that we are ethically, aesthetically, and politically bound to all other life in the planetary community, for better and worse. The COVID-19 era has shown us how extensively we are tied to planetary processes and human connectivity. It has further dramatized the dangers of staying siloed in our disciplinary and methodological mindsets. Instead of modeling solitary expertise or the superiority of this field or that for addressing the planetary, the authors in this volume take part in a transdisciplinary conversation. It is our contention that no common ground might be found otherwise. Instead of modeling the organization of the planetary after what has been done in the past—Nation- states, the continuation of neoliberal capitalism as usual (with tepid social and ecological safety measures tacked on), or returning to nationalism or parochialism—these authors seek a planetary alternative in which humans are embedded in, and a part of, the interconnected worlds that make up the planetary community.

Chapter by Chapter Analysis

Opening the volume, Mary Evelyn Tucker takes up the art of Confucian immanental naturalism, which might be described as discovering one's cosmological being amidst daily affairs. For the Confucian, the ordinary is the locus of the extraordinary, the secular is the sacred, the transcendent is in the immanent. What distinguishes Confucianism is an all-encompassing, cosmological context that grounds its world-affirming orientation for humanity. This is not a tradition seeking liberation outside the world, but rather one that affirms the spirituality of becoming more fully human within the world. The way of immanence is the Confucian way. According to Chris Chapple, such an immanent way of thinking is also part and parcel of Hindu and Jain thought. In his essay, he contends that the *New Cosmology*, referred to variously in the emergence of the language of the Anthropocene, the New Story, the Epic of Evolution, and Big History, hinges on the dawning realization that humans now play a central, geological role in the ongoing state of the planet, despite the breathtaking brevity of the presence of homo sapiens on Earth. Indian traditions provide immanent narratives that might hold insights and techniques useful for *new planetary thinking*. Rather than seeking meaning in externals, these systems of Indian thought lift up interior states of awareness as a bridge from the realm of woe to a place of greater well-being. Chistopher Ives, writing on another major tradition out of India, Buddhism, posits that although early Buddhists denigrated this world as the realm of suffering, and argued that nirvana transcends it, Mahāyāna Buddhists have generally taken a more positive view of nature and construed nirvana more immanently. We see this in Sōtō Zen thinker Dōgen's concept of *presencing* (*genjō*) and in the embodiedness and embeddedness of Zen monastic life. With its non-dual epistemology, interrelational worldview, and claims that truth is something immanent, the overall Zen approach accords with key elements of New Materialism(s).

Moving toward what we might broadly call Indigenous and Indigenous-influenced traditions, Elena Jefferson-Tatum examines ethnographic examples from a number of African and African diaspora religious/material contexts and demonstrates that African religions, or in other words, orientational traditions, exemplify an immanent metaphysics wherein "gods," "spirits," "persons," and so-called "things" all share, and participate in, the *cosmos-world*. Graham Harvey then articulates just what is "new" about neo-animistic thought. Citing the work of Irving Hallowell and what he learned among Ojibway and Anishinaabe, this new approach to animism contrasts with that of Edward Tylor who used the term to define religion as "belief in spirits." What this research offers to the rethinking of planetary belonging and interactivity begins with a challenge to Cartesian and other dualisms and pursues some radical and pluralist approaches to deep and pervasive relationality. If, as Hallowell reports, Anishinaabe recognize that "*some* stones are alive," we are invited to think more carefully about the practicalities of multispecies cohabitation and codependence in the formation, maintenance, and dissolution of place/communities, ecologies, or planetary living. John Grim might agree with this sentiment. In his essay, he argues for a focus on entanglements of meaning and giving in

Indigenous community governance, pragmatic thought, and verbal-oriented languages. *Entanglements* is presented as a way of talking about an immanent religiosity inherently manifest in Indigenous perspectives on a pervasive, living wholeness that generates the world. Cosmovisions are the stories that locate this living wholeness in the lifeways of the people. Materialism, then, does not refer to a perspective in which humans are separated from matter, understood as objectified or dead. Nor does materialism refer to a worldview of hyperconsumption. Materialism, grounded in the lands of Indigenous cultures, gives rise to philosophical perspectives embedded in cosmovisions.

To conclude the first section of the book, Catherine Keller calls for a panentheistic reading of Christian theology. While *pure immanence* may be incoherent (immanence signifies an "in," a within of something not strictly identical with itself), secular philosophies of immanence have been indispensable to the evolution of ecological theologies. An Earth-tuned Christian intercarnation links its ontological relationalism and ethical animacy with a new materialism. It hopes to undo the exceptionally incarnate, ergo disembodied, transcendence—so as to release ancient and novel stimuli for our planetary becoming. O'neil Van Horn explores a materialistic understanding of hope from within a Jewish theological tradition. Drawing on early Rabbinic and post-Shoah literature, Van Horn's piece offers a Jewish inflection of hope—*tiqvah*—as one possible mode for faithfully tending to the woes of climate catastrophe without resorting to nihilism. Hope entangles questions of immanence and transcendence, offering a conception of creativity as emerging mystically from the borderlands between. This understanding of hope neither forsakes the terrestrial nor abandons notions of the divine. It refuses to justify the violence of Anthropocene, even as it confronts those very instances of violence. Finally, wrapping up this section of the book, Terra Rowe argues that although modern Christianity has indeed rendered matter dead and inert, late nineteenth century accounts of oil in the U.S. complicate narratives of wholesale Protestant disenchantment. Relying on a *petroculture* analysis, early oil narratives convey a sense of oil as animated, agential, feminine, or divine. Consequently, in addressing climate change today, religious new materialisms must seek to do more than resurrect dead matter. Rigorously held binaries between life and death must be disrupted while critically attending to both the enchantments and disenchantments of matter. Remapping such binaries to a more relational sense of responsive animation might allow for broader recognition of and proper response to both climate change and the pervasive influence of oil, while opening new relational possibilities for envisioning the sacred.

Launching the second section of the book, Kevin Minister draws on new materialisms to make two proposals for shifts in the field of religion and ecology. The first new materialist proposal calls for a reorientation of sustainability away from ways of being that offer hope in a future that preserves life as we know it, toward ways of being that feel out present modes of life that create space for the desires foreclosed by the need for heteronormative, White supremacist, capitalist futures. The second new materialist proposal seeks a theoretical shift in the understanding of religiosity in the field of religion and ecology, from a world religions model to an interreligious studies model that focuses on the interactive, intersectional, and interpersonal nature of religion. Joerg Rieger's essay continues to address questions of justice. He argues that while reclaiming attention to the material broadly conceived is crucial in an ocean of academic, economic, and political idealisms, not all materialisms are created equal. Taking a closer look at what we are up against when addressing the ecological and environmental challenges of our age might help develop a deeper sense of where materialist perspectives are most needed, and where philosophical and theoretical work still needs to be done. What would happen if we refocused the attention of materialism and planetary thinking on the histories of struggle and social movements? Sarah Pike continues thinking about different possible futures, but is more concerned with *practices*, especially ritualized practices, which express and constitute immanent worldviews. Pike's installment is concerned with how planetary and materialist thinking emerges from ritualized relationships with the more-than-human world, especially focusing on contemporary Pagans (Witches, Druids, and others recreating pre-Christian European traditions), radical eco-activists, and ancestral skills practitioners. For these communities, what is sacred and meaningful is on the Earth and in the company of the Earth's other denizens; they foreground immanence and reject transcendence.

Carol Wayne White focuses on religious naturalism as one new materialist orientation offering insights into the ethical dimensions of planetary thinking. She contends that religious naturalism affirms inseparable ethical connections between humanity's relationality with other natural processes on the planet and humans' activities with each other. Specifically, using the concept of metaphysical perspectivism, White highlights religious naturalism's affirmation that humans' perspectives are included with and inflected by the perspectives of other existents in the universe. In a different way, Kimerer LaMothe argues that dance is also a source of new materialist ways of understanding the world. LaMothe's essay frames *dance* as a critical category for illuminating the constitutive role played by rhythmic bodily movement in generating and maintaining religious worldviews—especially those characterized by relational immanence. With help from a *philosophy of bodily becoming*, she provides a phenomenological account of how dancing educates people's senses to the ever-present reality of their own movement-making—their *kinetic creativity*. In the end, she contends that relational immanence alone will not change behavior unless coupled with practices of bodily movement that educate our senses to the movements of the natural world in and around us.

Framed by the concerns of new materialist and new animist cosmologies, Mary-Jane Rubenstein explores the "old" animist materiality of Giordano Bruno (1548–1600). In his dialogical reconfiguration of Aristotelian metaphysics, Bruno proclaimed the divine animacy of all things by virtue of their shared participation in a material world soul. Structured around a close reading of *De la causa, principio e uno*, Rubenstein's essay walks through Bruno's subtle, and occasionally hilarious, deconstruction of Aristotle's hierarchical categories—hierarchies that Bruno boldly attributes to a longstanding philosophical loathing of women. By recoding femininity as active rather than passive, and full rather than empty, Bruno's spokesman "Teofilo" ultimately affirms material beings as universally animate and matter itself as the very principle of creation. To round out the second section of the book, Kevin Schilbrack examines the recently revived interest in emergence theory. This theory is valuable for materialists because it avoids the dualisms that hinder our ability to explain the world. That is, the concept of emergence offers an account of living entities without introducing any mysterious entelechies or *élan vital*; an account of human entities without speaking of immaterial souls or minds, and an account of cultural entities that does not invoke some Hegelian Geist.

Beginning the third and final section of the book, Heather Eaton argues that the appeal, power, and presence of new materialisms have become irresistible, intellectual tidal swells of novel insights, disruptions, and creativity. She begins by outlining a few new materialist characteristics, then proceeds to focus on symbolic consciousness, and how these processes reveal infinite entanglements. The third portion of her piece considers evolution and the importance for any planetary thinking infused with insights from new materialisms. Karen Bray argues that the possibility of thinking planetarily requires much more than thinking alone; it also requires paying attention to the deep, precognizant habits, feelings, desires, hopes, and emotions that shape our embodied realities. She contends that in our exuberance about affect, New Materialisms, planetarity, and relationality, we not forget or ignore the fact that relations and affects are pharmakon-like: they are curse and cure. Perhaps we need to recognize the damage done in our attempt to heal in the face of the traumas of Imperial truth-claims and modernist ideas of our fundamental autonomy and disentanglement from planetary relations. We might recognize the damage our very relationality produces, or, as Ecstatic Naturalist Robert

Corrington has put it, we should not forget that the web is also a killing machine.

If the Anthropocene assumes a monolithic human subject rooted in European industrialization and universally acting upon the planet—a narrative that champions technical responses, ignoring religious discourse and the complex worlding relations between humans and the more-than-human, then Matthew Hartman maintains that we need another metaphor. He examines the alternative concept of the Capitalocene, which might be better equipped to focus on human/Earth relations, as well as the intertwining nexus of theology, capitalism, and colonialism to engender new planetary thinking and creative religious responses. Sam Mickey critiques the Anthropocene and its anthropology. For Mickey, object-oriented ontology (OOO) provides some fruitful alternatives. In particular, he follows the account of the ecological rendering of OOO in the work of Timothy Morton, including his notion of *hyperobjects*—entities that are massively distributed relative to humans; like global warming, global capitalism, the Internet, and Earth. Indicating how Morton's ecological iteration of OOO intersects with his understanding of Buddhist notions of emptiness, *karma*, and compassion, he presents OOO as an ally with other modes of planetary thinking, in solidarity with nonhumans.

Whitney Bauman's essay argues that the reductive and productive version of naturalism that we have come to associate with modern science (and the Anthropocene) is actually a relatively recent understanding of the natural world, especially in the biological sciences. It was not until WWII that the reductive and productive understanding of nature became the dominant, scientific assumption. Prior to that, there were many scientists who argued that nature was somehow alive, ensouled, and/or agential. Such understandings of nature not only placed humans as part of a living, natural world, but also provided sources of spirituality. Bauman examines some romantic scientists in dialogue with contemporary understandings of emergence theory and new materialisms, in order to construct a Critical Planetary Romanticism (CPR) for the Earth. Finally, Philip Clayton rounds out the volume by contrasting New Materialism with the mainline discussion of religion and science over the last fifty years. Given that humanity now faces the most devastating crisis in the history of our species, planetary thinking becomes the prime criterion for "science and..." discussions, whether of religion, values, ethics, or politics. Using emergent complexity, he focuses on NM as the exploration of living matter. This paradigm recognizes intention and value all the way "down" to unicellular organisms, and all the way "up" to the planet as a whole—Gaia as a single interconnected system of living, valuing matter. Thus *matter values* in both senses: living things are material agents acting in webs of value, and the resulting material systems matter for the fate of the planet. Emergent NM gives rise to a planetary ethic, and that ethic in turn gives rise to politics for a planet in crisis.

Although this volume does not have to be read as a single manuscript, we do feel like all the chapters are important for fleshing out the details of vibrant matter in religious discourse. In the end, we hope it will spawn many conversations— rather than being the definitive word on anything. We hope that the reader will find many ways to interweave their own thoughts and imaginings about possible futures that promote the thriving and flourishing of the entire planetary community.

Notes

1. The term is widely attributed to Hunter Lovins of the Rocky Mountain Institute. It gained more traction with an article by Thomas Friedman titled "Global Weirding Is Here," *New York Times*, February 17, 2010.

2. Joerg Rieger and Edward Waggoner, eds., *Religious Experience and the New Materialism: Movement Matters* (New York: Palgrave Macmillan, 2015) and Catherine Keller and Mary-Jane Rubenstein, eds., *Entangled Worlds: Religion, Science, and New Materialisms* (New York: Fordham University Press, 2017).

3. Graham Harvey, ed., *The Handbook of Contemporary Animism* (New York: Routledge, 2013).

4. Timothy Morton, "Queer Ecology," *PMLA*, Vol. 125 (2) (March 2010): 273–82 and Karen Barad, *Meeting the Universe Halfway: Quantum Physics and the Entanglement of Matter and Meaning* (Durham: Duke University Press, 2007).

5. Matthew Arthur, "Writing Affect and Theology in Indigenous Futures," in *Religion, Emotion, Sensation: Affect Theories and Theologies*, eds. Karen Bray and Stephen D. Moore (New York: Fordham University Press, 2020),187–205.

6. William Connolly, *A World of Becoming* (Durham: Duke University Press, 2010); Donna Haraway, *Staying with the Trouble: Making Kin in the Chthulucene* (Durham: Duke University Press, 2016).

7. Terrence Deacon, *Incomplete Nature: How Mind Emerged from Matter* (New York: W.W. Norton, 2012), 1–42.

8. Tomoko Masuzawa, *The Invention of World Religions: Or, How European Universalism Was Preserved in the Language of Pluralism* (Chicago: University of Chicago Press, 2005).

9. Gilles Deleuze and Felix Guattari, *A Thousand Plateaus: Capitalism and Schizophrenia* (Minneapolis: University of Minnesota Press, 1987).

10. Walter Mignolo, *The Darker Side of Western Modernity: Global Futures, Decolonial Options* (Durham: Duke University Press, 2011).

11. Gayatri Spivak, *Death of a Discipline* (New York: Columbia University Press, 2003).

12. On "common grounds" vs. foundations see: Laurel Kearns and Catherine Keller, "Introduction," in *EcoSpirit: Religions and Philosophies for the Earth* (New York: Fordham University Press, 2007), 1–20.

13. Stefano Harney and Fred Moten, *The Undercommons: Fugitive Planning and Black Study* (Colchester, England: Minor Compositions, 2013).

Confucianism as a Form
of Immanental Naturalism

Mary Evelyn Tucker

Introduction

It is clear, by all accounts, that we are living in one of the most challenging moments in human history: when environmental problems have escalated, and solutions seem illusive. How can Earth's life-support systems, which give us food and water, be preserved? Where can we find traction for sustainability? Clearly, we need science, policy, law, technology, and economics to solve these issues, but spiritual and ethical perspectives of the world's religions must also be brought to bear. And so it is, against great odds, that some Chinese are trying to reconfigure their assumptions of endless growth and extraction and find a path toward a sustainable future. Even the Chinese Belt and Road Initiative of massive infrastructure development is coming under scrutiny for its environmental impact.

Why now? The pressing answer is that pervasive pollution across China is putting the entire nation at risk. It has been sobering to watch China over the last four decades struggle to feed large numbers of people and have fresh water for drinking, irrigation, and hydroelectric power. Careening toward major public health issues and facing one hundred thousand environmental protests a year, the Chinese government and some of its people are trying to steer a different course.

In the last decade, the Chinese have realized the need to create not just a prosperous and technologically sophisticated society, but an "ecological civilization" based on the cultural and religious traditions of China. These traditions were nearly obliterated in the Cultural Revolution under Mao from 1966 to 1976. He sought to destroy the Confucian past and create a new socialist future for China, with devastating impact on both society and the environment. Four decades later, politicians, academics, journalists, and ordinary people are exploring how Confucian moral philosophy and ethical concerns can be resuscitated to help strengthen a sustainable future. Thus, a revival of China's religious traditions, especially Confucianism, is underway, with significant implications for environmental awareness. The Confucian classics are again being taught in schools after decades of being forbidden or ignored. Philosophy departments are organizing conferences on Confucianism. Popular culture is noticing this. Indeed, a non-scholarly book on the Confucian *Analects* by media professor Yu Dan has sold over ten million copies. The widespread reexamination of Confucian moral values and cultural tradition is noteworthy, and, perhaps, unprecedented in world history, after such an overt attempt to eliminate it during the Cultural Revolution.

This revival has played into the notion of *ecological civilization*, which is being promoted on various levels—political and academic. In November 2012, the government added the goal of "establishing ecological civilization" to the Chinese constitution. Numerous policy papers have been written on this and conferences have been organized on how to realize this long-range goal. While "ecological civilization" sometimes has lofty sounding ambitions and laypeople wonder about what role they might play, efforts such as this for a sustainable ecological society are noteworthy.

Academics are exploring Chinese traditions and have translated books about religion and ecology by Western scholars into Chinese. This includes the three volumes on Confucianism, Daoism, and Buddhism from the Harvard conference series on world religions and ecology that John Grim and I organized from 1996 to 1998. One of the leading scholars of the Confucian revival is Tu Weiming, formerly at Harvard, and now directing the Institute for the Advanced Studies in Humanities at Beijing University. He has been an inspiration to many on the role of Confucianism within modernity, especially in its relation to ecology.

Thus, amid immense challenges, a revival of Confucian values is growing. Confucian ecological philosophy and environmental ethics are emerging. The future of our planet may well depend on the pace of that growth—not in material wealth, but in moral values that return us all to the essential Confucian virtue of humaneness (*ren*). This virtue implies that people belong to a larger whole, the great triad of "Heaven, Earth, and Humans." Humans are part of the processes of the cosmos and of nature and are responsible for completing the triad. They can do this by creating the foundations for a flourishing future for the common good. No wonder there is a growing interest in a new Confucianism for contemporary China.

Without a doubt, there is something we in the West can learn from this rediscovery. Indeed, this book is dedicated to examining how new materialisms, in dialogue with the immanental dimensions of the world's religions and Indigenous traditions, may offer fresh perspectives on valuing nature. It is within this context that the *this-worldly* dimensions of Confucianism become important.

Immanental Naturalism

The art of Confucian immanental naturalism might be described as discovering one's cosmological being amidst daily affairs. For the Confucian, the ordinary is the locus of the extraordinary; the secular is the sacred; the transcendent is in the immanent. What distinguishes Confucianism is an all-encompassing cosmological context that grounds its world-affirming orientation for humanity. This is not a tradition seeking liberation *outside* the world, but rather one that affirms the spirituality of becoming more fully human *within* the world. The way of immanence is the Confucian way.[1]

The means of self-transformation is through cultivation of oneself in relation to others and to the natural world. This cultivation is seen in connection with a tradition of scholarly reflection embedded in a commitment to the value of culture and its myriad expressions. It aims to promote flourishing social relations, effective educational systems, sustainable agricultural patterns, and humane political governance within the context of the dynamic, life-giving processes of the universe.

One may hasten to add that, while subject to debate, aspects of transcendence are not entirely absent in this tradition; for example, in the idea of Heaven in classical Confucianism or the Supreme Ultimate in later Neo-Confucianism.[2] However, the emphasis of Confucian naturalism is on cultivating one's Heavenly endowed nature in relation to other humans and to the universe itself. There is no impulse to escape the cycles of samsaric suffering, as in Hinduism or Buddhism, or to seek other-worldly salvation as in Judaism, Christianity, or Islam. Rather, the microcosm of the self and the macrocosm of the universe are implicitly and explicitly seen as aspects of a unified, but ever-changing, reality.

The seamless web of immanence and transcendence in this tradition creates a unique form of spiritual praxis among the world's religions. There is no ontological split between the supernatural and natural orders. Indeed, this may be identified as one of the distinctive contributions of Confucianism, both historically and in its modern, revived forms. Thus, the term immanental naturalism is apt.

Confucian Immanental Naturalism

The cosmological orientation of Confucian naturalism has been described as encompassing a continuity of being between all life forms without a radical break between the divine and human worlds.[3] Heaven, Earth, and humans are part of a continuous worldview that is organic, holistic, and dynamic. Tu Weiming has used the term "anthropocosmic" to describe this integral relatedness of humans to the cosmos.[4] The flow of life and energy is seen in *qi* (material force or vital energy), which unifies the plant, animal, and human worlds, and pervades all elements of reality. The identification of the microcosm and the macrocosm in Confucian thought is a distinguishing feature of its cosmological orientation.[5]

Humans are connected to one another and to the larger cosmological order through an elaborate system of communitarian ethics. The five relations of society are marked, for example, by virtues of mutual exchange along with differentiated respect.[6] Reciprocity is key to Confucian ethics and the means by which Confucian societies develop a communitarian basis so that they can become a bonded "fiduciary community."[7] Moreover, the cultivation of virtue in individuals is the basis for the interconnection of self, society, and the cosmos. As P.J. Ivanhoe observes, the activation of virtue evokes response: "This mutual dynamic of *de* 'virtue' or 'kindness' and *bao* 'response' was thought to be in the very nature of things; some early thinkers seemed to believe it operated with the regularity and force of gravity."[8]

In all of this, Confucian naturalism aims at moral transformation of the human so that individuals can realize their full personhood. Each person receives a Heavenly-endowed nature and thus the potential for full authenticity, or even sagehood, is ever present. To become a noble person (*junzi*) is an achievement of continual self-examination, rigorous discipline, and the cultivation of virtue. This process of spiritual self-transformation is a communal act.[9] It is not an individual spiritual path aimed at personal salvation. It is an ongoing process of rectification so as to cultivate one's "luminous virtue."[10] The act of inner cultivation implies reflecting on the constituents of daily experience and bringing that experience into accord with the insights of the sages. The ultimate goal of such self-cultivation is the realization of sagehood, namely the attainment of one's cosmological being.[11]

Attainment of one's cosmological being means that humans must be attentive to one another, responsive to the needs of society, and attuned to the natural world through rituals, which establish patterns of relatedness. In the Confucian context, rituals were performed at official state ceremonials as well as at Confucian temples. However, the primary emphasis of ritual in the Confucian tradition was not liturgical ceremonies connected with places of worship (as in Western religions), but rituals involved in daily interchanges and rites of passage, intended to smooth and elevate human relations. For the early Confucian thinker, Xunzi, rituals are vehicles for expressing the range and depth of human emotion in appropriate contexts and in an adequate manner.[12] Rituals thus become a means of affirming the emotional dimensions of human life. They link humans to one another and to the other key dimensions of reality—the political order, nature's seasonal cycles, and the cosmos itself. Confucian rituals are seen to be in consonance with the creativity of the cosmic order.

Confucian naturalism, then, might be seen as a means of integrating oneself into the larger patterns of life embedded in society and nature. P.J. Ivanhoe describes this effort succinctly when he observes that the Confucians believed:

> . . . that a transformation of the self fulfilled a larger design, inherent in the universe itself, which the cultivated person could come to discern, and that a peaceful and flourishing society could only arise and be sustained by realizing this grand design. Cultivating the self in order to take one's place in this universal scheme describes the central task of life. . . .[13]

This is the basis for Confucian immanental naturalism.

Confucianism, then, is more than the conventional stereotype of a model for creating social order and political stability sometimes used for oppressive or autocratic ends. While Confucianism aimed to establish stable and harmonious societies, it also encouraged personal and public reform, along with the reexamination of moral principles and spiritual practices appropriate to different contexts.[14] This is evident in Confucian moral and political theory, from the early classical concept of the rectification of names in the *Analects* to Mencius' qualified notion of the right to revolution. It is also seen in the later Neo-Confucian practice of delivering remonstrating lectures to the emperor and, when necessary, withdrawing one's services from an unresponsive or corrupt government.

On a personal level, the process of self-cultivation in Confucian spiritual practice was aimed at achieving authenticity and sincerity through conscientious study, critical self-examination, continual effort, and a willingness to change oneself.[15] "Learning for oneself," not simply absorbing ideas uncritically or trying to impress others, was considered essential to this process.[16] Authenticity could only be realized by constant transformation, bringing oneself into consonance with the creative and generative powers of Heaven and Earth.[17] These teachings sought to inculcate a process in tune with the dynamic, cosmological workings of nature, affirming change as a positive force in the natural order and in human affairs. This process of harmonizing with changes in the universe can be identified as a major wellspring of Confucian naturalism, expressed in various forms of self-cultivation.

This focus on the positive aspects of change can be seen in each period of Confucianism as well as in its spread to other geographical contexts. Change in self, society, and cosmos was affirmed and celebrated from the early formative period, which produced the *Classic of Changes* (Yi Ching). Later, Han Confucianism emphasized the vitality of correspondences between the human and the various elements in nature.[18] Eleventh and twelfth century Sung Neo-Confucianism stressed the creativity of Heaven and Earth. Confucian naturalism, in all its diverse expressions, was seen in East Asia as a powerful means of personal transformation. It was a potential instrument of establishing social harmony and political order through communitarian ethics and ritual practices. It emphasized moral transformation that rippled outward across concentric circles rather than the external imposition of legalistic and bureaucratic restraints. It was precisely this point that differentiated the Confucian aspirations and ideals from those of the legalists, such as Han Fei Tzu, who felt humans could be restrained by law and changed by punishment.[19] It is a tradition that has endured for more than two and a half millennia in varied historical, geographical, and cultural contexts, and is still undergoing transformation and revitalization in its contemporary forms.[20]

Categories for the Study of Confucianism

It may be helpful to distinguish the various kinds of Confucianism so as to reframe the questions surrounding the emergence and manipulation of political ideologies and separate them from the spiritual dimensions of Confucian naturalism. At the same time, we can acknowledge the ambiguous nature of many religions or philosophies in their frequent appropriation for manipulative or distorted ends.

Gilbert Rozman has proposed several types of Confucianism in his analysis of the "Confucian heritage and its modern adaptation" in East Asia.[21] He suggests a hierarchical categorization following the stratification of East Asian societies. At the top is imperial Confucianism, involving ideology and ritual surrounding the emperor. While this has often been seen as authoritarian, reform Confucianism has been regarded as invoking principles of renewal or dissent to benefit society. This may intersect at times with elite Confucianism, which reflects the interests and concerns of the educated scholar-official class.

Next, there is merchant-house Confucianism, which includes business groups in both the pre-modern and modern period. Finally, there is mass Confucianism which embraces the vast numbers of peasants and ordinary citizens in East Asian societies.

Another classification of types of Confucianism has been made by Kim Kyong-Dong.[22] He speaks of the religious aspects in reference to ideas of Heaven and rituals of ancestor worship or veneration of the sages. He cites Confucianism as a philosophical system, which includes cosmology and metaphysics, theories of human nature, ways of knowing, and ethics. Confucianism, he notes, also embraces visions of governance based on political theories and social norms guiding human relations. Confucianism can be seen as a system of personal cultivation aimed at achieving inner equanimity and thus extending harmony to the world.

While Rozman's categories are quite apposite, they are limited by the perspective of social classification, and may be appropriated by those who see Confucianism only as a political ideology. Similarly, Kim's categories are fitting, but they separate philosophical aspects from social norms and personal cultivation. To evaluate the complex role of Confucianism as immanental naturalism from a more comprehensive perspective than social classifications or ideology, we will first identify four, various, broad categories of Confucianism as having had a significant impact on East Asian history and society. We will discuss Confucian naturalism and its expression in three distinctive forms, namely, communitarian ethics, modes of self-transformation, and ritual practices.

Let us first identify some broad descriptive categories of Confucianism:

1. *Political Confucianism* refers to state or imperial Confucianism, especially in its Chinese form, and involves such institutions as the civil service examination system and the larger government bureaucracy, from the local level to the various ranks of court ministers. In Korea, Confucian bureaucratic government was adapted in the Koryo dynasty and in 958 the civil service examination system was adopted as a means of selecting officials. Confucianism was further established as official orthodoxy under the Yi dynasty in 1392 and civil service examinations were inaugurated. In Japan, there were no civil service examinations, but Confucian ideas were used by the Nara government and in Prince Shotoku's "constitution" of 719, as well as in legitimizing the Tokugawa Shogunate and later the Meiji government's "Imperial Rescript on Education."

2. *Social Confucianism* alludes to what one might call family-based or human relations-oriented Confucianism. This involves the complex interactions of individuals with others, both within and outside of the family. It has been

described by Thomas Berry as a cultural coding and by Tu Weiming as a cultural DNA, or "habits of the heart" passed on from one generation to the next. These interactions both reflect and create the intricate patterns of obligations and responsibilities that permeate East Asian society. In Japan, for example, these patterns are expressed in concepts such as *on* and *giri* (mutual obligations and debts requiring repayment).

3. *Educational Confucianism* encompasses public and private learning in schools, in families, and by individual scholars and teachers. It refers, although not exclusively, to the curriculum of study of the Four Books—the *Great Learning*, the *Mean*, the *Analects*, and the *Mencius*—selected as the canon by Zhu Xi. This was used as the basis of the civil service examination system in China and Korea. Educational Confucianism incorporates the adaptation of that curriculum to other educational institutions and venues in East Asia. In Japan and Korea, for example, it includes the various schools set up both privately and by national and provincial governments, especially in Yi dynasty Korea and Tokugawa Japan. In addition, it refers to some of the moral training that continued to be part of the educational system in Korea and Japan in the twentieth century.[23] Educational Confucianism can be said to go beyond schools, institutions, and curriculum to include, at its heart, the notion of learning as a means of self-cultivation, an approach that is emphasized in the *Analects* and *Mencius*.[24]

4. *Economic Confucianism* describes business/company forms of Confucianism in the modern period and merchant-related Confucianism in the pre-modern period, especially in Qing China, Yi Korea, and Tokugawa Japan.[25] It includes the idea of familialism and loyalty as critical principles for the transmission of family-based Confucian values into organizational structures within the business community. This is particularly widespread in East Asia, especially in the last fifty years.[26] It also includes the transmission across the society of values often associated with Confucianism such as frugality, loyalty, and industriousness.

Confucian Naturalism: Organic Cosmology and Communitarian Ethos

Confucian naturalism is evident in its dynamic cosmology interwoven with communitarian harmony of society, moral cultivation of the individual, and public and private ritual expressions. This linking of self, society, and cosmos was affected through elaborate theories of correspondences.[27] Clifford Geertz's conceptual framework of worldview and ethos helps to further elucidate the interaction of Confucian cosmology and Confucian ethics.

Confucian naturalism, while by no means singular or uniform, is one that can be described as having an organismic cosmology[28] characterized as a "continuity of being"[29] within an "immanental cosmos."[30] There is no clear separation, as in the Western religions, between a transcendent, otherworldly order and an immanent, this-worldly orientation. As the *Mean* (*Zhongyong*) states: "The Way of Heaven and Earth can be described in one sentence: They are without any doubleness and so they produce things in an unfathomable way. The Way of Heaven and Earth is extensive, deep, high, brilliant, infinite, and lasting."[31]

Without an ontological gap between this world and another world there emerges an appreciation for the seamless interaction of humans with the universe. The Confucian cosmological worldview is one that embraces a fluid and dynamic continuity of being. In terms of ethos or ethics, this involves working out the deep interconnections of Heaven, Earth, and humans. This profound symbolic expression of the triadic intercommunion of an immanental cosmos is invoked repeatedly in both the Confucian and Neo-Confucian texts cited across East Asia.[32] As Tu Weiming notes, this cosmology is neither geocentric nor anthropocentric, but rather *anthropocosmic*.[33] In this sense, the emphasis is not exclusively on the divine or on humans, as is the prevailing model in the West. Rather, the comprehensive interaction of Heaven, Earth, and humans is what is underscored by the term anthropocosmic. The worldview of an organic cosmology creates a context for the intricate communitarian model of social ethics that distinguishes East Asian societies.

The mutual attraction of things for one another in both the human and natural worlds gives rise to an embedded ethical system of reciprocal relationships. The instinctive qualities of the human heart toward commiseration and empathy are what is nurtured and expressed through human relations and ritual practices. (*Mencius* 2, A:6.). The human is not an isolated individual in need of redemption by a personal God but is deeply embedded in a network of life-giving and life-sustaining relationships and rituals. As expressed in the *Great Learning*, within this organic universe, the human is viewed as a microcosm of the macrocosm where one's actions affect the larger whole, like ripples in a pond. There is a relational resonance of personal and cosmic communion animated by authenticity (*cheng*) as illustrated in the *Mean*.[34] The individual is intrinsically linked, via rituals, to various communities, beginning with the natural bonding of the family and stretching out to include the social-political order and embrace the symbolic community of Heaven and Earth.[35] Humans achieve their fullest identity as members of the great triad with Heaven and Earth. Within this triad Heaven is a guiding moral presence, Earth is a vital moral force, and humans are cocreators of a humane and moral social-political order.

The cosmological orientation of Confucian naturalism involves the recognition that humans are embedded in and dependent on the larger dynamics of nature. The fecundity of cosmic processes is not simply a static background but a fundamental context in which human life flourishes and finds its richest expression. To harmonize with the creativity and changes in the universe is the task of the human in forming one body with all things. This is the anthropocosmic vision that Tu Weiming has articulated:

> Human beings are . . . an integral part of the 'chain of being,' encompassing Heaven, Earth, and the myriad things. However, the uniqueness of being human is the intrinsic capacity of the mind to "embody" (*ti*) the cosmos in its conscience and consciousness. Through this embodying, the mind realizes its own sensitivity, manifests true humanity and assists in the cosmic transformation of Heaven and Earth.[36]

The ethical dimensions emphasize the communitarian nature of Confucian moral philosophy. The individual is always seen in relation to others, not as an isolated, atomistic individual. The embeddedness of a person in a web of relationships involves mutual obligations and responsibilities rather than individual rights. A series of relationships expressed as correspondences is established between the person, the family, the larger society, and the cosmos itself. Humans are considered to be in relational resonance with other humans as well as with animals and plants, the elements and the seasons, colors, and directions. Communitarian ethics thus involves an elaborate patterning that binds together the society and the cosmos. There is a rich moral continuity between persons and the universe at large.

Cosmology and Cultivation: Creativity of Heaven and Transformation of Humans

The Confucian organic cosmological order is distinguished by the creativity of Heaven as a life-giving force that is ceaselessly self-generating.[37] Similar to Whiteheadian process thought, the Confucian universe is seen as an unfolding, creative process, not a static, inert mechanistic system controlled by an absent or remote deity.[38] As a protecting, sustaining, and transforming force, Heaven helps to bring all humans to their natural fulfillment as cosmological. This is because humans are imprinted with a Heavenly-endowed nature, which enables them to transform themselves through self-cultivation.[39]

The ethos of this creative cosmology is one that encourages education, learning, and self-transformation. The optimistic view of humans as receiving a Heavenly-endowed nature results in a Confucian educational and family ethos, which ideally creates a value system for nurturing innate human goodness and the creative transformation of individual potential. This ethos is one that encourages a filial sense of repayment to Heaven for the gift of life and for a Heavenly bestowed nature. The way to repay these gifts is through ongoing moral cultivation for the betterment of self and society. The symbol or model that joins this aspect of the worldview and ethos together is the noble person (*junzi*), or the sage (*sheng*), who "hears" the will of Heaven and is able to embody it naturally in the ongoing process of learning and self-cultivation. The sage is thus the highest embodiment of the spiritual aspirations of the Confucian tradition.[40]

The self-transformational dimensions of Confucian naturalism entail various modes of personal cultivation. These include a variety of practices ranging from the aesthetic arts of music, poetry, painting, and calligraphy to spiritual disciplines such as quiet sitting, abiding in reverence, and being mindful when alone. One is encouraged to observe one's inner state of tranquility before the emotions arise (centrality) and to achieve the appropriate balance in expressing emotions (harmony).[41] One cultivates the inner life so as to return to the original mind-and-heart of a child, thus expressing the deepest spontaneities and clearest responses of the human to others and to the larger cosmos. Tu Weiming has described such cultivation as "self-transformation as a communal act."[42] One of the main concerns of Confucian naturalism is regarding moral and spiritual cultivation for the benefit of self, society, and the cosmos. This process of learning, reflection, and spiritual discipline is based on the understanding that to be fully human one aims to be a noble person (*junzi*) who is in harmony with the creative powers of Heaven and Earth and in accord with the larger human community. The goal of becoming a sage (*sheng*) is an over-arching goal of Confucian self-cultivation,[43] but always with the larger purpose to be of service to society and to participate in the cosmological processes of the universe. Confucian naturalism encompasses forms of practice that are attentive to the intersection of nature, community, and the self.

Vitalism of the Earth and Cocreativity of Humans: Cosmological Correspondences and Human Ritual

The creativity of Heaven in the Confucian cosmological worldview is paralleled by the vitalism of the natural world. From the early text of the *Classic of Changes* (*Yijing*), through the Neo-Confucian reappropriation of this text, the sense of the vitality of the natural world infuses many of the Confucian writings.[44] This vitality is understood as part of the seasonal cycles of nature, rather than as the developmental, evolving universe discussed by contemporary process philosophers and theologians. It is expressed in an elaborate series of correspondences (seasonal, directional, elemental) and rituals, which in Han Confucianism were seen as patterns suggestive of the careful regulation needed in the social and political realms.[45] This cosmological view of the integral cycles of nature reinforces an ethos of cooperating with those processes through establishing a harmonious society and government with appropriate ritual structures. The rituals reflect the patterned structures of the natural world and bind humans to one another, to the ancestral world, and to the cosmos at large.

The vital material force (*qi*) of the universe is that which joins humans and nature, unifying their worldview and ethos and giving humans the potential to become cocreators with the universe.[46] As Mencius notes, it is *qi* that unites rightness (ethos) and the Way (worldview), filling the whole space between Heaven and Earth.[47] The moral imperative of Confucianism, is to make appropriate ethical and ritual choices linked to the creative powers of the Way and thus contribute to the betterment of social and political order.

The Confucian worldview affirms change, as is manifest in the creativity of Heaven and in the vitality of Earth. In particular, the varied and dynamic patterns of cosmological change are celebrated as part of a life-giving universe. Rituals and music are designed to harmonize with these cosmic changes and to assist the process of personal transformation. Rituals help to join the worldview of cosmic change with the ethos of human changes in society, thus harmonizing the natural and human orders. Rituals and music are a means of creating grace, beauty, and accord. The natural cosmological structures of the Earth provide a counterpoint for an ethos of social patterns expressed in ritual behavior and music. Harmonizing with the universe in a cosmological sense is balanced by an ethos of reciprocal resonance in human relations and expressed in the patterned behavior of rituals.

The ritual dimensions of Confucian practice involve the ties that link these intersections. Individuals and groups are joined together for a larger sense of social harmony, political coherence, and cosmological relationship. In China, state rituals at the altars of Heaven and Earth were a central component of this category as were rituals at Confucian temples and educational institutions. In addition to these forms of public display, private rituals were conducted at ancestral altars, as well as within families, and even between individuals. These included rituals for rites of passage such as birth, marriage, death, and mourning. From the more public state ceremonies to the more private family interactions, these ritual structures became a means of reflecting the patterns of the cosmos, linking to the world of the ancestors, and binding individuals to one another.[48]

Conclusion

Confucian naturalism encompasses a dynamic, cosmological orientation that is interwoven with spiritual expressions in the form of communitarian ethics of the society, self-cultivation of the person, and ritual expressions integrating self, society, and cosmos. This tapestry of integration, which has had a long and rich history in China and other countries of East Asia, deserves further study. I trust these reflections will also point the way toward future forms of Confucian naturalism in new and creative expressions. In the context of the other essays in this book, I hope that the Confucian tradition may be a fruitful dialogue partner with others engaged in bringing forward various new materialisms. Our shared planetary future rests, in good measure, on this dialogue. Valuing nature as dynamic, emergent, and resonant will be a grounding, not just for sustainability, but also for the flourishing of the Earth community.

Notes

1. In this introduction, except where noted, we are using the terms Confucian and Confucianism to refer to the tradition in a broad sense without necessarily distinguishing between the early classical Confucian expressions and the later Neo-Confucian forms in China, Korea, and Japan. See Tu Weiming and Mary Evelyn Tucker, eds., *Confucian Spirituality*, 2 volumes (New York: Crossroad Publishing, 2003 & 2004). This paper was published in an earlier form in Donald Crosby and Jerome Stone, eds., *The Routledge Handbook of Religious Naturalism* (London & New York: Routledge, 2018).

2. As Tu Weiming puts it, "Despite the difficulty of conceptualizing transcendence as radical otherness, the Confucian commitment to ultimate self-transformation necessarily involves a transcendent dimension." *Confucian Thought: Selfhood as Creative Transformation* (Albany: SUNY Press, 1985), 137. This is not "radical transcendence but immanence with a transcendent dimension." *Centrality and Commonality: An Essay on Confucian Religiousness* (Albany: SUNY Press, 1989), 121. See similar arguments made earlier by Liu Shu-hsien, "The Confucian Approach to the Problem of Transcendence and Immanence," *Philosophy East and West*, Vol. 22 (1) (1972): 45–52. Roger Ames and David Hall have argued that the Confucian tradition, especially in its classical forms, does not focus on transcendence. See their books, *Thinking Through Confucius* (Albany: SUNY Press, 1987) and *Thinking from the Han: Self, Truth and Transcendence in Chinese and Western Culture* (Albany: SUNY Press,

1998). See also Roger Ames, "Religiousness in Classical Confucianism: A Comparative Analysis," *Asian Culture Quarterly*, Vol. 12 (2) (1984): 7–23.

3. Tu Weiming, "The Continuity of Being: Chinese Visions of Nature," in *Confucian Thought: Selfhood as Creative Transformation* (Albany: SUNY Press, 1985).

4. Ibid., 137–138.

5. See, for example, John Henderson, *The Development and Decline of Chinese Cosmological Thought* (New York: Columbia University Press, 1984), Henry Rosemont, ed., *Explorations of Early Chinese Cosmology* (Missoula: Scholars Press, 1984) and Charles LeBlanc, *Huai-nan tzu: Philosophical Synthesis in Early Han Thought* (Hong Kong: Hong Kong University Press, 1985).

6. The five relations are between ruler and minister, parent and child, husband and wife, older and younger siblings, and friend and friend.

7. Weiming, *Centrality and Commonality.*

8. P.J. Ivanhoe, *Confucian Moral Self Cultivation, second edition* (Indianapolis: Hackett, 2000), xii.

9. Weiming, *Centrality and Commonality*, 94–96.

10. This is from the *Great Learning* (*Daxue*). See Wm. Theodore de Bary and Irene Bloom, eds., *Sources of Chinese Tradition* (New York: Columbia University Press, 1999), 330.

11. For discussion of sagehood as the goal of Confucian spiritual practice see Rodney Taylor, *The Religious Dimensions of Confucianism* (Albany: SUNY Press, 1989), and Rodney Taylor, *The Cultivation of Sagehood as a Religious Goal in Neo-Confucianism: A Study of Selected Writings of Kao P'an-lung (1562–1626)* (Missoula: Scholars Press, 1978). For an insightful discussion of the various models of self-cultivation in the Confucian and Neo-Confucian tradition, see Ivanhoe, *Confucian Moral Self Cultivation.*

12. *Hsun Tzu, Basic Writings*, trans. Burton Watson (New York: Columbia University Press, 1963). The importance of ritual practice as critical to the Confucian spiritual path has been emphasized by Robert Neville. See, especially, his *Boston Confucianism* (Albany: SUNY Press, 2000). Edward Machle sees Xunxi's sense of ritual as reflecting a kind of cosmic dance. See *Nature and Heaven in the Xunzi: A Study of the Tian Lun* (Albany: SUNY Press, 1993).

13. Ivanhoe, *Confucian Moral Self Cultivation*, xiv.

14. Wm. Theodore de Bary, "A Reappraisal of Neo-Confucianism," *Studies in Chinese Thought: The American Anthropological Association*, ed. Arthur Wright, Vol. 55 (5), Part 2 Memoir No. 75 (December 1953).

15. Wm. Theodore de Bary, *Learning for One's Self: Essays on the Individual in Neo-Confucian Thought* (New York: Columbia University Press, 1991).

16. Ibid.

17. Weiming, *Centrality and Commonality.*

18. It is important to note that this ordering of cosmos and society can have both life-enhancing and life-constraining dimensions. When used as political ideology in the Han period, the record becomes more mixed.

19. The Confucians were, however, caught in matters of pragmatic politics of governance that often required not only an appeal to personal moral transformation and ritual practice as a means of restraint, but also recognized that law and punishment had their function, although as a secondary measure.

20. Many of the writings of Western Confucian scholars are being translated into Chinese as part of the renewed interest in Confucianism in China. These include works by Wm. Theodore de Bary, Tu Weiming, Roger Ames and David Hall, Robert Neville, John Berthrong, and the two volumes on *Confucian Spirituality* edited by Tu Weiming and Mary Evelyn Tucker.

21. Gilbert Rozman, ed., *The East Asian Region: Confucian Heritage and Its Modern Adaptation* (Princeton: Princeton University Press, 1991), 161.

22. Kim Kyong-Dong, "Confucianism and Modernization in East Asia," in *The Impact of Traditional Thought on Present-Day Japan*, ed. Josef Kreiner (München: Iudicium-Verl, 1996), 51–53.

23. Wm. Theodore de Bary and John Chaffee, *Neo-Confucian Education: The Formative Stage* (Berkeley: University of California Press, 1989).

24. De Bary, *Learning for One's Self*.

25. See, for example, the study by Tetsuo Najita, *Visions of Virtue in Tokugawa Japan: The Kaitokudo Merchant Academy of Osaka* (Chicago: University of Chicago Press, 1987).

26. Tu Weiming, *Confucian Traditions in East Asian Modernity (Cambridge: Harvard University Press, 1996)*. See also the articles on modernization and development by Ronald Dore, Tu Weiming, and Kim Kyong- Dong in *The Impact of Traditional Thought on Present Day Japan*, ed. Josef Kreiner (München: Iudicium-Verl, 1996).

27. This is especially evident in Han Confucian thought, but it continued to influence the later Neo-Confucian tradition as well.

28. Joseph Needham, *Science and Civilization in China, Volume 2* (Cambridge: Cambridge University Press, 1956), 291–293.

29. Weiming, *Confucian Thought: Selfhood as Creative Transformation*.

30. This is a term used by Roger Ames and David Hall in *Thinking Through Confucius*, 12–17.

31. Wing-tsit Chan, trans., *A Source Book in Chinese Philosophy* (Princeton: Princeton University Press, 1963), 109.

32. These include, among others, the *Book of Changes* (Third Appendix); the *Book of Ritual* (7th Chapter); the *Mean* (Chapter 22); Dong Zhongshu, *Luxuriant Gems of the Spring and Autumn Annals* (Chapter 44); the *Diagram of the Great Ultimate* of Zhou Dunyi; the *Western Inscription* of Zhang Zai; and the *Commentary on the Great Learning* by Wang Yangming. See these texts in *The Sources of Chinese Tradition*.

33. Weiming, *Centrality and Commonality*, 102–107.

34. See Chapters 22, 25, 26 of the *Mean* in *The Sources of Chinese Tradition*.

35. See Thomas Berry's article "Affectivity in Classical Confucian Tradition." in *Confucian Spirituality, Volume 1.*

36. Weiming, *Confucian Thought: Selfhood as Creative Transformation*, 132.

37. *Book of Changes*, Appendix HI 2:1/8. See also the chapter on "Creative Principle," in Hellmut Wilhelm, *Heaven, Earth, and Man in the Book of Changes* (Seattle: University of Washington Press, 1977). The Neo-Confucians frequently refer to the productive and reproductive forces of the universe (Ch. *sheng sheng*, Jp. *sei sei*).

38. John Berthrong, *All Under Heaven* (Albany: SUNY Press, 1994).

39. See the *Mean*, Chapter 1, in *The Sources of Chinese Tradition*.

40. Taylor, *The Religious Dimensions of Confucianism*.

41. See the *Mean* in *Sources of Chinese Tradition*, 333–339.

42. Weiming, *Confucian Thought: Selfhood as Creative Transformation*, 133.

43. Taylor, *The Religious Dimensions of Confucianism*; Taylor, *The Cultivation of Sagehood as a Religious Goal in Neo-Confucianism*.

44. Kidder Smith, Peter Bol, Joseph Adler, and Don Wyatt, *Sung Dynasty Uses of the I Ching* (Princeton: Princeton University Press, 1990).

45. See Rosemont, *Explorations in Early Chinese Cosmology*; Sarah Queen, *From Chronicle to Canon: The Hermeneutics of the Spring and Autumn According to Tung Chung-shu* (New York: Cambridge University Press, 1996); Robert Eno, *The Confucian Creation of Heaven* (Albany: SUNY Press, 1990); and John Henderson, *The Development and Decline of Chinese Cosmology*.

46. See Tu Weiming's use of the term cocreator in *Centrality and Commonality*, 70, 78, 98, 102, 106.

47. *Mencius*, 2A:2.

48. Rodney Taylor, ed., *The Ways of Heaven* (Leiden: Brill, 1986); Thomas Wilson, ed., *Genealogy of the Way: The Construction and Uses of the Confucian Tradition in Late Imperial China* (Stanford: Stanford University Press, 1995); Patricia Ebrey, *Confucianism and Family Rituals in Imperial China* (Princeton: Princeton University Press, 1991); and Patricia Ebrey, *Chu Hsi's Family Rituals* (Princeton: Princeton University Press, 1991).

Immanence in Hinduism and Jainism: New Planetary Thinking?

Christopher Key Chapple

Turning and turning in the widening gyre
The falcon cannot hear the falconer;
Things fall apart; the centre cannot hold;
Mere anarchy is loosed upon the world,
The blood-dimmed tide is loosed, and everywhere
The ceremony of innocence is drowned;
The best lack all conviction, while the worst
Are full of passionate intensity.[1]

The opening stanza of the William Butler Yeats' poem, "The Second Coming," calls out the intensity of the times. It also serves as a poignant reminder that the world continually passes through cycles of conflict and resolution. This poem was first published in *The Dial* exactly one hundred years ago. The year 2020 will be remembered as the lost year, the year of pandemic and West Coast fires, the year of renewed awareness that Black Lives Matter, an election year in the United States unlike any other. It may also be remembered as the year of a great awakening, perhaps even the catalyzing year for what Ray Kurzweil calls the *Singularity*. He suggests that this pivotal event will signal the merger of human technology with human intelligence, at which point "the universe wakes up" (Kurzweil, *The Singularity Is Near: When Humans Transcend Biology*, 21). However, this "waking up" is more like awakening into the dystopian dream in Lilly and Lana Wachowski's *The Matrix* or Octavia Butler's *The Parable of the Sower*.

We have learned that democracies are fragile, only as strong as the education and insight of the common voters. Human nature is fragile, all too susceptible to manipulation through misinformation, as demonstrated in Jeff Orlowski's *The Social Dilemma*. The lessons of history and psychology show, repeatedly, that times of consistent peace are rare. Wars and human suffering are the norm. What can be done? Let us begin with an examination of three realms that serve as metonymies for the human condition: the very small, the interpersonal, and the universe.

The Micro Sphere

Small things make a big difference. James Lovelock, when tasked by NASA in the 1960s to find life on other planets, made an interesting discovery. He compared our life-filled planet with its closest neighbors. Venus seemed to have too many gases and too much heat to support life.[2] Mars has too few gases and is much too cold, though traces of water can be found. Lovelock, in developing Gaia Theory, discovered that the signature feature of planet Earth, with its unique blend of atmospheric oxygen, nitrogen, and carbon dioxide, was in fact generated by life itself. Where can one find the proverbial hand of God? What started the creation process that culminated in life itself? Bacteria!

The Earth formed more than 4.5 billion years ago, and bacteria followed a billion years later. Bacterial processes gave birth to the many forms of life that flourish on the planet today. Bacterial microbes ate and digested rocks, absorbed carbon dioxide from the atmosphere which they deposited into limestone and other excreted minerals, and they released (essentially burped) nitrogen and oxygen upward, eventually crafting our current delicate balance of 78% nitrogen and 21% oxygen. James Lovelock describes this process as follows: "The air we breathe, the oceans and rocks are all either direct products of living organisms [think of the cliffs of Dover, just one gigantic pile of shells] or else they have been greatly modified by their presence . . . [O]rganisms . . . live with a world that is the breath and the bones and the blood of their ancestors."[3] Biological forces interacted with geological and atmospheric forces in a life-giving feedback loop. Not only were the gases perfect for the diversification and complexification of life, but the temperature gradient symbiotically supported these expressions of life—a gradient disrupted since global industrialization.

Microbes continue to play an integral role in the human drama. Billions of bacteria thrive in the gut of every human being. For hundreds of years before the technique was known in Europe, the elders of Africa and India understood the dynamics of using microbes as a defense against disease, specifically small-pox. On the advice of his house slave Onesimus, and after consulting widely with other enclaved people in Boston, Cotton Mather (1663–1728) explored the process of scraping a bit of pus onto a thorn and then into one's own skin. On June 6, 1721, Mather advocated that Boston's physicians begin widespread practice of this early form of vaccine. His advocacy met stiff resistance, akin to the "anti-vax" movement today.[4] Though this practice saved countless lives, it did not enter the mainstream for many years.

In three major ways, the Jain community of India has been attuned to the presence of bacterial and viral forms of microbial life for more than two thousand years. First, in the highly complex Jain bio-cosmology, bacteria and viruses occupy a genus niche above elemental life forms (particles of dirt, drops of water, flames, and gusts of wind) and below the more complex forms of trees and other plants. They have a name: *nigoda*. Like the elemental and vegetative forms of life, they respond to stimuli through touch, the most basic sense according to the Jain view. Second, the presence of *nigoda* in water is known to be problematic. Jain hygiene protocol requires that drinking water always be filtered through seven layers of fabric. Additionally, the Jains recommend boiling water to release any *nigoda* that might still linger. Third, Jains recognize that *nigoda* can be airborne. The face mask has been in use throughout Jain communities for hundreds of years, designed to prevent the inhalation of *nigoda* and small bugs, as well as to regulate and slow the force of the human exhale, through which the *nigoda* might be inadvertently shared with others.

This intimate knowledge of living forces at a microscopic level in Africa, India, and beyond during pre-modernity, helps decenter the notions of Eurocentric superiority. Returning to the discussion of Gaia, Dorion Sagan and Lynn Margulis have written that "Gaia, the meditation upon the Earth as a living being, can be seen to be part of a philosophical monism and ancient animism that regards the cosmos as living."[5] Informed by the Gaian worldview, a new cosmology has emerged, referred to as the *Anthropocene*, the New Story, the Epic of Evolution, and Big History. A realization has dawned that humans now play a central, geological role in the ongoing state of the planet, despite the breathtaking brevity of the presence of *homo sapiens* on the planet. Acknowledging the role of microbes in planetary evolution plays an essential role in grasping the implications of this new cosmology. By understanding the vital role of the small, we can understand the place of the middle (the human) within the large (the ecosystem).

The narrative leading up to the year 2020 can be summarized as follows: humans, traumatized by the Black Death, sought to overcome the threat of disease, leading to the discovery and implementation of scientific method and then to unparalleled technological and medical advancements. This has resulted in an exponential growth of the human population and myriad unforeseen consequences including frightfully efficient warfare, pollution of the soil, water, and air, and the gradual vitiation of the human bloodstream with hundreds of previously unknown chemical compounds. In a grand attempt to create paradise on Earth through manipulation of nature, Pandora's Box has been opened.

The Social Sphere

From the Buddha's pronouncement that all things suffer (*sarvaṃ duḥkham*) to Shakespeare's keen insight regarding human treachery, the world has long been known as a place of danger, fear, and uncertainty. In his play *No Exit*, Jean Paul Sartre famously declared "hell is other people" while describing three people trapped with one another in the dark recesses of hell. Law and religion emerged in the course of human history to regulate the dark side of human behavior. Nonetheless, humans consistently violate social norms due to uncontrolled negative emotions, especially greed, lust, and anger. Within small groups, this can lead to misunderstanding and conflict. On a larger scale, the failure to check these tendencies can lead to racial prejudice, "othering," hatred, and war.

The COVID-19 pandemic, the exposed raw nerve of America's racist legacy, and the growing pressures of climate change laid bare by increased hurricanes in the southeastern United States and increased fires in the Western states combined to cause a re-examination of civility—a call toward understanding the conditions that gave birth, initially, to the civilizing impulses that brought law and religion. Yet both remain disputed areas. Rather than moving toward a common good, the stresses to the micro, interpersonal, and macro spheres have opened new fissures in the realm of jurisprudence and theology. Blaming "the other side" sells newspapers, social media views, and television shows.

We must move beyond the binary of "us vs. them." The deaths of George Floyd, Breonna Taylor, and countless other Black Americans prompted the spontaneous outpouring of concern worldwide. Study groups have been formed to read about the history of enslavement and the apartheid-like conditions abetted by business, government, and religious institutions in the United States from the colonial period until 1965 and beyond. Ibram X. Kendi's *Stamped from the Beginning: The Definitive History of Racist Ideas in America* and Richard Rothstein's *The Color of Law: A Forgotten History of How Our Government Segregated America* have become best sellers. The discussion of these books, and the ideas they contain, by federal workers while on the job has been banned by executive orders issued by the President of the United States. A level of social engagement unseen since the 1960s has emerged, with the potential to transform civic life in America.

The current situation calls for the development of a new social ecology, one informed by togetherness rather than divisiveness, unity rather than division. Two core practices of Yoga, nonviolence (*ahiṃsā*) and holding to truth (*satyagraha*), each grounded in intimacy, proved helpful for Mahatma Gandhi as he set forth a method that threw off the shackles of global colonialism and inspired the Civil Rights movement of the 1950s and '60s in America. Howard Thurman, an African American pastor, studying in India, met with Gandhi in the 1930s to understand the principles of social analysis and direct action through non-cooperation. Nearly twenty years later, James Lawson, after being released from prison for refusing to serve in the Korean War, became a school-teacher in Nagpur, learning techniques of nonviolent resistance from Gandhians in Sevagram, Gandhi's Ashram. Lawson, at the urging of Martin Luther King, conducted workshops that trained Rosa Parks, John Lewis, and others as they prepared for the lunch counter sit-ins and integration of buses. With great bravery and courage, countless souls inspired by Lawson and King endured savage beatings and imprisonments, galvanizing a shocked America to make much needed change.

At the core of these accomplishments, a way of being in the world, celebrated in the *Bhagavad Gita*, provided inspiration: embodying the comportment of the sage, unmoved amidst the onslaught of resistance and hatred. Grounded in a philosophy of the inviolability of the indwelling spirit or Self (*ātman*), this text urges restraint.

These verses from the second chapter of the *Bhagavad Gita* inspired Gandhi daily:

54. Arjuna asks Krishna:
How can the person of steady wisdom be described,
that one accomplished in deep meditation?
How does the person of steady vision speak?
How does such a one sit and even move?
55. The Blessed One responds:
When a person leaves behind all desires
that arise in the mind, Arjuna,
and is contented in the Self with the Self,
that one is said to be steady in wisdom.
56. The person who is not agitated by suffering (*duḥkha*),
whose yearnings for pleasures have evaporated,
whose passion, fear, and anger have evaporated,
that sage, it is said, has become steady in vision.
57. One whose passions have been quelled on all sides

whether encountering anything, whether pleasant or unpleasant,
who neither rejoices or recoils,
such a person is established in wisdom.
58. And when this person can draw away from the objects of sense
by recognizing the senses themselves
like a tortoise who draws in all five of its limbs,
such a person is established in wisdom.
59. For some, the sense objects will recede
but the hunger remains within the body.
Having seen the Supreme, the flavor
and the hunger cease.
 60. Arjuna, even in the case of the resolute person
who has achieved some insight,
the rapacious senses
carry away the mind as if by force.
61. One who is able to apply restraint on all sides,
who is disciplined, intent on me, should sit
with the senses firmly under control.
Such a person is established in wisdom.
62. Fixation on objects
generates attachment.
Attachment generates desire.
Desire generates anger.
63. Anger generates delusion.
From delusion, mindfulness wanders.
From wandering mindfulness arises the loss of one's intelligence.
From the loss of intelligence, one perishes.
64. By giving up desire and hatred
even in the midst of the sense objects
through the control of the self by oneself,
a person attains peace.
65. This peace generates for that person
the end of all sufferings (duḥkha).
The one with a peaceful mind indeed
attains steady intelligence.
66. There is no intelligence if one is not disciplined.
Without discipline there is no meditation.
Without meditation there can be no tranquility.
Without tranquility, how can there be happiness?
67. When the mind is governed

by the wandering senses,
all wisdom goes away
like wind drives a ship on the water.
68. Therefore, O Arjuna of Mighty Arms,
when the senses are gathered inward on all sides
and directed away from objects,
that person is established in wisdom.
69. When it is night for all other beings,
the adept remains wakeful.
When those beings are wakeful
it is night for the sage who sees.
70. Just as waters continually enter the ocean
and yet it remains full, unmoving, and still,
so also, all manner of desires can enter but do not disturb
the one who has attained tranquility.
This is not so for those who desire desires.
71. The person who abandons all desires
moves about free from lust,
free from possessiveness, free from ego.
That person attains tranquility.
72. This is the godly state, Arjuna.
Having attained this, one is not deluded.
Staying in this even up until the time of death,
one reaches Brahma Nirvana.[6]

The *Bhagavad Gita* also inspired Nelson Mandela during his years of imprisonment on Robben Island.

James Lawson, who continues to train activists in Los Angeles and beyond, captures the essence of how to apply this strategy of staying tranquil while effecting social change:

> One of the things I lifted up, and still do in teachings and trainings . . .
> is that religion at its best tries to get human beings . . . to accept their
> fundamental humanity and take responsibility for the management
> and control of their anger, for their fear, for their animosities. Do not
> pretend it comes from somebody else. Develop a spiritual life. You can
> be in a very hostile situation, but you can still try to shape your own
> life. Life is a gift. And you can mold it.[7]

Self-control holds the key to being effective in the process of making change. To thwart desire, anger, and delusion through the steady practice of understanding oneself provides the great gift of allowing a goal and mission to become primary, stymying selfish ego and greed. This power moves mountains.

The *Yogabindu*, a sixth century Sanskrit Jain text, lauds the transformative power of Yoga practice:

> 37. Yoga is the best wish-granting tree.
> Yoga is the highest jewel of consciousness.
> Yoga is the source of *dharmas* (teachings and laws).
> Yoga brings one to the perfection of self.
> Thus it is the fire applied to the seeds of rebirth.
> It is the highest maturity of the aged.
> It lays waste to all sufferings.
> It is called the death of death.
> The sharp weapons of passion are always blunted
> when they encounter actions suffused with resolve
> in a mind protected with the defenses of Yoga.[8]

By holding to a higher vision, a person girded with Yoga endures adversity. Gandhi, Lawson, King, and Mandela demonstrated that persons of goodwill can affect lasting social change.

The Macro Sphere

Since the advent of Gaia Theory, a new, holistic vision of reality has emerged that looks at systems rather than discrete units. No event stands in isolation. Rupert Sheldrake referred to this web of interconnection as the "morphic field," described as follows in a *Scientific American* interview:

> Morphic resonance is the influence of previous structures of activity on subsequent similar structures of activity organized by morphic fields. It enables memories to pass across both space and time from the past. The greater the similarity, the greater the influence of morphic resonance. What this means is that all self-organizing systems, such as molecules, crystals, cells, plants, animals and animal societies, have a collective memory on which each individual draws and to which it contributes. In its most general sense this hypothesis implies that the so-called laws of nature are more like habits.[9]

This view of science invites a reconsideration of fixed materiality. By seeing all things in a continual process of unfolding and in-folding relationships, the cosmos simultaneously collapses and explodes.

Brian Swimme and Mary Evelyn Tucker have encapsulated the history of the cosmos in a remarkably succinct timeline, further truncated here:

13.7 billion years ago: the beginning . . . particles expanding away
13 billion years ago: first massive stars begin to emerge
12 billion years ago: 100 billion galaxies, including our Milky Way
4.5 billion years ago: Sun is born
4.45 billion years ago: Earth forms . . . atmosphere, oceans, and
     continents
4 billion years ago: first cells emerge
3.9 billion years ago: photosynthesis
2.3 billion years ago: first Ice Age
2 billion years ago: first cells with nuclei; first multicellular organisms
1 billion years ago: sexual reproduction
542 million years ago: flat worms
488 million years ago: Cambrian extinctions
480 million years ago: Supercontinent Gondwana
440 million years ago: Ordovician catastrophe
425 million years ago: jawed fishes appear; life moves ashore;
     Silurian Period
395 million years ago: insects
370 million years ago: Devonian catastrophe
350 million years ago: conifers, Carboniferous Period
256 million years ago: warm blooded reptiles
245 million years ago: Permian extinctions
235 million years ago: dinosaurs appear; Triassic Period of the
     Mesozoic
220 million years ago: Pangaea . . . all continents are joined
210 million years ago: first mammals; birth of the Atlantic Ocean
150 million years ago: birds, Jurassic Period
70 million years ago: primates emerge
65 million years ago: Cretaceous extinctions
55 million years ago: Cenozoic Era; Paleocene, Eocene, Oligocene,
     Miocene, Pliocene, Pleistocene Periods follow
6 million years ago: modern dogs (Miocene)
1.5 million years ago: Homo erectus
500,000 years ago: Clothing, shelter, fire, hand axes
200,000 years ago: Homo sapiens . . . human art, caves of South Africa
32,000 years ago: Musical instruments
18,000 years ago: Cave paintings in southern Europe

12,700 years ago: Sheep and goats tamed in Middle East

11,000 years ago: Southeast Asia: rice, water buffalo, pigs, chickens tamed

5,500 years ago: World population is 5–10 million people

20 years ago: World population reaches 6 billion

[Present: World population at 7.8 billion][10]

This remarkable sweep of the history of the Universe and humans reveals the relatively long existence of planet Earth (approximately as third as old as the Universe itself) and the decidedly short existence of the modern human (a mere second on the cosmic clock). It is also striking to note that we are in the midst of the sixth great extinction and the dawn of the Anthropocene, the name given to this period in history where human interference has taken on geological proportions in altering life on Earth.

Reviewing this timeline can help restore a sense of order and calm. The year 2020, the year 2001, and, before that, the assassinations of the 1960s, respectively delineate historical markers; preceded by the World Wars and countless prior wars that have defined and in many cases scarred succeeding generations. The human drama is miniscule when regarded from the perspective of Big History. Learning geology, the origins of plants, insects, fish, reptiles, birds, and mammals gives one a sense of perspective, perhaps even, in the words of Rachel Carson, "a sense of wonder." Hope can emerge from wonder.

Immanence and Intimacy as the Way

In *The Varieties of Religious Experience*, William James celebrates the interior life as the key to personal and societal betterment, "Religious rapture, moral enthusiasm, ontological wonder, cosmic emotion, are all unifying states of mind, in which the sand and grit of the selfhood incline to disappear, and tenderness to rule."[11] This insight helped inspire what Thomas Berry hailed as the most important spiritual movement of the twentieth century—Alcoholics Anonymous. Just as addicts worldwide gather daily in small groups to share their struggles, so too a similar recovery of a sense of wonder is needed to re-mediate human delusion, to reconcile anger, and to reconfigure an economy that feeds human greed at the expense of the Earth. Another word for James's "tenderness" might be intimacy. Recovery of intimacy will help address the realities of stress, anxiety, and depression that plague the post-modern human.

Action stems from a cosmological view. The prevailing consumer global economy arose from a desire to maximize human comfort, to avoid death, and to seek youth and beauty. In many ways, this effort has borne fruit. As noted by optimist Johan Norberg, "Poverty, malnutrition, child labor and infant mortality are falling faster than at any time in human history." (Norberg, *Progress: Ten Reasons to Look Forward to the Future*, 4) However, well-being and happiness cannot be measured exclusively by statistics. Along with the increase in life span and decrease in disease, various forms of depression and lack of purpose can be detected. The narrative of Aristotle's eudaemonia clashes with the experience of many modern humans, who struggle to find meaning in a world defined by status and possessions.

India provides a counter narrative that, at first glimpse, appears glum but might hold insights and techniques useful for "new planetary thinking." Rather than making happiness the ultimate goal, Buddhism, Hinduism, and Jainism posit freedom as the ultimate goal, asserting happiness (*sukha*) to be merely the flip side of discomfort (*duḥkha*). Rather than seeking meaning in externals, these systems of Indian thought lifted up interior states of awareness as the bridge from the realm of woe to the place of greater weal. Thomas Berry described the Yogic state of Samadhi, highly treasured in Hinduism, Buddhism, and Jainism, as follows:

> The world is no longer opaque. It comes together into an ordered cosmos . . . The one point of meditation enables the light of the self to permeate the universe. The point on which the mind is fixed becomes simultaneously center and circumference of reality. The whole is known simultaneously as the notes of a melody, successive in execution, are heard simultaneously by the mind. This status of the mind is sometimes referred to as the "cosmic mind." A new awareness floods the mind. (Berry, *Religions of India*, 1992, 102)

By examining the root cosmologies of the Hindu and Jain traditions, a different relationship with the physical world can be gleaned, one that values and emphasizes intimacy rather than distance, process rather than product or production.

Let us begin with the core Vedantic premise from the *Upanishads*: Tat Tvam Asi, or Thou Art That. This great sentence speaks to an underlying intimacy amongst all aspects of reality. Wherever one turns, there can be found a manifestation of the ultimate reality. One modern expression of this connection can be seen in the following song, *Maha Deva*, by Gurani Anjali Inti (1935–2001):

> These mountains, these valleys, the earth below.
> The sky, the birds, the breath in me.
> The waves of the ocean, the sound of the wind.

They tell me it's you. They tell me it's you. They tell me it's you.
These feelings, these senses, this heart of mine.
The babies, the aged, the hungry and the despised.
The sunrise in the morning, the dew on the grass,
They tell me it's you. They tell me it's you. They tell me it's you.

The song begins with an evocation of physical immensities: Earth, sky, ocean, and wind. It then turns inward to human affectivity, and then outward to concern for the dispossessed. It returns to an embrace of the morning, a hopeful observation of the new day dawning. This sense of planetary reconnection, hard-wired into the *Upanishads* and other early literature of India, such as the *Atharva Veda*, represents a distinctly Hindu approach for joining microcosm and macrocosm.

The Jain tradition offers another perspective on planetary thinking that differs in significant ways from the Hindu view. Whereas the *Upanishads* points to an underlying unity and the Yoga tradition works toward the attainment of the state of unity through the stilling of the mind, the starting point for Jainism is not a sense of totality but radical individuality. Each soul (*jīva*) has an inherent integrity, an eternality characterized by energy, consciousness, and bliss. Due to karmic lesions acquired over lifetimes, since beginningless time, the soul remains blinded to its higher potential. By entering the ethical path, one is able to systematically purge the karmic lesions that obstruct the soul. Karma consists of sticky, colorful particles that obscure knowledge and insight, which cause delusion, and present obstacles. The pathway to purification lies in the systematic observance of vows that minimize harm, lying, stealing, lusting, and covetousness. Jain ontology states that the soul can be found not only in humans but in gusts of wind, sparks of fire, flowing waters, plants, microbes, and rocks, stones, and soil, as well as in insects, reptiles, birds, marine life, and mammals. As argued by the late Nathmal Tatia and L.M. Singhvi in *Jainism and Ecology: Nonviolence in the Web of Life,* Jainism sets forth an inherently ecological worldview.

Sulekh C. Jain, in his recent book *An Ahimsa Crisis: You Decide,* provides updates and amendments to the Jain tradition, known for its strict monastic code and the encouragement of the lay community to adapt Jain principles to their home life and business undertakings. After explaining his conscious choice, as a male, to use the male pronoun, Jain writes, "If one is striving to be an ahimsak, he needs to be aware of whether, in satisfying his personal needs, he is involved in actions that lead to the suffering of any humans, animals, birds, fish, or insects . . . he must find out whether his home furnishings and flooring materials were made using child labor, sweatshops, or animal products (skins, fur, tusks, horns, antlers or bones from animals, or feathers from birds)." (Jain, *An Ahimsa Crisis*, 86)

In the style of the Quaker Query, Sulekh Jain poses several considered questions:

How many miles per gallon does his car get?
How much pollution does his company discharge into the air, land,
and water?
Does he use chemicals on his lawn?
Does he buy organic produce?
How much paper does he use?
How often does he use trains, airplanes, and other fuel-guzzling forms
of transport? (Jain, *An Ahimsa Crisis*, 87)

This self-examination extends from the personal to the social, with Jain urging one to guard against "any form of discrimination, racial profiling, unjust or discriminatory laws and regulations, in apartheid or slavery or untouchability, in crimes, terrorism, riots, unjust wars, or any other conflicts." (Jain, *An Ahimsa Crisis*, 87) Each year, at the end of August and the beginning of September, Jains worldwide engage in several days of penance and reflection. Sulekh Jain suggests that this is the best time to take full account of one's past behavior and set the course for the coming year.

In a refreshingly frank assessment, Sulekh Jain states that "We Jains rightly believe that several thousand years ago, we talked, preached, practiced and developed the whole concept of environmentalism. Yet the fact is that I find little actual environmentalism in the Jain community . . . I have seen Jains wasting food, water, electricity, using and discarding paper plates, polythene bags, and many other resources as if there is no tomorrow." (Jain, *An Ahimsa Crisis*, 230) He goes on to cite pollution caused by Jain textile mills in the city of Pali in Rajasthan and metal plating businesses in Delhi's Wazirpur industrial area. Despite the nobility of the monks and nuns, Jain daily practice in the aggregate has done little to advance the cause of environmentalism. Sulekh Jain makes a strong appeal for change, stating that "care for the environment is practice of ahimsa." (Jain, *An Ahimsa Crisis*, 235)

These traditions hold great promise as theoretical resources for both internal stabilization and for taking ethical action. Hindu Vedanta places the human person at the center point of the universe. Yoga can bring calm and a feeling of connection with the larger order of things. Jainism offers an impeccable code of ethics that, if applied judiciously, can inspire commitment to "do the right thing" and a long history of considering and reconsidering actions large and small.

Immanence, as explicated in the book edited by Loriliai Biernacki and Philip Clayton, *Panentheism Across the World's Traditions*, builds on the process theology of Charles Hartshorne, who posited that divinity does not dwell outside time, has consciousness, knowledge of the world, and in fact is "inclusive of the world: the world is contained within divinity." (Biernacki and Clayton, *Panentheism*, 3) This sensibility and sensitivity, when seen through the prism of Vedanta, Yoga, and Jainism, places the human simultaneously at the still point of rejoicing and celebrating the beauty of the world and upholding the tremendous burden of responsibility thrust upon the human species with the dawning of the Anthropocene; the grave task of altering human behavior to allow for the flourishing of all beings.

Applied Immanence

The world faces Herculean tasks: countering disease, overcoming systemic racism, and undoing the carbon load in the atmosphere. Each instance requires a questioning of human behavior. We need to interrogate the hubris that caused so many people to flout the advice of public health officials and refuse to wear masks during the COVID-19 pandemic. We need to acknowledge that part of the human psyche that seeks to own, exploit, and abuse other human beings. We need to drill down into the root causes of human greed; a toxic combination of human weakness, marketing geniuses, polymer engineers, social media, and a whole range of modern and post-modern factors that have contributed to rampant use of fossil carbon for production, transportation, and the explosion of unnecessary plastic items. These problems are systemic and must be addressed.

Returning to the dystopic narratives mentioned at the beginning of this chapter, Neo, in *The Matrix*, having seen through the false glitz of a video-created life to the harsh reality of human enslavement, decides to take action. In *Parable of the Sower*, Lauren Olamina creates her own religion, Earthseed, starting a back-to-the-land movement. James Lawson, as quoted earlier, urges us to, "Develop as spiritual life." Where can the needed change begin? Must this be a futuristic endeavor, a "singularity" as suggested by Kurzweil? Or might technologies already exist that can prepare the human being for an uncertain future?

To change systemic patterns, personal or societal, requires a systematic approach. The Sanskrit word for the cultivation of a spiritual life is *sādhana*, from the verb root *sādh* which means "to be successful, succeed, prosper."[12] Daily spiritual practice can help fortify any person with an inner strength that will be helpful in maintaining the emotional balance needed to endure. Additionally, as suggested by William James, an abiding, daily practice can help generate the "moral enthusiasm, ontological wonder, cosmic emotion"[13] needed to sustain steady analysis and provide corrective action.

Laura Cornell, founder of the Green Yoga Association, has suggested the adoption of a new eight-fold Yoga practice that would put into action the insights into what might be termed an incarnational or embodied presence of divinity within the world. She begins with core principles that inform Yoga practice, through which one brings the human body into alignment with the cosmos: *Jñāna Yoga*, which she defines as knowledge of the foundational interconnection between small and large, and *Bhakti Yoga*, a resulting reverential mindset, akin to James's sense of ontological wonder and cosmic emotion. She next outlines five pathways that arise from these core principles: developing a steady relationship with nature (*Āraṇyaka Yoga*) by practicing Yoga and meditation outdoors and scheduling regular hikes and beach visits; cultivating the physical and energetic body (*Haṭha Yoga*) by performing Yoga poses (*āsana*) and control of breath (*prāṇāyāma*) daily; training the mind (*Rāja Yoga*) by adherence to nonviolence (*ahiṃsā*) and holding to truth (*satyagraha*); developing a community of support (*Saṅgha Yoga*); and overcoming dualities (*Tantra Yoga*) through the integration of all paths. This leads to the final, most central practice of *Yoga Sādhana*: transformation of the world through action (*Karma Yoga*).[14] Note that this begins with analysis (*Jñāna Yoga*) and culminates with action (*Karma Yoga*).

Paolo Freire's move from conscientization to direct action serves as a model for the sort of radical systemic change needed. First, people must learn to assess their circumstances. In the present case, disease, systemic racism, and environmental degradation must be named and understood to move forward. Each must be "owned." Biology and medical science inform us about the nature of communicable disease. History and sociology teach us about the origins and ongoing manifestations of racism. Geology and meteorology explain the forces that have resulted in global climate change. Science is needed in every instance, underscoring Freire's prophetic voice: "For those who undertake cultural action for freedom, science is the indispensable instrument."[15] Recognition of human suffering and the suffering of the planet provide a clear call for action and change.

In Freire's words, "A true revolutionary project . . . is a process in which the people assume the role of subject in the precarious adventure of transforming and recreating the world."[16] Freire was writing about the overthrow of colonial power structures in Latin America, however our current predicament requires the undoing of the colonization-of-the-mind inflicted by corporate interests in maintaining the status quo—profit over people, commerce over the Earth. To accomplish this, there must be a radical change: "There can be no conscientization of the people without a radical denunciation of dehumanizing structures, accompanied by the proclamation of a new reality."[17] Just as Sulekh Jain unabashedly acknowledged the shortcomings of his own faith, so deeply committed to nonviolence, corrective action must also happen on three fronts: the personal, the inter-personal, and the transactional realm of commerce.

Conclusion

This chapter began with an examination of three spheres: micro, social, and macro. We have come to understand that human presence relies upon the small, individual decisions made in the marketplace. Consumer choices, writ large, are the bacteria generating waste and altering the atmosphere. Inequalities have long governed inter-group relations, resulting in abuse and enslavement. The toxicity of human waste, particularly in the form of pollution released into the land, water, and air, has put the planet itself in peril.

These troubled times call for a sober response that can only emerge from a place of shadow. Profound change requires an examination of self, society, and world. We need a new model of the good, one that considers the staggering destruction made real through the human effort to maximize comfort at the expense of what is reasonable.

We need to wake up from the darkness and welcome a new awareness:

The ones who worship ignorance enter into blind darkness.
The ones who delight in knowledge enter into darkness even greater
    than that!

Isa Upanishads, 9.

We need to once more honor and celebrate Mother Earth:

She is the form of the beautiful sky, of space itself.
She is seen as good fortune moving all that she holds.
She is the goddess of all these various worlds that she dances into being.
All cities and continents, mountains and islands,
hang on her agency as a string of gems around her neck.

Yogavāsiṣṭha, VI.II: 84:19–21

In the shadows, truths become known. By understanding the causes of exploitation, change can be affected. In embarrassment, purification happens. A catharsis undoes the doer, unglues the doing:

Darkness was concealed by darkness there, enveloped in chaos.
The heat of desire stirred. An unnamed one emerged from the void.

Rig Veda, X:129

Rather than rushing to produce, consume, and trash, humans need to take a pause, to reconsider every aspect of the cycle of human possibility. At some point, the proclamation will emerge: Enough is enough!

The purpose of creative vibration has been accomplished.
The outflow of the world ceases.

Yoga Sūtra, IV:32

Human grasping must be re-engineered.

Notes

1. William Butler Yeats, "The Second Coming," in *The Dial* (November 1920).

2. Recent detection of phosphine seems to signal the possibility of a form of life on Venus. See the research of David Grinspoon as summarized in *Scientific American*: https://www.scientificamerican.com/article/is-there-really-life-on-venus-theres-only-one-way-know-for-sure/.

3. James Lovelock, "The Gaia Hypothesis," in *Gaia, the Thesis, the Mechanisms and the Implications*, eds. Peter Bunyard and Edward Goldsmith (Cornwall: Wadebridge Ecological Centre, 1988), 38.

4. Ibram X. Kendi, *Stamped from the Beginning: The Definitive History of Racist Ideas in America* (New York: Bold Type Books, 2016), 73.

5. Dorion Sagan and Lynn Margulis, "Gaian Views," in *Ecological Prospects: Scientific, Religious, and Aesthetic Perspectives*, ed. Christopher Key Chapple (Albany: SUNY Press, 1994), 5.

6. Translation by author.

7. James Lawson, 2013.

8. Translation by author.

9. https://blogs.scientificamerican.com/cross-check/scientific-heretic-rupert-sheldrake-on-morphic-fields-psychic-dogs-and-other-mysteries/.

10. Brian Thomas Swimme and Mary Evelyn Tucker, *Journey of the Universe* (New Haven: Yale University Press, 2011), Appendix: Timeline, 119–131.

11. William James, *Varieties of Religious Experience* (New York: Crowell-Collier, 1961), 225. First published 1902.

12. Monier Monier-Williams, *A Sanskrit-English Dictionary Etymologically and Philologically Arranged with Special Reference to Cognate Indo-European Languages* (Oxford: Clarendon Press, 1899), 1200.

13. James, *Varieties.*

14. Laura Cornell, "Green Yoga: Contemporary Activism and Ancient Practice: A Model for Eight Paths of Green Yoga," in *Yoga and Ecology: Dharma*

for the Earth, ed. Christopher Key Chapple (Hampton: Deepak Heritage Books, 2009), 168.

15. Ibid., 515.

16. Paulo Freire, "Cultural Action and Conscientization," *Harvard Educational Review*, Vol. 69 (4) (Winter 1998), 511.

17. Ibid., 514.

Mountains Preach the Dharma: Immanence in Mahāyāna Buddhism

Christopher Ives

Early Buddhists tended to regard the material realm as permeated with suffering and *nibbāna* as something unconditioned and hence transcendent of "dependent arising" and nature, while Mahāyāna Buddhists, especially in East Asia, have generally taken a more positive view of matter and nature and construed *nirvāṇa* more immanently. This is evident in such doctrines as buddha-nature, suchness, and the non-duality of *nirvāṇa* and *saṃsāra*. Facets of this Mahāyāna orientation find expression in Sōtō Zen thinker Dōgen's *Sansuikyō* (Mountains and Waters Sutra) and in the embodiedness and embeddedness of Zen monastic practice. The importance of this for environmental ethics should not, however, be overstated.

"Immanence" implies that "the ultimate" is not transcendent of the world of matter, though it may have been, or in some respects may still be, transcendent. That ultimate reality can, of course, be conceptualized in various ways: as "the sacred," "the Real," "the Eternal One," "Yahweh," "Allah," "Brahman," "the Great Spirit," "God," "gods," or "Truth" (or, à la process philosophy, as multiple ultimates, whether God and creativity or some other combination). For Buddhism, several candidates emerge. Some might argue that ultimate reality for Buddhists is *nibbāna/nirvāṇa*. Another option is the Dharma, or perhaps the Buddha. Others might lift up dependent arising (Pali *paṭicca-samuppāda*, Sanskrit *pratītya-samutpāda*; also translated as "dependent origination") or emptiness/emptying (Skt. *śūnyatā*). In his treatment of views on ultimate reality across the plurality of religious traditions, John Hick focuses on "the Nirvana of Theravada Buddhism and the *Sūnyatā* of Mahāyāna Buddhism"[1] as Buddhist names for "the Eternal One."

Early Buddhist sources portray *nibbāna* as something unconditioned and hence an exception to the dependent arising that characterizes our material world, the world into which we are born and reborn, the realm of *saṃsāra*. This early Buddhist view of *nibbāna* and *saṃsāra* appears in the "inspired utterance" (*udāna*) passages in the *Khuddaka Nikāya* in the Pali canon, one of which quotes the Buddha as saying, "There is, monks, an unborn, unbecome, unmade, unconditioned. If, monks, there were no unborn, unbecome, unmade, unconditioned, no escape would be discerned from what is born, become, made, conditioned."[2]

As indicated by this statement, *nibbāna* is transcendent, not immanent, and the world in which we find ourselves—with things that are "born, become, made, conditioned"—is a realm from which we need to extricate ourselves. As David McMahan puts it, "The monks and ascetics who developed the concept of dependent origination and its implications saw the phenomenal world as a binding chain, a web of entanglement, not a web of wonderment,"[3] and early Buddhist texts in the Pali language advocate, not engagement, but "*disengagement* from all entanglement in this web,"[4] based on a feeling of disgust (*nibbidā*) with the world.[5] Thanissaro Bhikkhu expresses this view when he writes that "all interconnectedness is essentially unstable, and any happiness based on this instability is an invitation to suffering. True happiness has to go beyond interdependence and interconnectedness to the unconditioned."[6]

This stance has implications for the early Buddhist view of nature. David Eckel writes, "The natural world functions as a locus and an example of the impermanence and unsatisfactoriness of death and rebirth. The goal to be cultivated is not wildness in its own right but a state of awareness in which the practitioner can let go of the 'natural'—of all that is impermanent and unsatisfactory—and achieve the sense of peace and freedom that is represented by the state of *nirvāna*. One might say that nature is not to be dominated but to be relinquished in order to become free."[7]

It is important to note here that, as indicated by Eckel's reference to the "state" of *nirvāna*, most early Buddhists viewed *nirvāna* not as some "Eternal One" or analogue to "God," but as a state of being, and to some Buddhists, while *nirvāna* is not entangled in this dependently arising world of matter, it is not wholly transcendent of it either. Specifically, the term *nirvāna* derives from a Sanskrit root meaning "extinguishment," and in this case what gets extinguished are the fires of detrimental states of mind (*kilesa*), such as the Three Poisons of ignorance, greed, and ill will. What "nirvana" connotes is less a metaphysical presence than a psychological absence—the absence of certain mental states. And since those mental states are the cause of suffering, when one has extinguished them, one has brought about a cessation of suffering, which, as the third of the Four Noble Truths, is another way of referring to *nirvāna*. Along these lines, Paul Williams writes, "Nirvāna . . . is *not* 'the Buddhist name for the Absolute Reality' (let alone, God forbid, 'the Buddhist name for God'). Nirvāna is . . . an occurrence, an event (not a being, nor Being),"[8] an occurrence that happens here in this world.

As a result, even though some Buddhists have viewed *nirvāna* as a literal escape from this world to another realm, whether an actual place or some sort of immaterial, disembodied, blissful state of consciousness, this view is not hegemonic. Christopher Gowans writes, "It may be said . . . that in early Buddhism suffering is not an essential feature of the natural world as such, but of our unenlightened way of experiencing the world. Moreover, enlightenment is not an escape from the natural world, but a non-attached way of living in it (as exemplified by the life of the Buddha)."[9] In this respect, as *nirvāna* denotes an awakened way of living, it is very much here in this world and, hence, "immanent." This view of *nirvāna* as immanent is developed by Nāgārjuna (c. 150–250 CE) and other Mahāyāna Buddhist thinkers, who argue for the nonduality of *nirvāna* and *samsāra*.

If we shift from soteriological to metaphysical ultimates, a prime candidate for ultimate reality in Mahāyāna Buddhism is, as Hick correctly flagged, *śūnyatā*, emptiness or emptying. Though thinkers like Masao Abe have reified this construct by rendering it with an uppercase E ("Emptiness"), *śūnyatā* is not a transcendent, noumenal *real*. Rather, it refers to the mode in which reality—including the material realm in and around us—is happening. That is to say, every thing or, better yet, every *event* is empty of any independent, unchanging essence (or self-nature, Skt. *svabhāva*), and it exists, and can only exist, interrelationally, as it is affected by and affects other events in the ever-changing process we call reality (or nature). In this respect, *śūnyatā* is partly a restatement of the earlier Buddhist doctrine of dependent arising.[10]

Astute readers might counter that all such claims about *nirvāna* and *śūnyatā* aside, Mahāyāna Buddhist philosophy posits other forms of ultimacy in a more transcendent direction. The Mahāyāna tradition is replete with cosmic buddhas and bodhisattvas, such as Avalokiteśvara, Amitābha, Maitreya, and the sun buddha Vairocana (also known as Mahāvairocana), living in transcendent pure lands, heavens, and celestial mansions. The standard six forms of rebirth include gods (*deva*), and the Mahāyāna features myriad rituals, prayers, and incantations for entreating, thanking, placating, and in some cases destroying transcendent beings.

As Mahāyāna Buddhism found its way from India to East Asia and encountered Daoism and Shintō, certain cosmic buddhas were brought down to Earth.[11] In the Japanese esoteric Buddhist tradition of Mikkyō, for example,

"The masters of Mikkyō worked out elaborate mystical correspondences, or homologies, between . . . [the] hidden macrocosmic realm of dharma and the microcosm of our apparent world. Everything here was somehow an expression of something there. . . . Such correspondences held not just in our human practices of body, speech, and thought; they were spread throughout the world around us, mapped onto the landscape: every direction, every time, every physical element, every color and shape belonged to a particular buddha and had a mystic meaning in the sacred structure of the whole."[12]

The immanence here is not simply a matter of buddhas like Vairocana being mapped onto the landscape. The *Avatamsaka Sūtra* conveys the story of Sudhana, a seeker who has "a vision of the entire cosmos within the body of the Buddha Mahāvairocana" and becomes one with this deity, resulting in "the identification of a person with a being who is the universe itself or with the underlying reality of things."[13] This identification is fostered ritually through the use of mandalas, mantras, and mudras,[14] both to transform the participant's actions and to deepen the identification with immanent cosmic forces and, ultimately, Vairocana. A major practice in Japanese esoteric Buddhism is "plugging yourself into the mandala and getting into motion, visualizing the sacred realms, embodying the sacred beings, speaking the sacred sounds of the buddhas—enacting through ritual performance the spiritual life of the cosmic buddha body."[15] As someone who has engaged in this practice, Gary Snyder writes, "The point is to make intimate contact with the real world, real self. *Sacred* refers to that which helps take us (not only human beings) out of our little selves into the whole mountains-and-rivers mandala universe."[16]

A similar approach to immanence is set forth by Dōgen (1200–1253), the founder of Sōtō Zen in Japan. In his "Mountains and Waters Sutra" (*Sansuikyō*, which we can also translate as "Landscape Sutra"), one fascicle of his magnum opus, the *Shōbōgenzō*, Dōgen makes a statement that is typically translated as "The mountains and waters of the present are the actualization of the Way of buddhas long ago" (而今の山水は、古佛の道現成なり). The character *dō* (道) or "Way" here can also be translated as "words" or "speak,"[17] so we could also render this line as "The mountains and waters of the present are the actualization of the words of buddhas of long ago," or even combine the two translations: "The mountains and waters of the present are the actualization of the Way spoken of by buddhas long ago."

But in what sense are mountains and waters the words of ancient buddhas, the Way about which they spoke? Dōgen scholar Carl Bielefeldt comments, "The mountains and rivers around us right now are a sutra, a revelation of the eternal truth taught by Buddhism."[18] That is to say, "the landscape itself is a sutra, teaching us the meaning of the dharma."[19] This notion of non-sentient things like mountains and rivers teaching us appears in another *Shōbōgenzō* fascicle, titled *Mujō-seppō* (無情説法), "the insentient preaching the Dharma." But what is the Dharma, this Way, spoken of by buddhas of long ago? What, exactly, are the mountains and waters preaching, and in what exact sense are they doing so?

Dōgen's statement that mountains and waters are actualizing or preaching something has lent itself to varying interpretations. Bielefeldt mentions one: "This is an old Chinese literary tradition, which uses 'mountain and waters' to set off such polarities as 'yang' and 'yin,' 'male' and 'female,' 'hard' and 'soft,' 'solid' and 'fluid,' and so on."[20] Snyder affirms this view of mountains and waters as representing yang and yin, the two modes of energy in the cosmos: "Mountains . . . have mythic associations of verticality, spirit, height, transcendence, hardness, resistance, and masculinity. For the Chinese they are exemplars of the 'yang': dry, hard, male, and bright. Waters are feminine: wet, soft, dark, 'yin' with associations of fluid-but-strong, seeking (and carving) the lowest, soulful, life-giving, shape-shifting."[21] Snyder also assimilates mountains and waters to a pairing that is central to Mahāyāna Buddhism: "Mountains and Waters are a dyad that together make wholeness possible: wisdom and compassion are the two components of realization."[22]

Bielefeldt also argues that what mountains and waters are preaching is a lesson about impermanence: "Dōgen himself is particularly interested here in the contrast between 'permanent' and 'changing': what is ancient and always so, like the teaching of the buddha; and what is ever presently going on, like the practice of the buddha. Since, according to the buddha's teaching, what is always so is that things are ever changing, the contrast is built right into the Buddhist landscape."[23] This interpretation appears sound, for in *Sansuikyō*, Dōgen quotes a Chinese Zen master, Furong Daokai (1043–1119), who wrote the line, "The blue mountains are constantly walking." To Bielefeldt, this line refers to the fact that "the entire landscape [is] nothing but swarms of very small happenings. And nothing in between: no substance, no essence, no soul, no self."[24] In this respect, though we do not usually recognize it, mountains are constantly teaching us about impermanence, for they, too, are always eroding, crumbling, *changing*.

Even though we can justifiably interpret Dōgen as arguing that the "Way" or "words" here is a set of truths about yin/yang, impermanence, and wisdom and compassion, his expression, 道現成, "Way/words/speak actualization," deserves further scrutiny. Crucial here is the Japanese expression *genjō* (現成), which is usually rendered as "actualization" or "realization." One way to work with this expression is to offer another possible rendering: the neologism "presencing," which yields "presencing of the Way spoken by buddhas long ago."[25]

This rendering offers the benefit of alluding to such related concepts as "presenting itself," "in the present," and "with presence."

We can go a step further and, rather than translate *dō* (道) as Way or even as words, render it as the verb "speak" or, more exactly, the verbal expression "spoken of." This generates another possible translation: "The mountains and waters of the present are the presencing spoken of by buddhas long ago." In this rendering of the line, my focus is on the intransitive presencing itself, not some Dharma, Way, or words that are being actualized transitively by the mountains and waters. Gary Snyder seems to hold a similar focus when he writes, "The 'Mountains and Waters Sutra' is called a sutra not to assert that the 'mountains and waters of this moment' are a text, a system of symbols, a referential world of mirrors, but that this world in its actual existence is a complete presentation, an enactment—and that it stands for nothing."[26]

But what, exactly, is this presencing that is happening with the mountains and waters? What, exactly, does *genjō* connote here? Zen teacher Shohaku Okumura writes that when functioning as a verb,[27] *genjō* means to "manifest," "actualize," or "appear and become," and "As a noun it refers to reality as it is actually happening in the present moment,"[28] to "reality actually and presently taking place."[29]

Helpful for our purposes here, *genjō* appears in another fascicle of the *Shōbōgenzō*, which includes *genjō* in its title: *Genjōkōan* (現成公案). This title gets variously translated as "realized kōan,"[30] "the kōan realized in life,"[31] "the kōan realized in everyday life,"[32] "actualizing the vital point,"[33] "manifesting absolute reality,"[34] "presencing absolute reality,"[35] and "spontaneous realization of Zen enlightenment."[36] One way to distill this is through the translation, "absolute reality presencing itself."

But how might we experience "absolute reality presencing itself"? Dōgen offers us a clue in *Genjōkōan*, where we encounter several lines that are frequently quoted in writings on Zen epistemology:

> To convey the self to the myriad things to confirm them is delusion,
> for the myriad things to advance and confirm the self is satori
>   [awakening].

and

> To study the Buddha Way is to study the self,
> to study the self is to forget the self,
> to forget the self is to be confirmed by all things.

What Dōgen seems to be saying here is that, ordinarily, the mind reaches out to the objects it experiences, discriminates them, and evaluates them positively or negatively as we feel attraction to them or repulsion from them, based on our past experiences and mental constructs about them. Through seated meditation (*zazen*), however, one can "forget" oneself or let go of thinking and reactivity, in what Dōgen calls "the dropping off of mind and body." Now open and spacious, we can be filled or "confirmed" by what is usually grasped as an experiential object "out there"—a mountain, a cold stream, fragrant honeysuckle, or the sound of a warbler.

Zen masters usually claim that in such moments it is not me and the mountain, me and the call of the songbird, but just the mountain, just the bird call, happening in their suchness (Skt. *tathatā*). Though it may last only for a moment and no words can capture it, we might be inclined to exclaim, "In that moment, I was the mountain, and the mountain was me." Although the experience may be mediated at a preconscious, pre-reflective level by concepts (thank you Kant), in that moment of experience there is no thinking, self-reflective "me" apart from other things, but simply raw, immediate experience. Along these lines Sōtō Zen teacher Taigen Dan Leighton construes suchness as "simply this immediate present reality,"[37] as "just this": "the simplicity and immediacy of reality here now, beyond human conceptualizations."[38]

What Dōgen seems to be conveying with his exposition of *genjō* or presencing is how our judging, reacting, and grasping mind quiets down and we are calmly present, "presencing" ourselves. Our open, spacious mind can then be filled by the myriad things as they "advance," as they manifest themselves just as they are, as they presence themselves in their suchness, with their distinctive flowering, chirping, or gurgling. In this presencing of ourselves and our being filled by mountains and birds as they presence themselves, any dualistic sense of me "in here" and mountains and rivers "out there" drops away and what remains is the *presencing of reality*. Expressed differently, we awaken to "absolute reality presencing itself."

In this non-dual mode of experience we extricate ourselves from the notion that truth exists on a level transcendent of what is happening here, and from the damage this notion can wreak, whether denigration of embodiment and nature, disparagement of certain feelings and modes of knowing, or privileging of religious and social elites who supposedly have special access to that which is transcendent. With this type of experience, we also realize we are part of a larger, interconnected, ever-changing system; expressed differently, we realize we are embedded in nature *as* nature. With this epistemology, Zen can help extricate us from entanglement in dualism, from pernicious "transcendence," from monistic and atomistic perspectives, and from the sense of being separate or estranged from "nature."

Immanence holds sway in one other facet of Zen: the affirmation of embodiedness as the locus of this awakening that is taking place here and now in
the body, not in some disembodied high state of consciousness, an other-worldly
nirvana, a Pure Land, or a Heaven. As many Japanese Buddhists would say,
one awakens or "becomes a buddha" (成仏 jōbutsu) "right here in this body"
(即身 sokushin). Going even a step further in the direction of radical immanence, Dōgen tweaks the expression, "right here in this body becoming a
buddha" (即身成仏 sokushin-jōbutsu) and talks of "this very body is the buddha"
(即身是仏 sokushin-zebutsu).

This embodiedness of Zen is brought into stark relief for people who experience Zen practice for the first time. They are usually struck by the physicality
of zazen meditation as they sit in full- or half-lotus position, focused on the act
of breathing from the hara, the center of gravity in the belly, an inch or two
below the navel. Often the physicality gets insistent, with pain in the knees
and upper back.

Dōgen advocates pouring oneself (gūjin) into the act of breathing when on
the cushions and into all other acts as we give ourselves completely to the action
at hand, whether chanting, doing walking meditation, performing rituals, or
doing such tasks around the monastery as cooking meals, pulling weeds, or
ladling onto fields the contents of the latrine.[39] This orientation extends beyond
the monastery to the range of actions in the daily life of the laity, and it has
generated such oft-quoted statements as "Chop wood, carry water," "Go wash
your bowls," and "A day of no work is a day of no eating"—or, we might say, "A
day of no work is a day of no practice, and hence a day of no awakening." Zen
teacher Ruben Habito encapsulates this as "a constant return to that primordial
experience of awakening to one's true self in the ordinary events of life, such
as looking at a flower or chopping wood or carrying water."[40]

This valuation of the materiality of the body and the range of quotidian
actions we do with our bodies extends to the physical objects involved in those
actions in and around the monastery (or one's home). Zen is replete with ex-
hortations to appreciate the beauty of objects—bells, bowls, gongs, tatami mats,
robes, cushions—in their distinctiveness (again, their suchness). And Zen emphasizes "using completely the nature of things" (物の性を尽くす mono no shō
o tsukusu): respecting, valuing, and getting the most out of each thing by using
it as long—and in as many ways—as possible. As the shirt loses buttons, sew
on new ones. As needed, patch the elbows and stitch frayed hems back into
place. And when the collar splits, use the shirt as a rag.

The embodiedness of Zen life and related valuation of materiality is com-
plemented by the tradition's embeddedness. The Zen tradition views humans,
monasteries, and homes as embedded in nature. We see this in the preference

for tucking monasteries into mountain ravines and forests. In fact, Zen monasteries are often referred to as mountains (山) or forests (林). This embeddedness finds expression in Zen-inspired landscape paintings by artists like Sesshū (1420–1506), with buildings, and their inhabitants, depicted as blending into the landscape, into nature, rather than towering over it or dominating it; and with mist and indistinct forms conveying the permeability of the boundaries between "things," between humans and the landscape. The value Zen places on being embedded in one's physical locale, in one's particular local environment, appears to be one of the factors that led Gary Snyder to advocate reinhabitation, the act of studying and living fully in one's place, in one's watershed, in one's bioregion.

What I have sketched in the above attempt to understand Dōgen's complex standpoint in relation to immanence, however, may offer only limited resources for environmental ethics.[41] Granted, Dōgen seems to value nature as a teacher of Buddhist truth, whether concerning impermanence, presencing, or wisdom and compassion. Like other Zen figures, he values nature as a good place for practice, views people as embedded in it (and dependent upon it), and ascribes aesthetic value to it. The practices that he advocates free us from entrenchment in dualism, subvert the human sense of existing over and against nature, and foster identification with, if not valuation of, the more-than-human world as an interrelational, organic system. He values the body, even equating it with a buddha. His soteriology focuses on the here-and-now, not on a transcendent postmortem realm, and he is engaged in a value system and set of monastic practices that in key respects are "green" (though beyond the scope of this essay).

These epistemological, metaphysical, aesthetic, and normative dimensions of Dōgen's approach notwithstanding, the immanence in his system, the actualizing or presencing, may not offer much for environmental ethics. Even though it affirms the material realm and is fully immanent, the actualizing of the Dao/Way or the presencing of things in their suchness presumably applies to all things, not only mountains and rivers but smokestacks spewing carbon and nuclear reactors melting down. And it is not readily apparent how Dōgen, were he alive today, would make a distinction between such things. At the very least, any distinction that he might make would likely be made on some other basis, not the mere fact of immanence or presencing. Perhaps he would make a distinction based on the degrees to which the different things derive from and exacerbate detrimental mental states, get people stuck in dualistic discrimination and self-fixation, or in some other way cause suffering rather than liberation. Simply put, the immanence in Dōgen's system is a necessary, but not sufficient, condition for a Zen environmental ethic.

Notes

1. John Hick, *God Has Many Names* (Philadelphia: The Westminster Press, 1982), 53.

2. Bhikkhu Bodhi, trans., *In the Buddha's Words: An Anthology of Discourses from the Pali Canon* (Boston: Wisdom Publications, 2005), 366.

3. David L. McMahan, *The Making of Buddhist Modernism* (New York: Oxford University Press, 2008), 153.

4. Ibid., 154.

5. Ibid., 155.

6. Thanissaro Bhikkhu, "Romancing the Buddha," *Tricycle*, Vol. 12 (2) (Winter 2002): 112; cited by McMahan, 181.

7. David Malcolm Eckel, "Is There a Buddhist Philosophy of Nature?" in *Buddhism and Ecology: The Interconnection of Dharma and Deeds*, eds. Mary Evelyn Tucker and Duncan Ryūken Williams (Cambridge: Harvard University Press, 1997), 137.

8. Paul Williams, *Buddhist Thought: A Complete Introduction to the Indian Tradition* (New York: Routledge, 2000), 48.

9. Christopher W. Gowans, *Buddhist Moral Philosophy: An Introduction* (New York: Routledge, 2014), 284.

10. As John Cobb points out in a critique of Hick's theory of religious pluralism, "Emptiness is not an object of worship for Buddhists" and "it is not illuminating to insist that Emptiness and God are two names for the same noumenal reality." *Beyond Dialogue: Toward a Mutual Transformation of Christianity and Buddhism* (Philadelphia: Fortress Press, 1982), 43.

11. Bernard Faure, *The Fluid Pantheon: Gods of Medieval Japan*, 2 volumes (Honolulu: University of Hawaii Press, 2015).

12. Carl Bielefeldt, "Buddhism," in *Mountains and Waters without End* (unpublished), 9.

13. McMahan, *The Making of Buddhist Modernism*, 158.

14. These diagrams, chanted verses, and hand positions concern the mind, mouth, and body as the loci of the three types of action: mental, verbal, and bodily.

15. Bielefeldt, "Buddhism," in *Mountains and Waters without End*, 9.

16. Gary Snyder, *Practice of the Wild: Essays by Gary Snyder* (San Francisco: North Point Press, 1990), 94.

17. This multivalency generates the first line of the *Daodejing*: 道可道、非常道, which can be translated as "The Dao that can be spoken is not the constant Dao."

18. Carl Bielefeldt, "Dōgen's Shōbōgenzō Sansuikyō," in *The Mountain Spirit*, eds. Michael Charles Tobia and Harold Drasdo (Woodstock: The Overlook Press, 1979), 38.

19. Carl Bielefeldt, "The Mountain Spirit: Dōgen, Gary Snyder, and Critical Buddhism," *Zen Quarterly*, Vol. 11 (1) (1999): 21.

20. Bielefeldt, "Buddhism," in *Mountains and Waters without End*, 5.

21. Snyder, *Practice of the Wild*, 101.

22. Ibid.

23. Bielefeldt, "Buddhism," in *Mountains and Waters without End*, 5.

24. Ibid., 6.

25. I am not alone in working with this rendering. Though focusing on Dōgen's use of *genjō* in spots other than *Sansuikyō*, Gereon Kopf, *Beyond Personal Identity: Dōgen, Nishida, and a Phenomenology of No-Self* (Surrey: Curzon Press, 2001), 235 ff. and Brett Davis, "The Presencing of Truth: Dōgen's Genjōkōan," in *Buddhist Philosophy: Essential Readings*, eds. William Edelglass and Jay Garfield (New York: Oxford University Press, 2009), 255, have used "presencing" as well.

26. Snyder, *The Practice of the Wild*, 113.

27. In Japanese, *genjō suru*.

28. Shohaku Okumura, *Realizing Genjōkōan: A Key to Dōgen's Shōbōgenzō* (Boston: Wisdom Publications, 2010), 14–15.

29. Snyder, *The Practice of the Wild*, 113.

30. Carl Bielefeldt, *Dōgen's Manuals of Zen Meditation* (Berkeley: University of California Press, 1990), 249.

31. Hee-Jin Kim, *Dōgen Kigen: Mystical Realist* (Tucson: University of Arizona Press, 1987), 76.

32. Steven Heine, "Introduction: Dōgen Studies on Both Sides of the Pacific," in *Dōgen: Textual and Historical Studies*, ed. Steven Heine (New York: Oxford University Press, 2012), 3.

33. Taigen Dan Leighton, *Just This is It: Dongshan and the Practice of Suchness* (Boston: Shambhala Publications, 2015), 39.

34. Francis H. Cook, *Sounds of Valley Streams: Enlightenment in Dōgen's Zen* (Albany: SUNY Press, 1988), 65.

35. Francis H. Cook, "Dōgen's View of Authentic Selfhood and Its Socio-ethical Implications," in *Dōgen Studies*, ed. William R. LaFleur (Honolulu: University of Hawaii Press, 1985), 137.

36. Steven Heine, "What Is on the Other Side? Delusion and Realization in Dōgen's 'Genjōkōan,'" in *Dōgen: Textual and Historical Studies* (New York: Oxford University Press, 2012), 42.

37. Leighton, *Just This Is It*, 127.

38. Ibid., 34. As Gary Snyder puts it, "For those who would see directly into essential nature, the idea of the sacred is a delusion and an obstruction: it diverts us from seeing what is before our eyes: plain thusness. Roots, stems, and branches are all equally scratchy. No hierarchy, no equality." *Practice of the Wild*, 103.

39. Zen uses the expression *samu* to refer to this doing of tasks as a form of practice.

40. Habito, *Buddhism and Ecology*, 170.

41. Environmental ethics in two senses: a person's guidelines for living ecologically and Environmental Ethics as an intellectual endeavor rooted in the discipline of Philosophy.

Africana Sacred Matters: Religious Materialities in Africa, the Caribbean, and the Americas

Elana Jefferson-Tatum

Offering examples from a number of African and African Diaspora religious traditions, specifically those ancestrally-rooted in the geopolitical and socio-culturally constructed space of present-day West and Central Africa, this chapter illuminates a variety of *sacred matters*—namely, embodiments and material-izations of the invisible and yet the materially and naturally present. These *matters* thus reveal the operation of a metaphysics that is fundamentally *this* worldly. The "sacred" here does not presuppose a binary: there is no profane world that is its corollary opposite. Nor does this conception of the sacred nec-essarily presuppose the existence of a being or entity that is ontologically su-perior (e.g., the normative Christian conception of "God"). Rather, these sacred matters demonstrate the interrelationship between the metaphysical, the ma-terial, and the natural in Africana orientational worlds, an ontological universe wherein "gods," "spirits," "persons," and so-called "things" all share and par-ticipate in the *cosmos-world*. I use this concept of cosmos-world to revive the originary meaning of the Greek concept of cosmos in its widest sense—the conception of the world and universe as an organized, interwoven ecosystem. By using this concept, I am proposing that cosmology and geography, rather than being separate domains, are necessarily coexistent, overlapping, and in-terdependent. Africana orientational worlds are then materially vibrant and ontologically complex geo-cosmological spheres in which *any* and *all* bodies—human, animal, vegetal, mineral, and organic (as well as seemingly inorganic)—can be and become receptacles and material embodiments of the sacred. The sacred thus includes a variety of bodies and beings, as well as various forms of temporal-functional preeminence—a variety of wisdoms, knowledges, powers, and other forms of energic, natural, and material authority.[1] In this chapter, I argue that on the whole, Africana orientational worlds exemplify a non-transcendental metaphysics wherein nature, "spirits," matter, and "humanity" fundamentally coexist and commune, and wherein *matter* is the essential "stuff" of religion.

Sacred Matters: From Ewe Gorovodu to Cuban Santería

In the 1998 publication, *Possession, Ecstasy, and Law in Ewe Voodoo*, Judy Rosenthal recounts the philosophical musings of Fo Idi, an Ewe Gorovodu priest: "We Ewe are not like Christians, who are created by their gods. We Ewe create our gods, and we create only the gods that we want to possess us, not any others."[2] This Indigenous theorist is explicitly asserting that there is a fundamental ontological difference between his Vodun world—wherein creation is an open-ended and widely accessible capacity through which the divine is made and re-made present, and the normative Christian cosmos—wherein creation is often the sole enterprise of God. Detailing how to "make Nana Wango," the great Grandmother Crocodile, Amouzou, another Gorovodu priest, explains:

> . . . cowries representing crocodile skin need to be placed on the statue. Plants are placed inside Wango's head. Wango replaces (is) the person. She must be treated with respect; one must never point and call her "wood."[3]

Nana Wango materialized, through the making and activating of her body-statue, is not dead matter. Rather, her material presence challenges the parameters of normative subjectivity and the presumed division between "materiality" and the "divine," and between the "sacred" and the "natural." This Vodun world is not a cosmos in which creator and creature are inescapably juxtaposed. In this cosmos-world, the *vodun* are not otherworldly entities. This is not Rudolf Otto's "wholly other" or Mircea Eliade's "the sacred."[4] This conceptualization of the divine does not presuppose a numinous, transcendent, or supernatural entity set apart from the brute materiality of life. Nor does it imply an ontological reduction—a diminution of the sacred into a lesser vessel or entity. Rather, through their materialization, these sacred matters find their rightful form, and their full embodiment and significance within community—a cosmos-world that includes both the visible and the invisible, both humanity and nature. The *vodun* reveal themselves as immanent metaphysical beings, intimately interwoven into the lifeworld of nature. Divinity, nature, and materiality are intimately interrelated and eternally tied.

Furnishing another apt example of the kinship between creation, craftsmanship, and spiritual materialization in pre-colonial Kongo society, the *nkondi* (pl., *minkondi*) was a class of *nkisi*, (pl. *minkisi*), what Wyatt MacGaffey later aptly terms "bodies of power"[5] specifically created and manufactured to "hunt."[6] Describing their various uses, MacGaffey provides the following descriptions, collected circa the 1920s from two informants:

> When a carrying-band has been cut and those who did it will not admit to doing so for fear of a lawsuit, the owner, not knowing who the guilty party is, will have the bit that remains confided to Lunkanka [a type of *nkondi*], that he may seek them out and inflict sickness upon them.
>
> If they wish to enter into brotherhood with a stranger to the clan, they fetch Minkondi, which are large and awesome *minkisi* (*mibafueti vumina*). They lick the spikes and drive them into the Nkondi, one for each side. Then they sacrifice two chickens and pour the blood on the spikes to strengthen them, so that whosoever should attack his fellow by witchcraft will be killed with his whole clan by the furious *nkisi*, which drives them mad and burns them with fire.
>
> If a man has been ailing for some time, with bad dreams, but does not know who is responsible, he attaches hairs from his head, or finger nails, or a little piece from his loincloth, wraps them in a rag and has the *nganga* [priest] attach them to Lunkanka, saying, "Look upon me, for I have come to put this relic in your body; it is my entire self I have come to put in safekeeping."[7]

Through the craft of creating *nkisi*, of concretizing "spirit" (e.g., the *bakulu* ancestors), *bisimbi* (local nature beings), *minkuyu* (ghosts), and other powerful entities, the *nkondi*, the "hunter" *nkisi*, is empowered to pursue a wrongdoer, to seal, bind, and enforce a contract, and even to protect a member of the community. As MacGaffey explains, rather than merely being a ritual "object," the crafted and activated *nkisi* becomes a "person"—namely, a social being—and a "body of power," a potent entity and mystical technology able to yield social, political, and religious power and authority.[8] MacGaffey later argues, the *nkisi* was not unlike other "bodies of power," or other socio-metaphysical technologies (or, even technologists) such as; *bangnaga* (ritual specialists), *mfumu* (chiefs), and *bankdoki* (persons with unique qualities and powers),whose roles and titles were at times interchangeable given their similar access to and use of specialized knowledge and power.[9] While Kongo *minkisi* and many of their counterparts are no longer overtly present in contemporary Congolese communities and societies, their "camouflaged successors" (e.g., the Haitian *wanga*, African American *gris-gris*, Afro-Brazilian *calundú*, etc.) continue to provide a central means for both individuals and communities to make, access, and utilize their knowledge and potency for personal and communal well-being.[10] As "bodies of power," to again borrow MacGaffey's concept, they are exemplars of an immanent metaphysics in which materiality is the fundamental agent and primary technology through which the sacred is known, accessed, and experienced.

A diaspora creolization of the Kongo *nkisi*, in Cuban Palo Monte Mayombe (also referred to as Reglas de Palo), the *nganga*—a new world materialization of spirit in the vessel of a clay pot, gourd, or iron cauldron—is a microcosm through which humans, nature, and the dead are intimately interconnected.[11] Describing this *nganga*-centered cosmos, art historian Judith Bettelheim elaborates:

> The word *nganga* . . . refers to the pot itself, the power of the pot, or the owner of the pot. It is a world in miniature, and when one is initiated into the final level in Palo, one receives a personal *nganga*. Thus, initiation makes *nganga*—both priest and pot.[12]

While in pre-colonial West Central Africa *nganga* generally referred to a ritual specialist (or, more precisely, an Indigenous physician-priest), in Cuba *nganga* specifies an entire cosmos centered around a potted spirit and force.[13] In this nature-centered ontological universe, the *nganga* is simultaneously matter, power, and priest. Hence, like its Kongo counterpart, the *nganga* is also a "body of power," materializing and actualizing other "bodies of power," namely, the living, the dead, and the priest. No *tata* or *yaya* (Palo priest/priestess) can work without the power of their own *nganga*. Through "making" *nganga*, both power and priesthood are created, activated, and materialized. *Nganga* is a microcosm that embodies and materializes the interconnectedness between power and matter, spirit and nature, and death and life. The Palo *nganga* is thus an immanent materialization and concretization of the sacred and, therefore, another apt expression of the concrete, this-worldly nature of the metaphysical in Africana cosmos-worlds.

Equally illustrative of an eco-ontological cosmos in which an ongoing process of creation and re-creation is a primary means of rousing, materializing, and concretizing spirit, the anthropologist Patricia de Aquino, in 1995, transcribed the following narrative by a Candomblé initiate:

> I was shaved [i.e., initiated] in Salvador for Osala, but I had to seat Yewa and Mother Aninha [my spiritual godmother] sent me to Rio because at the time Yewa was already an endangered *Orisa*, so to speak. There were many who no longer knew the *oro* [that is, "the words and rites"] of Yewa.

I am from Oba, Oba is almost dead already because no one knows
how to seat her, no one knows the craft, so I came here because I was
shaved here, and they are not going to forget the *awo* [that is, the
"secrets"] for making her.[14]

In this Candomblé cosmos-world, we see that the *orisa*[15] Yewa and Oba are
literally "made" and "seated" onto the *ori* ("head") of the devotee. The "craft"
of making Candomblé is, then, a process by which the *orisa* are embodied and
concretized via the physical head. To provide an analogous illustration, in 1958
when Oseijeman Adefunmi, the future founder and *oba* (king) of Oyotunji
Village in Sheldon, SC, was "making *ocha*" (*kariocha*)—undergoing initiation
into Cuban-based Santería (also known as Reglas de Ocha)—he described his
bewilderment to Carl Hunt, one of the first chroniclers of the North American
Yorùbá movement, who recounted the following:

The shock did not come until after the ceremony because [Adefunmi]
could not see everything while he was being initiated. He could see
the bringing in of goats, rams, and pigeons to make up the various
gods and the blood being poured in and on objects. But he could not
see what was in the pot given him containing his gods . . . "I remember the shock after the initiation rituals were over and we were preparing to leave the Temple and return to America, our God father told us
to open our pots and see what we had and there was *nothing in them
but stones.* And I thought, Oh My God! You mean I spent two thousand dollars for some stones to take home to my wife; and there was
embarrassment and shock at the same time." Although this was a set-
back, it proved to be only momentary. He was too deeply involved, and
the enthusiasm of the crowd was so great when they were leaving the
Temple with their singing and chanting, that he realized he had found
what he had been seeking.[16]

Adefunmi would come to comprehend and fully embody this Yorùbá cosmos-
world wherein seemingly mere stones, "made" and consecrated with the blood
of sacrificed offerings and other natural elements, became *otanes*—*orisa* stones,
the primary material presence of the *orisa* ("deities," "gods," "spirits"). As sacred
matters, like the Palo *nganga* and even the consecrated *ori* ("head") of a Candomblé initiate, these *otanes* are immanent embodiments of often invisible
and yet materially present and powerful metaphysical beings. Building on my
foregoing argument, I reason, therefore, that these Africana sacred matters
epitomize a cosmos-world in which nature, matter, and the divine are fundamentally bound and mutually interdependent.

An Africana Immanent Metaphysics

In this preliminary sketch, I have delineated six ethnographic vignettes: the Ewe Gorovodu cosmos wherein Nana Wango is "made" and materialized through a consecrated body (i.e., a statue); the Kongo *nkondi*, a *nkisi* that materializes its "hunter" capacities and powers to pursue, protect, and bind; the Palo *nganga* through which power, the living dead, and the priesthood are made and materialized; the Candomblé "making" and "seating" of the *orisa* Ywa and Oba onto the *ori* (head) of an initiate; and the Santería initiation of Oba Oseijeman Adefunmi and the "making" of his *otanes*, his *orisa* stones. While providing only brief illustrations, these narratives demonstrate the ontological complexity and deep material-natural engagement of Africana orientational worlds. These examples elaborate Africana cosmos-worlds wherein immanence—the full spectrum of nature's reality and materiality—is the primary medium of the divine expression of the metaphysical. From this material- and nature-centered perspective, it becomes apparent that within Africana-sacred landscapes the metaphysical is made *real*, *material*, and even *active* through the natural. Through the materialization of the *vodun* Nana Wango, through the making and activating of *nkondi*, through the manufacturing of a *nganga*, through *kariocha* ("making *Ocha*") via the seating (*asiento*) of Yewa and Oba onto the *ori* ("head") of an initiate, and through concretizing the presence of the *orisa* through the *otanos* (*orisa* stones). These material illustrations, exemplify an Africana ontological universe in which a wooden statue, an ordinary clay pot, the head of an initiate, and seemingly unremarkable stones can equally *be* and *become* receptacles, embodiments, and concretizations of the sacred and of the immanently metaphysical.

By the metaphysical, I am referring to the primordial kinship between the visible and materially concrete. The often invisible and yet still, to borrow Kwasi Wiredu's concept, "quasi-material"—those "forces," "energies," and "powers" that often defy spatial, geographical, temporal, and ontological norms and thus are not concretized or fixed into any particular body, place, or time.[17] While these entities and powers often remain imperceptible to the common eye (especially, when unattached to fully concretized forms), within Africana metaphysics they are nonetheless materially real and present. Despite being transitory and fluid in nature, these vibrational forces remain interwoven and, even to a certain degree, indistinguishable from the fully concrete and material (e.g., rivers, oceans, stones, and various other sacred bodies), and thus are better understood as "quasi-material," rather than immaterial in nature. Given its Greek origins, metaphysics has often been interpreted as either that which is beyond the senses or as the abstract and transcendent. Yet, both conceptualizations are too narrow to adequately capture the material-ontological complexity of many Africana orientational worlds. As Nana Wango and the *orisa otanes* demonstrate in Gorovodu and Santería, the metaphysical is not beyond the senses. Rather, the *vodun* and *orisa* make themselves known and sensed through their transitory bodies (a sculpture, a devotee, or even a stone) and during their often transient and temporal incarnations; through the boom of voices, the clapping of hands, the sweat of brows, and the beating of feet as their "horses"[18] dance. The metaphysical in the Africana world is anything but not sensed.

In Africana sacred worlds, the metaphysical is also not transcendent or, more precisely, not beyond the natural, material world. While scholars often label African "gods" and "spirits" as "supernatural," these entities and forces are neither otherworldly nor transcend nature. Instead, nature is their abode; it is the landscape and medium of their familiar bond with the visible world. Exemplifying the intimacy between the natural and the metaphysical, José Bedia, a Palo priest and artist, described Palo as follows:

Palo teaches that there are spiritual forces within the natural world. Our lives depend on them. The religion offers ways to relate to those forces and bring their energy to play more directly in our lives. Prayer in Palo can be to go off alone to some natural place and leave a simple offering. Prayer can be collecting small bits of nature—particular roots or plants or stones or water or animals—and combining them in sacred recipes which release certain energy.[19]

In Palo, the *mpungu* ("spirits") are not set off from nature. Instead, they are *in* and part and parcel of nature. It is directly through devotional engagement with nature that the *palero* (Palo initiate) comes to effectively know and utilize its "sacred recipes" and "energy." Likewise, defining Santería as an "earth-centered religion," Miguel A. De La Torre states:

In a very real sense, Santería is a terrestrial religion, firmly rooted in the earth. While Western religions tend to emphasize a heavenly place, or stress the placement of the stars and planets to determine the course of human events (remember that a shining star over Bethlehem was said to signify the birth of Jesus), Santería is shaped and formed by earth-centered forces of nature . . . Its rituals utilize earthly things— stones, herbs, water, plants, trees, soil, seashells, and so on all become sacred objects. Seldom do devotees of Santería look to the stars or to the movement of planets to determine the will of the gods; rather, they look to the earth for the answers to all conceivable questions.[20]

This terrestrial-centered ethos has meant that in many Africana religious cultures, divine forces and energies are understood as operative and active within *this* world. There is, often no otherworldly domain—no realm set apart from the realities of this world. Rather, the divine realm is a parallel and overlapping cosmos that is constantly concerned with and engaged in this-worldly concerns, problems, and experiences.

Traditions like Haitian Vodou, that explicitly maintain a connection between the celestial and the terrestrial, as embodied, for instance, in the cosmic unity between Danbala (the wise serpent) and Ayido (the rainbow) who together form the "cosmic egg," the "innerworld" of the terrestrial Danbala and the "outerworld" of the celestial Ayido, are still *not* otherworlds.[21] Elucidating the complex cosmological orientation of Haitian Vodou, Patrick Bellegarde-Smith and Claudine Michel explain:

> Life is about passages, but it is also a constant exchange between deities and human spirits that results in optimal life opportunities, which ultimately lead to never-ending life cycles and reincarnation. We are our own ancestors . . . Matter/flesh and mind/spirit are one in a closed system, a cosmos from which we arose but which we create in tandem.[22]

Creation itself is testimony to the fundamental interrelatedness and interdependence of the spiritual and the material, of the Earthly and the terrestrial. Not only is it that "we are our own ancestors," we are our own gods. Through reincarnation, literally everything is recycled and remade—everything is interconnected. Mirroring this cosmic-nature of interrelatedness, Bellegarde-Smith and Michel affirm:

> Vodou embraces nature and the larger cosmos that validates our existence as part of the material world. We live in a biocentric universe in which we create the universe and are created by it. We exist as organic beings in line with all other energetic forces, including spirits. So part of us is divine as well. Danbala and Ayida serve as archetypes that remind us how to be divine but also earthly.[23]

In this sense, while the relationship between humans and "spirits" might, as Bellegarde-Smith and Michel propose, be described as an intimate interrelationship between "innerworlds" and "outerworlds" the divine world is still not *other* than or apart from. It is *part of* our sensorial experience and *part of* the natural, material reality of life. The sky and the moon are as present as the earth and the ocean—they are both part and parcel of the cosmic kinship of nature.

If nature and the cosmos are, in fact, understood as kin and inherently inseparable, rather than relating to metaphysical beings as otherworldly and wholly other entities, it is preferable, at least in many Africana worlds, to conceptualize spirits and divine authority figures in sociological terms. In this respect, metaphysical beings (e.g., ancestors, deities, and even witches) are related to as social beings existing within a shared community; as fathers and mothers, grandfathers and grandmothers, uncles and aunts, and husbands and wives.[24] By and large, even if and when Africana "deities" and "spirits" may be considered transcendent in some respect, I would still suggest that these entities are generally only transcendent in the same manner as a monarch might be perceived as transcendent (functionally and/or socially above and beyond) in comparison to their subjects. For instance, in Brazilian Candomblé, the *orisa* are known to emerge in full pageantry invoking ancient motifs of royalty. Describing the regal character of the *terreiro* (sacred grounds) and the *barracão* (dance pavilion), which are central to the procession of the *orisa*, Joseph A. Murphy notes:

> The architecture of the *terreiro* with its central court and constellations
> of outbuildings reflects the ideal organization of the Nagô kingdoms.
> The *barracão* is the court of the king of the nation of the *terreiro's*
> founder whether this is reckoned by blood descent or by initiation.[25]

Then, describing the procession of the *orisa*, Murphy states, "The *orixás* of candomblé [sic] are remote royal personages, affectless in their procession before the community."[26] As cosmic royalty, in Candomblé, the *orisa* are ideals of the well-lived life. Yet, even still, the *orisa* remain terrestrially rooted and are intimately connected to nature and the Earth. They are neither otherworldly nor entirely immaterial entities. They kiss, dance, shake hands, praise, and scold their beloved devotees.

This conception of the metaphysical does not presuppose a cosmos-world bifurcated into material and immaterial, or even physical and spiritual domains. Spirit, nature, and materiality are always present and interrelated. As Kwasi Wiredu suggests, "A people can be highly metaphysical without employing transcendental concepts in their thinking, for not all metaphysics is transcendental metaphysics."[27] Taking Wiredu's philosophical proposition further, I propose that, by and large, Africana orientational worlds embody and actualize an immanent metaphysics—an orientation to the metaphysical and yet naturally and materially present that necessitates an on-going, intimate interrelationship between the innerworlds of the visible and the concrete and the outerworlds of the invisible and yet quasi-material.

Conclusion: Immanent Religiosities and The Question of Africa in the Study of Religion

Yet, while explicitly rejecting Western categories of thought—polytheism, fetishism, animism, totemism, shamanism, and the list goes on—that have sought to make sense of, or "classify and conquer" the religious Other,[28] this Africana immanent metaphysics is not, what Graham Harvey and others in this volume might describe as a "new animism;" a scholarly approach that seeks to reclaim and revolutionize the very colonialist category that rendered Indigenous peoples into religious Others in the first place.[29] Rather it is the assertion and affirmation that Africana metaphysics are a normative onto-epistemology in which the sacred, matter, and nature are coexistent, intersectional, and ontologically equivalent.

Africana sacred matters explicitly calls into question transcendence as normative and inherent to the metaphysical and the religious. Rather than simply exploring the question, "What is *religious* about *Africa?*" (with a set of pre-given "religious" expectations), I have, instead, attempted to inquire: What is *African* (or, more precisely, *Africana*) about *religion*, the *religious*, and the *metaphysical?*[30] To this end, I am reminded of the words of my great mentor and teacher, Bernard Adjibodou, a Hogbonou court official and the Spiritual Chief of the town of Adjarra in Porto-Novo, Benin, who explained, "The philosophy is that African people never saw *god* when they looked up into the sky and looked around, and so . . . they took soil, leaves, and other substances of nature to make their *vodun*."[31] This Indigenous theorist is positing, that *vodun, orisa, lwa,* and other similar, effectual, ontological entities cannot be separated from the matters and materialities of nature and the natural world. Rather, these powerful entities are made concrete, perceptible, and perceivable through their materialization into empowered shrines, statues, devotees, and other organic and even seemingly inorganic bodies—whether a wooden *vodun*, an activated *nkondi*, a *nganga* pot, a consecrated *ori*, or even *otanes*.

Returning to the powerful words of the Ewe Gorovodou priest who stated, "We Ewe are not like Christians, who are created by their gods. We Ewe create our gods, and we create only the gods we want to possess us, not any others," I propose that Africana material-orientational worlds are ever unfolding of creation and re-creation, of becoming and being. While similar to other forms of, what Whitney Bauman, Karen Bray, and Heather Eaton (see their essays in this volume) refer to as, "immanent religiosities," *Africana sacred matters* are still more than merely "icons," "theophanies," or even "hierophanies."[32] Perhaps most similar to Catherine Keller's Panentheism (see, "Amorous Entanglements: The Matter of Christian Panentheism") in which God "emerges in relation," I would argue that Africana metaphysics is still not fundamentally theistically-centered nor singular (or, even plural, for that matter) but rather, conceptualizes the cosmos-world. To borrow Ogbonnaya's critical theological concept, as a "communotheism," or, better yet, a *communo-cosmos*—namely, an interwoven community, extended family, and network of kin.[33]

Revisiting the questions I posed, perhaps the central query is not, as I suggested, "What is *religious* or even *metaphysical* about *Africa* and *Africana* orientational worlds?" Rather, if we take seriously what *Africana sacred matters* contribute to the study of religion, then in the words of a Vodun priest, the central questions in our field might be: "What do you create? And, what possesses you?" From a process theological perspective, Keller might still argue that in the process of being and becoming we create God/god (and I would not disagree), but we are also the creators of anti-social and anti-communal forces and institutions (e.g., colonialism, racism, incarceration, White supremacy, xenophobia, the planetary crisis, capitalism, police brutality, and COVID-19). These *ajogun*, these "anti-gods," these anti-social "warlords," (to adopt a Yorùbá Indigenous concept) are also "gods" of our own making.[34] Thus, Joerg Reiger (see, "Which Materialism, Whose Planetary Thinking?") is right to ask: "What are we up against?" The answer is not simply neoliberal capitalism, as he would suggest, but rather all of the anti-social forces and institutions, all of the *ajogun*, that we continue to feed and to which we literally give power and life itself. So, to the question, "What are we to do?" I agree that we must "do rituals" (Pike, "Rewilding Religion for a Primeval Future"), "dance" (LaMothe, "Dancing Immanence: A Philosophy of Bodily Becoming"), and "eat and kill better together" (Bray, "Gut Theology: The Peril and Promise of Political Affect"), but we must also stop creating and often unintentionally feeding *gods* that no longer serve and nourish our diverse planetary communities.

Notes

This chapter was inspired by an undergraduate course of the same name that I designed and have taught at Hobart & Smith Colleges and Tufts University in which students are introduced to African and African diaspora religious traditions through a focus on art and material culture (including, dance, ritual, divination systems, warfare technologies, etc.).

1. For a further exploration of this notion of temporal functional superiority, please see A. Okechukwu Ogbonnaya, *On Communitarian Divinity: An African Interpretation of the Trinity* (New York: Paragon House, 1994), 29–85, and Kọ́lá Abímbọ́lá, *Yorùbá Culture: A Philosophical Account* (Birmingham, UK: Irókò Academic Publishers, 2006), 70–71. Proposing the concept of "communotheism" as

a means of conceptualizing how African entities function collectively, Ogbonnoya theorizes that instead of these entities being classified according to their ontological substance that there is a hierarchy of responsibilities. Likewise, Kọ́lá Abímbọ́lá argues against the notion of "hierarchal dogmatism" and demonstrates that Yorùbá cosmology can either be organized according to an existential or functional hierarchy (26–29, 59–60).

2. Rudy Rosenthal, *Possession, Ecstasy, and Law in Ewe Voodoo* (Charlottesville: University of Virginia Press, 1998), 45.

3. Ibid., 67.

4. See for example, Rudolf Otto, *The Idea of the Holy* (New York: Oxford University Press, 1958) and Mircea Eliade, *Patterns in Comparative Religion* (Lincoln: University of Nebraska Press, 1996).

5. Wyatt MacGaffey, *Kongo Political Culture: The Conceptual Challenge of the Particular* (Bloomington: Indiana University Press, 2000), 13.

6. *Nkondi* literally means "hunter" and was also commonly referred to as *mbwa* meaning "dog." Wyatt MacGaffey, "The Personhood of Ritual Objects: Kongo 'Minkisi,'" *Etnofoor*, Vol. 3 (1) (1990): 52.

7. Wyatt MacGaffey, "Complexity, Astonishment and Power: The Visual Vocabulary of Kongo Minkisi," *Journal of Southern African Studies*, Vol. 14 (2) (January 1988): 201.

8. MacGaffey, "The Personhood of Ritual Objects," 51, 54; MacGaffey, *Kongo Political Culture*, 13.

9. MacGaffey, *Kongo Political Culture*, 13.

10. MacGaffey, "The Personhood of Ritual Objects," 49. For examples of African diaspora counterparts to the Kongo *nkisi*, see for example: Carolyn Marrow Long, *Spiritual Merchants: Religion, Magic, and Commerce* (Knoxville: University of Tennessee Press, 2001); James H. Sweet, *Recreating Africa: Culture, Kinship, and Religion in the Afro-Portuguese World, 1441–1770* (Chapel Hill: University of North Carolina Press, 2003); and Karen McCarthy Brown, "Making Wanga: Reality Construction and the Magical Manipulation of Power," in *Religion and Healing in America*, eds. Linda L. Barnes and Susan S. Sered (New York: Oxford University Press, 2005), 173–194.

11. Judith Bettelheim, "Palo Monte Mayombe and Its Influence on Cuban Contemporary Art," *African Arts*, Vol. 34 (2) (Summer 2001): 36–37.

12. Ibid., 37.

13. Ibid., 36.

14. Bruno Latour, "Fetish-Factish," *Material Religion*, Vol. 7 (1) (2015): 46 (emphasis added); Patricia de Aquino (Personal Communication).

15. This Yorùbá word is also often transcribed as *orixá, orisha, orisa* (with or without a tonal dot under the "s" to signify the "sh" sound), and with all its tonal marking as *òrìṣà*.

16. Qtd. in Tracey Hucks, *Yoruba Traditions and African American Religious Nationalism* (Albuquerque: University of New Mexico Press, 2014), 90–91; Carl

Hunt, *Oyotunji Village: The Yoruba Movement in America* (Washington, DC: University Press of America, 1979), 27.

17. Kwasi Wiredu, *Cultural Universals and Particulars: An African Perspective* (Bloomington: Indiana University Press, 1997), 35–36.

18. In some Africana religious cultures, such as Santería and Vodou, the person being "mounted" by a deity or spirit is referred to as the "horse" (or, in French, "cheval") of the *orisa* or *lwa* who is temporarily materialized through the devotee's body.

19. Qtd. in Bettelheim, "Palo Monte Mayombe," 37; Elizabeth Hanly, "Symbol Truths," *South Florida* (April 1994): 52.

20. Miguel de la Torre, *Santeria: The Beliefs and Rituals of a Growing Religion in America* (Grand Rapids: Eerdmans Publishing, 2004), 14.

21. Patrice Bellegarde-Smith and Claudine Michel, "Danbala/Ayida as Cosmic Prism: The Lwa as Trope for Understanding Metaphysics in Haitian Vodou and Beyond," *Journal of Africana Religions*, Vol. 1 (4) (2013): 470–471, 479.

22. Ibid.: 468.

23. Ibid.: 470.

24. Robin Horton, *Patterns of Thought in Africa and the West* (Boston: Cambridge University Press, 1997), 369.

25. Joseph M. Murphy, *Working the Spirit: Ceremonies of the African Diaspora* (Boston: Beacon Press, 1994), 187.

26. Ibid., 187.

27. Wiredu, *Cultural Universals and Particulars*, 87.

28. For an examination of this topic in the study of African religions in particular, see David Chidester, *Savage Systems: Colonialism and Comparative Religion in Southern Africa* (Charlottesville: University Press of Virginia, 1996); David Chidester, "Classify and Conquer: Friedrich Max Müller, indigenous traditions, and imperial comparative religion," in *Beyond Primitivism: Indigenous Religious Traditions and Modernity*, ed. Jacob K. Olupona (New York: Routledge, 2004), 71–88; and David Chidester, *Empire of Religion: Imperialism and Comparative Religion* (Chicago: University of Chicago Press, 2014).

29. Though Graham Harvey acknowledges and, to some degree, grapples with the pitfalls of the "new animism" concept (noting, for instance, his on-going debate with Jacob Olupona), he fails to fully understand why reclaiming such a concept continues to reinforce Western (Christian) religion as the norm and the traditions of Black, Brown, and Indigenous peoples as religiously Other.

30. Newell S. Booth, "An Approach to African Religions," in *African Religions: A Symposium*, ed. Newell S. Booth (New York: Nok Publishers, Ltd., 1977), 1–11. Booth introduces this problematic in the study of African religious cultures noting that while anthropologists, on the one hand, in their attempt to capture the whole of African *culture* often lose sight of the *religious*, historians of religion, on the other hand, in their pursuit of the phenomenon of *religion*, often obscure what is specifically *African*. While Booth's initial definition of this problematic was still

flawed in assuming an inherent separation between "religion" and "culture," which I reject (note my use of "Africana religious cultures"), his query importantly suggests a deeper examination of the relationship between "religion" and "Africa" in the study of religion.

31. Bernard Adjibodou, conversation with author, January 7, 2014.

32. Eliade, *Patterns in Comparative Religion*, 29. According to Eliade, a hierophany sets up a problem and paradox wherein the sacred and the profane come together. Yet I am, alternatively, arguing that in an Africana immanent metaphysics no such paradox actually exists.

33. Ogbonnaya, *On Communitarian Divinity*, 28, 30.

34. For an examination of *ajogun* in Yorùbá religious culture and philosophy, see Abímbọlá, *Yorùbá Culture*, 49. Specifically, he notes that the "eight warlords of the Ajogun are: Ikú (Death), Àrùn (Disease), Òfò (Loss), Ègbà (Paralysis), Ọràn (Big Trouble), Èpa (Curse), Èwọn (Imprisonment), and Éṣe (Afflication).

We have always been animists . . .

Graham Harvey

Animism has been significantly revisited (to use Nurit Bird-David's term[1]) by a host of scholars, from multiple disciplines, in recent years. A new animism has arisen—not as a new religion or cultural movement but as a scholarly approach. It is a "turn" towards those who conceive of themselves as participants, with other than human persons, in a larger cosmic society[2]—not "in general" but always in terms of relationships which enhance human capacities to "feel, think, and imagine" the world communally.[3] Engagement with this scholarly turn has led increasing numbers of people to identify as animists, and some animist networks or movements have been established. While these provide further contexts for research and debate, they are not the main focus of this article or the principle reference of the term "new animism."

In this ongoing debate, the term "animism" is deployed strategically, in full awareness that it has become a weapon used against many of the Indigenous[4] and other people whose knowledges, practices, and lives have inspired the new animism, new materialism, and other turns. Indeed, the new animism entails explicit challenges to colonialist, dismissive, and derogatory engagements with Indigenous peoples and their knowledges. In part, the new use of "animism" gains confidence from the successes of the reclamation of other contentious terms (e.g., "gay" and "native"). Such reclamations and resistances contribute to profound challenges to dominating ideologies, ontologies, and practices— and thereby to increased respect and justice.

By reflecting on Bruno Latour's provocative assertion that, "We have never been modern,"[5] the following section is intended to show where the title of this essay comes from. A summary of the new animism project so far will be followed by an evocation of the constitutive relationality of the world and of the coevolving and symbiotic nature of all reality. In dialogue, these sections identify a friction between the efforts to be(come) "Moderns" by separating humans from the world and the animistic efforts to live respectfully with/in larger-than-human relationships. Although Anna Tsing identifies friction as enabling movement (e.g., wheels are ineffective without friction against road surfaces),[6] the project of Modernity seductively invites us to glide over matters that deserve more sustained attention and more, rather than less, participatory engagement. I propose here that confronting Modernity's "nature" concept provides the necessary traction to shift from efforts to be(come) Modern towards efforts to recognise our (human and scholarly) participation in the larger-than-human world.

Put differently, while we are necessarily (ontologically) participants in an animate world, continuously and constitutionally interacting with all existences, we have often been misled into joining the human separatist movement named Modernity. Re-placing ourselves as located kin, as symbionts, as coevolvers, and as citizens in a larger polity *could* improve our scholarly activities and ambitions. An examination of the contrast between Modern and animist ontologies and epistemologies will be aided by reflection on the promulgation of and resistance to the Modernist notion of "nature." This ends with an examination of the emphasis on obligations and immediacy among the relations; which and who make the real, but threatened, world.

Never Having Been Modern . . .

Ever since Bruno Latour asserted that "we have never been modern," he has often used "Moderns" as a label for people who identify with or contribute to a particular way of being human and/or a particular ideological project.[7] While we might not do so deliberately—all of us alive today are susceptible to the disenchanted temptation to adhere to the Modern project. Modernity, in this schema, is the project of separating some elements of reality from others. Although "modern"—like "premodern" and "postmodern"—presents a semblance of temporality, it is not a neutral descriptor of an era. Even prior to Latour's capitalized usage, the word "modern" promoted a colonialist contrast between cultures and peoples. The term is a flag for value judgements at the heart of a purity system at least as vigorous and complex as those known as Kashrut and Halal. There is nothing natural or necessary about the separations structuring Modernity. Indeed, at the heart of Latour's argument is the assertion that "natural" is a marker labeling one part of the world which Moderns seek to construct. It is opposed to "culture," in such a way that to be a Modern (cultural) human is to be separated from the (natural) rest of the world, or separate enough to be able to use the world as if it were inert and inconsequential, and as if we were unburdened by reciprocal obligations of larger-than-human kinship. That academia is implicated in Modernity's project, perhaps even spearheading it, is evidenced by the structural separation of the "natural sciences" from the "social sciences" or by the use of "Science" in contrast with "the Humanities."

Latour thinks that we have never been Modern because we have never fully enacted the divisions required of us. We continuously participate in networks of actors which are neither purely natural nor purely cultural. One of Latour's early examples was the hole in the ozone layer which resulted from interactions between human and larger-than-human processes in a world without impermeable boundaries. He has also argued that scientific laboratory practices demonstrate the failure of Moderns to separate humans from the rest of life.[8] In his many publications and lectures, he argues that there is no "society" but only myriad changing interactions between all sorts of beings, objects, and ideas that together form the shifting patterns of "actor-networks." His argument draws significantly on Indigenous[9] and other perspectives and knowledges which posit the vital role of relational interaction in making and unmaking the coevolving assemblage we call the world.

The failure to be(come) Modern is rooted in the fact that "we" have not succeeded in renouncing our non-Modern affective relationships with beings of other species. We scholars have certainly made efforts to become Modern— including maintaining boundaries between academic disciplines (and those "natural" and "social" sciences)—but these very efforts illustrate that we are not (yet) fully Modern. Even as we try to construct a Modern world we are caught between transgressing and policing boundaries that fail to discover a world in which humans are separate from our other-than-human kin or in which there is a "nature" separate from the constructions that reputedly evidence "culture." Equally, we have failed to achieve the objectivity modeled on that imputed to a transcendent deity, instead finding ourselves attracted to, and enthused by, research and teaching projects that willfully present themselves to us.

It should be clear that Latour's title and argument are deliberately provocative. If we have never been Modern, it is not for want of trying. When Latour sets out how Modernity's proposed borders and divisions fail to structure reality, some of his readers might respond with more vigorous efforts to enforce or enact them. Alternatively, some might celebrate the failure of Modernity and be encouraged in pursuing diverse other-than-Modernisms—Animism can be one such way. However, my title ("We have always been animists . . .") is as deliberately provocative as Latour's. If Latour's work shows that we have never been fully Modern despite significant efforts, my title is intended to indicate that within an inherently animate (relational) world, some people are trying to be more fully animistic. The following section sets out my understanding of the ways in which new animism research presents animism as a deliberate practice or, as Tim Ingold says, a way of being alive *to* the world "characterized by a heightened sensitivity and responsiveness, in perception and action, to an environment that is always in flux, never the same from one moment to the next."[10]

New and Old Animisms

The new animism might be briefly encapsulated as a focus on ontologies and lifeways in which the world is assumed to be full of persons, most of whom are not human, all of whom are kin, some closer than others, and all of whom deserve respect. The assumption is foundational to those efforts to interact respectfully which are elaborated as "traditions" or "cultures" shaped by obligations and responsibilities. Indeed, the pervasive Indigenous use of the term "respect"—glossed by Mary Black as "careful and constructive" engagement[11]— points to inculcated and practised attention to specific persons and to locally appropriate etiquettes of relationship.

Elsewhere, I have devoted some attention to the "old animism," an approach most usefully associated with Edward Tylor (first professor of anthropology at Oxford University).[12] For the current purpose, it is enough to note that Tylor used "animism" to label systems of belief in souls, spirits, or metaphysical entities which he argued was definitive of religion.[13] Within Tylorian discourses, animism is a misinterpretation of observations and experiences which seem to suggest intentionality, agency and/or other modes of specifically human ability in the wider world. More succinctly, this animism (which glosses "religion" and characterizes all religions) is the attribution or projection of human likeness or capacities where they are unwarranted.

It is vital to emphasise the point that the new animism is a still evolving and diverse assemblage of scholarly approaches. The phenomena that new animism scholars aim to understand, debate, make familiar, and/or bring into provocative dialogue with other ontologies and epistemologies are not new. Multiple Indigenous and other communities engage with the world around them in the explicitly and deliberately relational ways that may be labeled "animism." Many have done so for millennia. Many have been derided for their animism—and for subsets of animism such as: "fetishism," in which artifacts are treated as persons; "totemism," in which clans or families embrace other-than-human kin; and "shamanism," in which myriad more-or-less invisible actors require diplomatic negotiation. Both "materialism" and "spiritism" have also been used as accusations against animists who might dare to value ritual performances; to contest an emphasis on (human) mind or spirit while also imagining a proliferation of interactive, but not always visible, agents who cocreate reality.

In Chickasaw author Linda Hogan's wise reflections on scholarship about animism, "we call it *tradition*," she explains that she has been thinking about:

> . . . what this newly accepted area of study means to those of us whose cognitive and spiritual worlds are already created by our rivers, mountains and forests. For those who have always prayed with, to, and for the waters, and known our intimate relatives, the plant people, the animals, insects, and all our special relations, the field of animism is a belated study. It has not gone unnoticed that without these relationships, a great pain and absence has been suffered by humanity, an absence and loss we ourselves have felt as a result of the determinations of the Western mind to separate us from our homelands, and which has created great destruction to the living body of the continent. We know our own pain as we have been forced, often through violent means by the governing politics, to take up values vastly different from our own.[14]

In short, "tradition" should be enough to sum up the varied, locally appropriate efforts to engage respectfully with kin among the larger-than-human community. It points to tested etiquettes, ceremonies, attitudes, protocols, obligations, and other aspects of life that are offered to succeeding generations to improvise and improve on.

Both tradition and animism are well (re)presented in Hogan's many creative and scholarly publications. Her novels and essays are infused with what Nigerian scholar of literature Harry Garuba calls, "animist materialism." He summarizes his discussion of "animist realist" literature and culture(s) by saying that he has tried to "sketch out the logic of animist thought and then focus on what I believe is its primary characteristic: the continual re-enchantment of the world."[15] He shows that:

> A recurrent theme in accounts of the meeting between traditional ways of life and modernity is the clash of cultures and the agony of the man or woman caught in the throes of opposing conceptions of the world and of social life. In these narratives a binary structure is usually erected, and within this world the agonistic struggle of the protagonist is drawn . . . What may be much closer to the reality is that animist logic subverts this binarism and destabilizes the hierarchy of science

over magic and the secularist narrative of modernity by reabsorbing historical time into the matrices of myth and magic.[16]

In explicit contrast with colonialist denigration of animist logic, materialism, and cultural performances (ritual, artistic, or narrative), Garuba "visibly and enthusiastically" champions the recognition that "an animistic understanding of the world applied to the practice of everyday life has often provided avenues of agency for the dispossessed in colonial and postcolonial Africa."[17] He indicates the necessity of understanding such animism if Africa and its diasporas are to be understood.

Two matters are crucially interwoven into Garuba's argument: (a) the continuous nature of re-enchantment and (b) the need to rethink the encounter between "tradition" and "modernity." Animist realist novels (and other animist discourses) do not always flag "myth and magic" as exceptional or peculiar. Everyday life, politics, economics, subsistence, entertainment, and other areas of life are thoroughly and expansively relational, and thus enchanted beyond the narrow human obsession of Modernity. The pervasive assumption is that members of the larger-than-human community *might* be interested in human acts and affairs—and should be approached or avoided on that basis. Rocks, plants, animals, drinking vessels, and others are active participants in the assemblages that become recognizable as local cultures or "the way things are." Thus, there can be animistic efforts to be(come) (more) involved in the globalized world without rejecting the animist traditions that may make life possible or bearable.

Turning to those Moderns, meanwhile, it is notable that the despisers of animistic ways of relating to the world continuously fail to de-animate their own relations. They talk to dogs, cats, cars, and computers. They fetishize money, wedding rings, and other symbolic regalia. They bow to invisible forces like "the market"—even as they make conflicts out of the material forms of goods and labor. They give power to totemic notions such as "race" and "class." Perhaps all of these deserve elaboration as entanglements in which "Moderns" continue to work to be(come) Modern by struggling against their continuing animism. But an alternative possibility emerges in which never having been Modern brings an exciting possibility of celebrating enchantment and relationality—and, within this, of scholarship that reimagines the roles "religion" might play.

I turn now to a broader sense of animism: summarizing some of the ways in which humans and all beings (indeed, all existences) *are* relations, constituted by interactions with kin and others among all species, whether or not we wish to do so, whether or not we celebrate doing so.

Kinship in a Relational World

In the larger-than-human world that struggles (and often fails) to resist the Modern human separatist project, we are all kin (remembering that relatives do not always like each other). All of us, even if we are trying to be(come) Moderns, and even if we are not making efforts to be(come) animists, are already animistic in that we live relationally. Much like Monty Python's eponymous "not the messiah / naughty boy" Brian, the promoters of Modernity encourage us to be individuals. But not only is our individuality a communal effort (with all sorts of social structures encouraging individuation), we continuously interact in ways that demonstrate what McKim Marriot and Marilyn Strathern called "dividuality."[18] We are "dividuals," more-or-less fluidly moving between different ways of interacting with other dividuals to whom we are, for a while at least, family, colleagues, shoppers, carers, observers, participants, opponents, and/or celebrants. In no single one of these relationships are all possible facets of our being or our moving-through-the-world revealed or enacted. We are partible and partial beings,[19] engaging more-or-less immediately with specific partners.[20] Perhaps here we see the masks / faces / characters of classical Mediterranean "person" concepts—with no allusion to dissembling or deceit but, rather, to the dynamic presentations of relationality.

In fact, our relationality is more deeply woven than merely interactivity. Not only have we coevolved with species as diverse as grains and canids, our relations with myriad bacteria make us symbionts or perhaps holobionts.[21] Bacterial DNA exceeds "human" DNA on and within "our" bodies.[22] They collaborate with our mothers and other relations to aid our growth before and after birth. They make nutrients available from the food that we (and they) consume. They aid our dissolution into nutrients useful for other life—or would if we were not so obsessed with individualized, sterile, and separatist death. Our relationship with bacteria provides the perfect example of what Lynn Margulis identified as "symbiogenesis,"[23] the standard process by which evolution forms everything relationally.

Other recent terminological and theoretical interventions could aid recognition and even celebration of the normality of relationality. They include Eduardo Viveiros de Castro's "multinaturalism" and "perspectivism,"[24] both flowering from the stock of Amazonian Indigenous enacted knowledges.[25] These require a rethinking of Modernity's perception of a universal identity of "nature" over the fecundity of plural "cultures." Isabelle Stengers' "cosmopolitics"[26] invites us to re-view the complex negotiations, conflicts, and diplomacies (politics) in which all existences (cosmos) are implicated. Karen Barad's "agential realism"[27] enriches understanding of the constitutive inter- and (more significantly) intra-activity performed by, among, and as, all phenomena from quantum particles to (human) social inequality. Elizabeth Povinelli's "geontologies"[28] composts the division between *bios* and *geos*—or lively biology and inert mineral geology—to insist that being in place(s) creates compulsions or carries obligations. Then, if we follow Donna Haraway's call to "stay with trouble" as we negotiate our way among the tentacular processes and practices of the "Chthulucene"[29] and Anna Tsing et al's *Arts of Living on a Damaged Planet*, we might (re)encounter the captivating, pulsing "shimmering" of generative life to which Debbie Bird Rose[30] was introduced by her Aboriginal Australian teachers. All these, and other ways of understanding encounters with(in) the larger-than-human community, present us with provocative insights into the deeply constitutive relationality of all existences and deserve further consideration for what more they might contribute to the "new animism" and "new materialism" projects.

Using "Nature" as a Lever

Both Modern separatism and animistic relationality form mediums in which we engage with others (whether or not we celebrate such entanglements). If we have never been completely Modern it is in part because we have never completely ceased to be animists, always identifying ourselves by our relations and obligations. We remain enchanted by our subjectivities. Simultaneously, some of us struggle to enact our animistic ambitions in the face of Modernity's dominating restructuring of reality, which encourages us to find our (separate, individualized, and often "inner") selves. In either situation, our inherent relationality sometimes becomes evident to us. Sometimes we respond by increasing our individuating efforts. Sometimes we respond by celebrating our dividuality. The notion of "nature" implicates—indeed engineers—this tension.

If, as Latour suggests, the project of Modernity cannot succeed—because the relationality of reality resists individuality and separatism—its promoters deploy "nature" as a tool to construct or reinforce their world-making. "Nature," we are encouraged to accept and assume, is a resource from which matter can be extracted, in which recreation can be pursued, or which might inspire decorative cultural products. These include TV "natural history" documentaries, information boards at "nature reserves," and rhetoric about the therapeutic or spiritual value or sacrality of "natural places." The removal of humans from places that can then be imagined as "pristine nature"—sometimes as a stage in a longer genocide and sometimes only in a willful act of ignoring farm labor—paradoxically cultivates a notion of "nature" that can be sold to or consumed by cultured connoisseurs of the "nonhuman" or "primitive." In all these cases, despite appearances, "nature" requires considerable effort to manufacture and to mask as something "natural" and separate from humans and our cultures.

Academia is a participant in this deceit, continuously elaborating distinctions between scholarship about humans (the arts, humanities, and social sciences) and the "hard sciences" which engage with "nature." However, things are changing. Botanists are making clear the ways in which plants communicate,[31] and ethographers are making clear the ways in which animals require humans to interact attentively.[32] The "new animism" and the "new materialism"—both braided into the "ontological turn" and the "turn to things"—are increasingly vociferous in challenging the de-animation of the larger-than-human world. More powerfully yet, the rise of multi- and trans-disciplinary Indigenous Studies—conducted in the glare of ongoing colonialism and imperious Modernity—is insistent that the world remains thoroughly relational and that this diverse and rich assembling of relationships should be central to theorizing, teaching, and other academic pursuits. Importantly, Indigenous Studies scholars do not romanticize relationality but recognize and re-cognize conflict and cannibalism as modes of relationship that are just as creative as totemic and familial relations.

Nonetheless, Modernity's "nature" is vigorous in continuing to lever de-animation and separatism into popular consciousness and discourse. For instance, when self-identified animists, Pagans, and others talk about the "celebration of nature" (or "Nature") they undermine their efforts to contest human separatism or to compose discourses or rites of respect. This is particularly evident when the celebrated "nature" is rural or allegedly "wild" rather than immediately accessible in our biodiverse urban spaces and homes. "Nature" at its most ever-present, we might imagine, is the air that we breathe and the weather we move through.[33] If so, it is everywhere and nobody (no body, no thing) is separate. Therefore, more than semantic quibbles are involved in contesting the term.

Given the extremity of the unfolding climate disaster and mass extinction, Modernity's "nature" is a term of terror that must be contested. Even if humans have not become separate from the world in reality—indeed, cannot become separate—Modernity's "nature" has enabled, provoked, and propelled the immense damage still being done to all parts of the planet. Animism cannot be about finding an alternative word that does a similar job. It is about undercutting the assumptions of inanimacy and separatism and about injecting relationality into discourses and practices that matter.

The anthropologist Irving Hallowell asked an Anishinaabe elder and medicine man, Kiiwiich Alec Keeper, a question that is cited in most "new animism" publications. Hallowell writes:

> Since stones are grammatically animate, I once asked an old man: Are
> *all* the stones we see about us here alive? He reflected a long while and
> then replied, "No! But *some* are."[34]

Again, this might appear to be about language but is rich in its implications and value. As Hallowell notes, "This qualified answer made a lasting impression on me" and reinforced his understanding that in conversations with such people, "We are confronted with the philosophical implications of their thought, the nature of the world of being as they conceive it."[35] The "world as they conceive it" is also the world that they compose as they interact, seeking to be carefully and constructively respectful of kin and companions. It is also the world in which all humans live. It is not a world in which it is necessary to ask, "are stones alive?" because that is the Modernist question. It is a question about differences and identities; about *geos* and *bios*; about "our culture" and "theirs." It is not a question about relations. But note, Hallowell asked "are *all* the stones we see about us here alive?" That is, he asked about nearby stones, ones in whose presence the conversation took place. His phrasing was open to the relationality that became the subject of the discussion developing from Kiiwiich Keeper's wonderfully enigmatic answer. The elder's answer requires more words, in stories and rites, in greeting etiquette and gift exchange, to reveal that his world was one in which knowledge about stones involves those stones. As Povinelli learnt in dialogue with the Karrabing film collective and other Aboriginal Australians, the point of collecting, recording, and sharing knowledge of Dreaming places and relations is:

> Not merely to gain knowledge but to keep an arrangement in place by
> the activity of using it in a specific way. Knowledge about country
> should be learned, but abstract truth is not the actual end of learning.
> Learning—knowing the truth about place—is a way to refashion bodies
> and landscapes into mutually obligated bodies.[36]

During fieldwork with the "Reassembling Democracy" project,[37] I met a Sámi man by a river surrounding the Riddu Riddu festival site. We had a casual but serious conversation about the river, which seemed likely to flood the surrounding land. When I expressed concern, he not only explained that this was a result of climate change (mountain snow melting faster than ever before) but that the rapid flow and cold temperature of the water would prevent salmon and trout entering the river from the fjord. If they did not manage that soon, he said, they would fail to breed and thus disappear from the river forever. Whether his analysis was correct or not, it is his concern for the well-being of the fish (and other river dwellers) that has made a lasting impression on me. He might also have worried about human subsistence and Coastal Sámi culture, but what he talked about was the shame of what human consumerism has done to other-than-human kin such as the fish and the waters.

Animism, then, is not a confusion of life and death, animacy and inanimacy. It is not a projection of human-likeness on to a passive and inert world of uncultured nature. It is not about extending the rights of "persons" to things ("natural" or artifactual). Neither is it about identifying all existences under the label "persons." Rather, animism is about relatives and interactions. As Bird-David and Naveh insist, animism is about specificity and immediacy.[38] It is about actual engagements between relatives of whatever species. There is no categorical separation between kinds of being, no limit to the possible or potential ways of interacting and intra-acting, but locally encouraged traditions of appropriate etiquette and engagement. In the more-than-human world[39] there is no "nature" except that of Modernity, which interrupts relationality perhaps forever. Failure to understand this not only damages scholarship by imagining a world different to the one we are evolving in, but also threatens to do yet more damage to the relationships that are the real world.

Notes

Ideas for this article were first aired in a presentation in November 2019 in Harvard University's lecture series, "Matter and Spirit: Ecology and the Non-Human Turn." I am grateful to Charles Stang and colleagues at the Center for the Study of World Religions for inviting me. It was also a privilege and pleasure to spend time with the Center's Animism Study Group. I am grateful to Molly Kady for accompanying me, and for her advice on a draft of this article.

1. Nurit Bird-David, "'Animism' Revisited: Personhood, Environment, and Relational Epistemology," *Current Anthropology, Vol.* 40, Supplement (1999): S67–S91.

2. A. Irving Hallowell, *The Ojibwa of Berens River, Manitoba*, ed. Jennifer S. H. Brown (Fort Worth: Harcourt Brace, 1992), 80.

3. Isabelle Stengers, "Reclaiming Animism", *e-flux, Vol.* 36. http://worker01.e-flux .com/pdf/article_8955850.pdf. (2012), (accessed. January 16, 2020): 9. I am also grateful to Jacob Olupona for pushing me to justify using a term ("animism") that has been deployed aggressively against many Indigenous people and cultures. Our conversation is unresolved, ongoing, and vital.

4. The capitalization of "Indigenous," "Modern," and some other words is strategic and points to the hard work they undertake in carrying a heavy freight of

meaning in contested debates. Briefly, both are projects of world-(re)making: Indigeneity has to do with First Nations while Modernity has to do with human separatism. See Graham Harvey, "Performing Indigeneity and Performing Guesthood," in *Religious Categories and the Construction of the Indigenous*, eds. Christopher Hartney and Daniel J. Tower (Leiden: Brill, 2017), 74–91 and Miguel Astor-Aguilera and Graham Harvey, "Introduction: We have never been individuals," in *Rethinking Personhood: Animism and Materiality*, eds. Miguel Astor-Aguilera and Graham Harvey (New York: Routledge, 2018), 1–12.

5. Bruno Latour, *We Have Never Been Modern* (New York, Harvester Wheatsheaf, 1993).

6. Anna L. Tsing, *Friction: An Ethnography of Global Connection* (Princeton: Princeton University Press, 2004).

7. Latour, *We Have Never Been Modern*. Also see Gurminder K. Bhambra, *Rethinking Modernity: Postcolonialism and the Sociological Imagination* (Basingstoke: Palgrave Macmillan, 2007).

8. Bruno Latour, *Reassembling the Social: An Introduction to Actor-Network-Theory* (New York: Oxford University Press, 2005) and Bruno Latour, *An Inquiry into Modes of Existence: An Anthropology of the Moderns* (Cambridge: Harvard University Press, 2013).

9. Latour often cites these second-hand, e.g., acknowledging Eduardo Viveiros de Castro's work rather than that of Davi Kopenawa—noted later. See Zoe Todd, "An Indigenous Feminist's Take on The Ontological Turn: 'Ontology' Is Just Another Word for Colonialism," *Journal of Historical Sociology*, Vol. 29 (1) (2016): 4–22.

10. Tim Ingold, "Rethinking the Animate, Re-Animating Thought." *Ethnos*, Vol. 71, (1) (2006): 10.

11. Mary B. Black, "Ojibwa Power Belief System" in *The Anthropology of Power*, eds. Raymond D. Fogelson, and Richard N. Adams (New York: Academic, 1977), 141–51.

12. Graham Harvey, *Animism: Respecting the Living World*, second revised edition (London: Hurst, 2017) and Paul Tremlett, Liam Sutherland, and Graham Harvey, eds., *Edward Tylor, Religion, and Culture* (London: Bloomsbury, 2017).

13. Edward B. Tylor, *Primitive Culture*, 2 Volumes (London: John Murray, 1871).

14. Linda Hogan, "We call it *tradition*," in *The Handbook of Contemporary Animism*, ed. Graham Harvey (New York: Routledge, 2013), 17–18.

15. Harry Garuba, "Explorations in Animist Materialism: Notes on Reading/Writing African Literature, Culture, and Society," *Public Culture*, Vol. 15 (2) (2003): 261–85, 284.

16. Ibid., 270

17. Ibid., 285

18. McKim Marriott, "Hindu Transactions: Diversity Without Dualism," in *Transaction and Meaning: Directions in the Anthropology of Exchange and Symbolic Behavior*, ed. Bruce Kapferer (Philadelphia: ISHI, 1976), 109–42, and Marilyn Strathern, *The Gender of the Gift: Problems with Women and Problems with Society in Melanesia* (Berkeley: University of California Press, 1988).

19. Ernst Halbmayer, ed., "Debating Animism, Perspectivism and the Construction of Ontologies," *Indiana, Vol.* 29 (2012): 9–23.

20. Nurit Bird-David and Danny Naveh, "Relational Epistemology, Immediacy, and Conservation: Or, What Do the Nayaka Try to Conserve?" *Journal for the Study of Religion, Nature, and Culture, Vol.* 2 (1) (2008): 55–73 and Nurit Bird-David, "Persons or relatives? Animistic scales of practice and imagination," in *Rethinking Personhood: Animism and Materiality, eds.* Miguel Astor-Aguilera and Graham Harvey (New York: Routledge, 2018), 25–34.

21. Lynn Margulis and René Fester, *Symbiosis as a Source of Evolutionary Innovation* (Cambridge: MIT Press, 1991).

22. Scott F. Gilbert, "When 'personhood' begins in the embryo: avoiding a syllabus of errors," *Birth Defects Research, Part C, Vol.* 84 (2) (2008): 164–73.

23. Lynn Sagan, "On the origin of mitosing cells," *Journal of Theoretical Biology, Vol.* 14 (3) (1967): 255–74.

24. Eduardo B. Viveiros de Castro, "Cosmological Deixis and Amerindian Perspectivism," *Journal of the Royal Anthropological Institute, Vol.* 4 (1998): 469–88 and "Exchanging Perspectives: The Transformation of Objects into Subjects in Amerindian Ontologies," *Common Knowledge Vol.* 10 (3) (2004): 463–84.

25. Eduardo B. Viveiros de Castro, "The Crystal Forest: Notes on the Ontology of Amazonian Spirits", *Inner Asia, Vol.* 9 (2007): 13; citing Davi Y. Kopenawa, "Sonhos das origens," in *Povos indígenas no Brasil (1996–2000)*, ed. Carlos A. Ricardo (São Paulo: ISA, 2000); and Bruce Albert and Davi Kopenawa, *Yanomami, o Espírito da Floresta* (Rio de Janeiro: Centro Cultural Banco do Brasil / Fondation Cartier, 2004). Also see Davi Y. Kopenawa and Bruce Albert, *The Falling Sky: Words of a Yanomami Shaman* (Cambridge: Harvard University Press, 2013).

26. Isabelle Stengers, "The Cosmopolitical Proposal," in *Making Things Public: Atmospheres of Democracy, eds.* Bruno Latour and Peter Weibel (Cambridge: MIT Press, 2005), 994–1003; *Cosmopolitics I & II* (Minneapolis: University of Minnesota Press, 2010 and 2011); and "Reclaiming Animism," *e-flux, Vol.* 36 (2012), http://worker01.e-flux.com/pdf/article_8955850.pdf. (accessed January 16, 2020).

27. Karen Barad, *Meeting the Universe Halfway: Quantum Physics and the Entanglement of Matter and Meaning* (Durham: Duke University Press, 2007).

28. Elizabeth A. Povinelli, *Geontologies: A Requiem to Late Liberalism* (Durham: Duke University Press, 2016).

29. Donna Haraway, "Anthropocene, Capitalocene, Plantationocene, Chthulucene: Making Kin," *Environmental Humanities, Vol.* 6 (2015): 159–65 and *Staying with the Trouble: Making Kin in the Chthulucene* (Durham: Duke University Press, 2016).

30. Deborah B. Rose, "Shimmer: When All You Love is Being Trashed," in *Arts of Living on a Damaged Planet: Ghosts and Monsters of the Anthropocene, eds.* Anna Tsing, Heather Swanson, Elaine Gan, and Nils Bubandt (Minneapolis: University of Minnesota Press, 2017), G51–G63

31. Robin W. Kimmerer, *Braiding Sweetgrass: Indigenous Wisdom, Scientific Knowledge, and the Teachings of Plants* (Minneapolis: Milkweed, 2013).

32. Vinciane Despret, *What Would Animals Say if We Asked the Right Questions?* (Minneapolis: University of Minnesota Press, 2016).

33. Tim Ingold, "Rethinking the animate, re-animating thought," *Ethnos, Vol.* 71 (1) (2006): 9–20;. and "Earth, sky, wind, and weather," *Journal of the Royal Anthropological Association Vol.* 13 (1) (2007): S19–S38.

34. A. Irving Hallowell, "Ojibwa Ontology, Behavior, and World View," in *Culture in History,* ed. Stanley Diamond (New York: Columbia University Press, 1960), 19–52.

35. Hallowell, "Ojibwa Ontology," 20; also see Maureen Matthews and Roger Roulette, "'Are all stones alive?': Anthropological and Anishinaabe approaches to personhood," in *Rethinking Personhood: Animism and Materiality,* eds. Miguel Astor-Aguilera and Graham Harvey (New York: Routledge, 2018), 173–92.

36. Povinelli, *Geontologies,* 157.

37. Led by Jone Salomonsen and funded by the Norwegian Research Council.

38. Nurit Bird-David, "Persons, or relatives?," 25–34; Danny Naveh and Nurit Bird-David, "Animism, conservation and immediacy," in *The Handbook of Contemporary Animism,* ed. Graham Harvey (London: Routledge, 2013), 27–37; and Bird-David and Danny Naveh, "Relational Epistemology," 55–73.

39. David Abram, *The Spell of the Sensuous: Perception and Language in a More-than-Human World* (New York: Vintage, 1996).

Indigenous Cosmovisions and a Humanist Perspective on Materialism

John Grim

"An economy that grants personhood to corporations but denies it to more-than-human-beings: this is a Windigo economy."[1]

"The original common condition of both humans and animals is not animality but rather humanity. The great mythical separation reveals not so much culture distinguishing itself from nature but nature distancing itself from culture: the myths tell us how animals lost the qualities inherited or retained by humans. Humans are those who continue as they have always been: animals are ex-humans, not humans ex-animals. In sum, the common point of reference for all beings of nature is not humans as a species but rather humanity as a condition."[2]

Beginning with Animating Language

A fundamental insight, evident in a vast majority of Indigenous languages, is that they are primarily based on verbals rather than nouns—verbal languages evoke and engage a living, dynamic world.[3] This broad statement about such diverse, Indigenous cultures affirms the possibility that the material world speaks, rather than simply being an objective, inanimate world waiting to be named or "nouned."[4] Verbal languages can also be said to anticipate, engender, and cultivate that sense of a living world. One learns the language of this living materialism by close listening and careful observation over time.[5] This agrees with what the Citizen Potawatomi botanist, Robin Wall Kimmerer, describes as a *grammar of animacy*:

To whom does our language extend the grammar of animacy? Naturally, plants and animals are animate, but as I learn, I am discovering that the Potawatomi understanding of what it means to be animate diverges from the list of attributes we all learned in Biology 101. In Potawatomi 101, rocks are animate, as are mountains and water and fire and places. Beings that are imbued with spirit, our sacred medicines, our songs, drums, and even stories, are all animate. The list of the inanimate seems to be smaller, filled with objects that are made by people. Of an inanimate being, like a table, we say, "*What* is it?" And we answer *Dopwen yewe*. Table it is. But of an apple, we must say, "*Who* is that being?" And reply *Mshimin yawe*. Apple that being is.[6]

Increasingly, many Indigenous peoples are speaking up for those living, endangered worlds that speak to them. Academics who study Indigenous religions and ecology are reporting on that resistance to environmentally damaging projects not simply as political resistance for natural, or spiritual entities. Rather, Indigenous environmental activism is increasingly framed as *protection*. This is a protection that interweaves a material and spiritual ensemble well-known among Indigenous cultures as lifeways, or lifeworlds,[7] that communicates itself in cosmovisions, namely, a holistic vision of acknowledged realities. An underlying point of this protection is interdependence: humans protect that which gives them life and which, in turn, humans give back in the diverse ways about which people tell stories. These traditions narrate *cosmovisions* with the reverence, respect, and reciprocity given to a living sacred expression of breath, words, and spiritual presences. This is a fundamental disposition in the immanent religiosity of Indigenous lifeways.

Around the world, Indigenous peoples are claiming, through their environmental activism, that they are part of communities of entangled beings that other worldviews separate into water, land, and biodiversity. In protection of that "entanglement," Indigenous communities object to extractive mining such as: the proposed Pebbles gold mine in Alaska; the endangerment of water by the Dakota Access Pipeline by Hunkpapa and other First Nations activists; the fracking of deep-rock gases in New York State by the Haudenosaunee Environmental Task Force; and the transport of fossil fuels through the Coastal Gaslink Pipeline across their lands by Wet'suwet'en peoples in British Columbia. These actions may not always halt these projects, but the activism flows from a long-term commitment to protect Native kinship with our entangled worlds.

Indigenous youth and elders come together to protect water against damming, and to speak against "wilderness" parks designed to exclude them. They are bringing their traditional religious and spiritual relations with land and biodiversity into contemporary environmental action as protection for those with whom they form the material world. It is this creative entanglement of materialism that Candace Slater calls the images and stories of *encante*: an enchanted place.[8] A "new materialism" is well-served by attending to this Indigenous materialism of *encante*. So also, the term, *cosmopolitics*, has emerged to speak of these forms of political ecology motivated by Indigenous cosmovisions.[9] This contemporary protection/resistance/resilience entanglement of Indigenous political sovereignty also manifests an ancient attention to prior environmental agreements.[10]

Indigenous Environmental Activism

Ancient rituals celebrate treaties, or agreements, between Native peoples and local land and biodiversity. Echoes and statements of affirmation of these agreements can be heard in such complex ceremonials as the Anishinaabe *Midewiwin*,[11] Tsistsistas *Masaaum*,[12] and Haudenosaunee *Thanksgiving Address*.[13] Many of these rituals renew the early movements of peoples and their request to animals and lands for permission to enter, live, and sustain themselves in a region.[14] The stories told during these rituals often constitute governance systems, legal relationships, and cosmological narratives—in total, a materialism of environmental survivance.[15] Not simply surviving on the land and animals, but an integral ecology that manifests an immanent religiosity of reverence, reciprocity, and respect among living beings here on Earth and throughout the cosmos.

With colonization by European powers, these integral ecologies were observed, subverted, and intentionally fragmented. But survivance—the entangled nurturing of living narratives—continued among many Indigenous communities. In some, such as the Kayapo in Amazonian South America, rituals celebrating relations with the cosmic powers in the natural world were used to resist developmental projects such as dams, water diversion, and forest removal for outsider agriculture.[16] In other instances, Indigenous rituals have been brought into alliance with non-Native environmental resistance. For example, the Sacred Pipe has been used by Anishinaabe and Lakota elders in solidarity with non-Native environmentalists as both fought extractive mining projects in northern Wisconsin,[17] northern Minnesota, and North and South Dakota.[18] The Lummi master carver, Jewel James, along with other Indigenous elders of the Northwest have brought lineage poles, often called totem poles, into environmental activism from 2015 to the present.[19] These alliances demonstrate a contemporary turn in ancient Indigenous cosmovisions, namely, recognizing survivance in the conjunction of social justice with environmental justice, or cosmopolitics.[20]

Tragically, some forms of conservation biology have led environmentalists to favor the removal of Indigenous peoples in the name of preserving habitats for endangered species. Yosemite and Yellowstone parks in the United States were cleared of Native peoples by this type of thinking.[21] In the forested regions of the Congo in Africa, in Amazonian South America, and in the forests of Southeast Asia, a "fortress conservation" mentality favored non-human species over resident Indigenous people.[22] However, many Indigenous communities resisted these land grabs in the same ways they had resisted lawless seizures of their lands by invading settlers. Gradually Nation-states, such as Gabon and Chad in Africa and Guyana in South America, as well as many global conservation organizations, acknowledged these land-grabs and undertook the designation of "protected areas" that pointedly referred to Indigenous peoples as more-than-equal players in protecting wildlife. Indigenous resilience in cosmopolitical struggles has been directly acknowledged as a sign of the enduring strength of their cosmovisions.[23]

Cosmovision

Cosmovision is presented as an experiential and conceptual narrative among differing Indigenous peoples for entering into communication with a world that speaks to them. Rather than simply an "origin story," cosmovision places a community in the wholeness of its vision of reality. Cosmovision provides a name for that "family resemblance" among core narratives evident in many Indigenous traditions, describing living relationships with land, biodiversity, and celestial formations. For Indigenous peoples, cosmovision allows beings in the world to tell their own stories; and, in the telling, manifest living presences in the world. These are narratives that do not separate the cosmos into good or evil, cultural or natural, human or non-human. While recognizing differences, Indigenous narratives and rituals strive for dynamic balances, and vital interactions.[24]

Cosmovisions are not simply metanarratives that impose a fixed truth. A cosmovision is not one true story, or oral scripture, to which individuals and communities commit themselves as forms of dogma or belief. Cosmovision indicates a path that unfolds in the telling; and that telling arises from the lips of the world. Cosmovisions live primarily in Indigenous oral forms that are plural and porous. Consquently, that orality is thoroughly entangled in place, personhood, and perspective as the world speaks to Indigenous visionaries. Traditional narratives orient the many, living, small-scale societies, called

Indigenous, in their changing relations and daily exchanges with the social organizations of biodiversity, with the flow of local ecologies, and the communications from distant stars.[25]

Indigenous

Often, the term Indigenous, is used to designate local, small-scale societies that have a distinct shared language, kinship system, and stories of their relationships with local lands and biodiversity. Many of these integrated societies remember their emergence or assembly from separate groups that coalesced. They came together as the agreement of differing peoples who created their shared language as they shaped their social coherence. Hybridity is often at the heart of Indigenous cosmovisions, reflecting language accretions and participation by local biodiversity in shaping the people. By the political reality of the colonialist period from the sixteenth to the twenty-first centuries, many of these societies were overwhelmed.

During this encounter period, Indigenous societies became increasingly dominated by non-Native religious narratives, the economic practices of settler-colonialism, and the politics of emerging Nation-states. Metanarratives of Manifest Destiny, and accompanying market economies, implemented capitalist agendas of acquisition and extraction. Karl Marx observed the "metabolic rift" of capitalist accumulation theories in which both nature and labor were removed from their sustainable context.[26] Sadly, communist regimes took a similar path as capitalist democracies—towards Indigenous nations plagued with misunderstanding, neglect, and forced cultural change. The history of relations between socialist and communist countries and Indigenous peoples is full of typically tragic narratives. Impositions of colonialist, imperialist, and economic agendas resulted, for example, in the Soviet Union's oppression of traditional shamanic practices,[27] state collectivization and subversion of ancient community subsistence practices,[28] and the environmental degradation of homelands.[29]

Throughout the Americas, the subversion of Native peoples often resulted in alienation and self-loathing as epidemics killed off community elders. The Pequot writer, William Apess, educated in Western normative education, expressed this despair in 1829:

> We form our opinions of the Indian character from the miserable
> hordes that infest our frontiers. These, however, are degenerate beings,
> enfeebled by the vices of society, without being benefited by its arts of
> living. The independence of thought and action, that formed the main

pillar of their character, has been completely prostrated, and the whole moral fabric lies in ruins . . . The forest, which once furnished them with ample means of subsistence, has been leveled to the ground— waving fields of grain have sprung up in its place; but they have no participation in the harvest; plenty revels around them, but they are starving amidst its stores; the whole wilderness blossoms like a garden, but they feel like the reptiles that infest it.[30]

Were Apess himself not a native Pequot, from the Connecticut region, it would be convenient to read this passage as simply describing the Indian as victim of the oppressive colonial regime. By the early nineteenth century, the New England states had thoroughly undermined Native lifeways. William Apess does critique some Native individuals as alienated and self-destructive, but what he deplored more deeply, was a nation living with abundance while totally ignoring its role in the loss of Native health, dignity, and purpose. Ancient cosmovisions celebrated a life of circular giving in reciprocity with the land, while spiritual starvation accompanied the centuries-long holocaust of Indigenous peoples in the Americas.

During this era, Indigenous cosmovisions were marginalized, erased, exploited, and co-opted by dominant actors.[31] Today, Indigenous leaders speak out with more force, with many allies in mainstream communities, realizing the limits of thoughtless development. Regenerating Indigenous voice, often associated with *indigeneity*, is clearly evident in the 2007 *United Nations Declaration of the Rights of Indigenous Peoples* (UNDRIP).[32] Yet, even this promising document locates Indigenous Nations in the sovereignty of Nation-states rather than a larger vision of an immanent and shared cosmological heritage of a living world. The shadow pattern of colonial domination and dehumanization, identified with the Christian doctrine of discovery, continues to validate the taking of Native lands and the marginalization of Native voice.[33]

The terms, Indigenous and indigeneity, are not unilaterally used by Native scholars. Some African, and South and Southeast Asian scholars, consider "Indigenous" as a colonialist carry-over. This term is seen as continuing the degradation of peoples and places. It does so, some protest, by presenting these communities as frozen in some "traditional" past. Some scholars also suggest that the term Indigenous confuses historical encounters in which local peoples were displaced, for example, in the Americas, with other regions where no such widespread displacement occurred, for example, in Africa, India, or Southeast Asia.[34] Several Nation-states especially object to the term, Indigenous, as an outsider perspective, separating out certain peoples as authentic or original to a place and excluding others who have been in country for extended periods of time. In some areas, such as the Americas and the Pacific region, Indigenous is an accepted term referencing Native societies and their diverse sovereign voices in distinction from settler immigrants who arrived over the past five hundred years.[35]

Lifeway as a Way of Knowing

Acknowledging these differences, Indigenous is used here to indicate peoples originally located in place before colonial expansion. They have often been voiceless in historical and contemporary economic, political, cultural, and environmental forums. Cosmovision enriches the voices of Indigenous elders by framing their views in a holistic perspective that is overtly recognized by Native peoples, especially in the American hemisphere. Indigenous ways of being-in-the-world are signaled by the term *lifeway*.[36] Moreover, phrases such as "Indigenous knowledge" and "traditional environmental knowledge" also open pathways into understanding lifeway as a way of knowing. As one Indigenous scholar has said:

> Perhaps the closest one can get to describing unity in Indigenous knowledge is that knowledge is the expression of the vibrant relationships between people, their ecosystems, and other living beings and spirits that share their lands . . . All aspects of knowledge are interrelated and cannot be separated from the traditional territories of the people concerned . . . The purpose of these ways of knowing is to reunify the world or at least to reconcile the world to itself. Indigenous knowledge is *the way of living* within contexts of flux, paradox, and tension, respecting the pull of dualism and reconciling opposing forces.[37]

That the immanental character of cosmovision is something woven into cultural life is what is meant by lifeway. Indigenous knowledge (IK) and traditional environmental knowledge (TEK) are terms that describe ways of listening and observing the materialism of the world embedded within cosmovision.[38]

Indigenous Humanism as a Perspective on a Materialism of Giving

What emerges in the current attention to Indigenous religions and ecology is a larger understanding of Indigenous rituals and oral narratives as a lifeway that acknowledges the personhood of materialism. By means of cosmovisions, Indigenous communities narrate and locate themselves in respectful sustenance relations with dimensions of their homelands, animals, and plants as persons.[39]

Indigenous humanism locates the human person within a community of other-than-human persons. In this sense, Indigenous humanism is not centered on the human-as-separate but rather on a hominizing perspective of persons-in-the-world seeing and being seen as humans.[40] The term, *anthropocosmic*, suggests the deep embeddedness of Indigenous cosmovisions in the flow of the world. This surfaces in the environmental protection of Indigenous peoples in ways that can be quite different than scientific ecology or political environmentalism. The phrase "Indigenous humanism" suggests that these micro-macro interactions are affirmed in their cosmovisions of a giving world.

First Nations scholars in a study of the Ojibway peoples' exchange with animals, observed that:

> . . . if non-foragers [humans] "gave gifts to the foragers without receiving gifts of food in return, they would shame not only the foragers but also themselves." Is it not also the case that the members of the Whitesands Indian Band would bring shame upon themselves if they stood by and did nothing while the habitat of those other-than-human persons with whom they exchange gifts is threatened or destroyed? After all it is through the exchange with gifts that one maintains one's membership in Ojibway society. Are not these other-than-human persons with whom they exchange gifts members of that society and entitled to the same respect and help accorded to any other member of the community? There is, we suggest, a moral obligation to protect the habitat of the moose, the beaver, the muskrat, and the lynx; the habitat of geese, ducks, grouse and hare, not just because members of the band wish to continue hunting and trapping, but because these other-than-human persons are also members of Ojibway society."[41]

The authors of this ethical statement interpret the reciprocal exchanges of humans with animals in the context of the moral obligation required between fellow members of a shared world. Their understanding of societies of different beings extends into the natural world in which gift-giving marks their relationships. Narrating the origins and procedures of this material gift-giving is cosmovision. Narrating cosmovisions are forms of Indigenous environmental ethics acknowledging the entanglements of their lifeways into protection of that giving world.

The contemporary environmental protection of Indigenous peoples, concerning environmental projects seen as harmful to lifeway relationships, has entered into national and international news. In fact, this activism is as old as the resistance to colonial subversion. Indigenous leaders resist such environmental disasters as tar sands petroleum extraction in Alberta, Canada even as these projects give jobs to members of their own communities. There are the seemingly endless pipelines proposed across Indigenous lands, and the allure of huge financial windfalls that tempt small coastal Indigenous communities to allow terminals to process and ship coal to foreign countries. Awareness of these conflicts and ambiguities awakens outsiders to the complexity of Indigenous protection / resistance / resilience.

Questions surface for sympathetic outsiders who ask: How did cosmovisions preserve Indigenous knowledge through such extended periods of cultural oppression and depravation? Similarly, we now find new perspectives arising from questions about the ways in which humans became alienated from their surrounding worlds, into a materialism of commodification which is causing environmental devastation. One new academic approach, environmental humanities, asks both how humans dissociated themselves from the context of life and what a reconnection might look like.

Indigenous humanism provides perspectives on a materialism that is neither relativism nor essentialism. Diverse forms of Indigenous humanism bring ancient, ecological wisdoms expressed in gratitude, practiced in ritual cycles of giving, along with historical memories of resistance to the thoughtless destruction of the communities of life. Indigenous humanism brings awareness of relational solutions to environmental challenges that are embedded in the nature of their cultural agreements with material reality itself.[42] Cosmovisions affirm humans, along with biotic and abiotic life, as having an entangled future. Still, many questions remain unanswered: Can non-Native humans see the sixth extinction of biodiversity in this age of the anthropocene as a loss of that which is giving life? Can we hear both scientists and Indigenous elders when they offer empirical observations regarding climate change? Can humans feel what the material worlds are saying?

Notes

1. Robin Wall Kimmerer, *Braiding Sweetgrass: Indigenous Wisdom, Scientific Knowledge, and the Teachings of Plants* (Minneapolis: Milkweed, 2013), 376. "Windigo" is a cannibal monster in Anishinaabe stories who eats uncontrollably.

2. Philippe Descola, *La nature domestique*, cited in Eduardo Viveiros de Castro's, "Cosmological Deixis and Amerindian Perspectivism," *The Journal of the Royal Anthropological Institute*, Vol. 4 (3) (September 1998): 469–488.

3. Gary Witherspoon, *Language and Art in the Navajo Universe* (Ann Arbor: University of Michigan Press, 1977).

4. Julie Cruikshank, *Do Glaciers Listen: Local Knowledge, Colonial Encounters, and Social Imagination* (Vancouver and Toronto: University of British Columbia Press, 2005).

5. Eduardo Kohn, *How Forests Think: Toward an Anthropology Beyond the Human* (Berkeley: University of California Press, 2013).

6. Kimmerer, *Braiding Sweetgrass*.

7. Tim Ingold, *Perception of the Environment: Essays in Livelihood, Dwelling, and Skill* (New York: Routledge, 2000).

8. Candace Slater, *Entangled Edens: Visions of the Amazon* (Berkeley: University of California Press, 2001).

9. Willis Jenkins, Mary Evelyn Tucker, and John Grim, eds., *Routledge Handbook of Religion and Ecology* (London and New York: Routledge, 2013).

10. Indigenous Environmental Network: ien.org.

11. Sound of the drum. See "North America: Native ecologies and cosmovisions renew treaties with the earth and fuel indigenous movements," in *Routledge Handbook of Religion and Ecology*, eds. Willis Jenkins, Mary Evelyn Tucker, and John Grim (London and New York: Routledge, 2013), 138–147.

12. Masked animal dance. See Karl Schlesier, *The Wolves of Heaven: Cheyenne Shamanism, Ceremonies, and Prehistoric Origins* (Norman: University of Oklahoma Press, 1987).

13. Tom Porter, *Kanatsiohareke: Traditional Mohawk Indians Return to Their Ancestral Homeland* (Greenfield Center: Bowman Books–Greenfield Press, 1998).

14. John Grim, *The Shaman: Patterns of Religious Healing among the Ojibway Indians* (Norman: University of Oklahoma Press, 1983).

15. Gerald Vizenor, *Manifest Manners: Narratives on Postindian Survivance* (Lincoln: University of Nebraska Press, 1999).

16. Darrell Posey, *Indigenous Knowledge and Ethics: A Darrell Posey Reader*, ed. Kristina Plenderleith (New York & London: Routledge, 2004).

17. Al Gedicks, *The New Resource Wars: Native and Environmental Struggles against Multinational Companies* (Boston: South End Press, 1993).

18. *Idle No More Manifesto*: idlenomore.ca/manifesto; *Indigenous Environmental Network*: ien.org.

19. https://totempolejourney.com.

20. Jace Weaver, ed., *Defending Mother Earth: Native American Perspectives on Environmental Justice* (Maryknoll: Orbis Books, 1996).

21. Justin Farrell, *The Battle for Yellowstone* (Princeton: Princeton University Press, 2015).

22. Mark Dowie, *Conservation Refugees: The Hundred-Year Conflict between Global Conservation and Native Peoples* (Cambridge: MIT Press, 2009).

23. Jesse Mugambi, "Africa: African heritage and ecological stewardship," in *Routledge Handbook of Religion and Ecology*, eds. Willis Jenkins, Mary Evelyn Tucker, and John Grim (London and New York: Routledge, 2013), 109–119.

24. Thomas King, *The Truth About Stories: A Native Narrative* (Minneapolis: University of Minnesota Press, 2003).

25. Timothy McCleary, *The Stars We Know: Crow Indian Astronomy and Lifeways* (Prospect Heights: Waveland Press, 1996).

26. Karl Marx and Frederick Engels, *Collected Works, Volume 37* (New York: International Publishers, 1975), 732–33.

27. Vilmos Dioszegi, *Tracing Shamans in Siberia* (New York: Humanities Press, 1968), 14.

28. Rane Willerslev, *Soul Hunters: Hunting, Animism, and Personhood Among the Siberian Yukaghirs* (Berkeley: University of California Press, 2007).

29. Piers Vitebsky, *The Reindeer People: Living with Animals and Spirits in Siberia* (New York: Houghton Mifflin Company, 2006).

30. William Apess, *On Our Own Ground: The Complete Writings of William Apess, A Pequot*, ed. Barry O'Connell (Amherst: University of Massachusetts Press, 1992), 61.

31. Vine Deloria, "Trouble in High Places: Erosion of American Indian Rights to Religious Freedom in the United States," in *The State of Native America: Genocide, Colonization, and Resistance*, ed. M. Annette Jaimes (Boston: South End Press, 1992), 267–290.

32. UNDRIP: http://www.un.org/esa/socdev/unpfii/documents/DRIPS_en.pdf.

33. Steven Newcomb, *Pagans in the Promised Land: Decoding the Doctrine of Christian Discovery* (Golden: Fulcrum Publishing, 2008).

34. Ken S. Coates, *A Global History of Indigenous Peoples: Struggle and Survival* (Basingstoke, UK: Palgrave Macmillan, 2004).

35. Christian Erni, ed., *The Concept of Indigenous Peoples in Asia: A Resource Book* (Copenhagen, Denmark, and Chiang Mai, Thailand: International Work Group for Indigenous Affairs and Asia Indigenous Peoples Pact Foundation, 2008).

36. Similar to Tim Ingold's use of *lifeworld*: Ingold, *Perception of the Environment.*

37. Marie Battiste and James Henderson, *Protecting Indigenous Knowledge and Heritage* (Saskatoon, SK: Purich Publishing, 2000), 35.

38. Fikret Berkes, *Sacred Ecology: Traditional Ecological Knowledge and Resource Management* (Philadelphia: Taylor & Francis, 1999).

39. Melissa Nelson, ed., *Original Instructions: Indigenous Teachings for a Sustainable Future* (Rochester, VT: Bear and Company-Inner Traditions, 2008).

40. Eduardo Viveiros de Castro, "Cosmological Deixis:" 469–488.

41. Dennis McPherson and J. Douglas Rabb, *Indian from the Inside: A Study in Ethno-Metaphysics* (Thunder Bay, Ontario: Lakehead University, Centre for Northern Studies, 1993), 90.

42. Akwesasne Notes, ed., *Basic Call to Consciousness* (Summertown: Native Voices, 2005).

Amorous Entanglements: The Matter of Christian Panentheism

Catherine Keller

The Incarnation has designated the single event in which transcendence and immanence fused in a body. Its singularity has tragically backfired. It soon became the exception that proves the rule—of a transcendent sovereignty over the world, an ontotheological over-and-above, condescending to intervene here and there while summoning its, no, *His*, subjects to join Him eschatologically in an ultimate transcendence of Earthly matter. Bound for Heaven, no longer Earth-bound, Christians were to leave the ecologies and economies of the planet to the Lord; more specifically, to his lordly minders.

I have not summarized the history of Christianity. I have offered a caricature of the betrayal of a messianic moment, a surrender to Caesar, massive in its institutions, religious and secular, insidious in its parasitism of the very flesh and blood of its messiah. I am tempted to counter the Incarnation with *inter*-carnation. To pit a relational materialization of immanence against the exit strategy of transcendence.[1] But such a binarism cannot help but perform its own exceptionalism. The nickname intercarnation, as syntagma and as symbiosis, remains immanent to a particular globalized history. It feeds on the Christian narratives of the particular incarnation. Since it performs its own christology it cannot refuse responsibility for the capitalist/climate apocalypse that a christological exceptionalism has done much to fund. That Incarnation has invested itself in a dynamic, modern set of exceptionalisms: national, racial, sexual, and capitalized; all folding into the ever-renewed *human* exceptionalism that now poses the primary problem for the human future. The present garish form of American exceptionalism has resulted in our truly exceptional failure to manage the materiality of COVID-19. In this essay, I will explore the political theology of that exceptionalism regarding the tense relation of the new materialism's sheer immanence with panentheism's traces of a troubled transcendence.

Beyond and beneath every exceptionalist one-off, the intercarnation signals, not mere opposition, but an up-close juxtaposition to the narrative of the single incarnation. In that one has always carried a potentiality immanent to Christianity and exceeding, which is to say *transcending*, its sovereign exclusivisms: the possibility of a radical redistribution of divinity, and of the entangled materiality, old and new, of its "Heaven and Earth." The very 'in' of the incarnation signals its interactivity with all flesh, all materiality. Such a cosmos of entanglement endlessly precedes the life of Jesus, even as it exceeds its every iteration. The captures and incorporation of incarnation as *Christology, Inc.* have always required imperial conversion, and its liberation to interdependence with all bodies requires something very like a political theology of the Earth.[2]

The exception, *excipere*, morphs, in such a political theology, into the *inception*. The divine appears all-in *(pan-en)*. A come down from traditional transcendence. It is not coming down to immanence that stakes its radicality on the moment of a post-Christian take-out—a transcendence of transcendence. It appears too ecopolitical, too impure in its movements for that. In this time of accelerating climate warming and Earth waste, there is no time to squander such resonant resources as the prophetic exposé of the empires of oppression. The ancient possibility of a creation-honoring justice, of a revolutionary love, still trickle around the planet from the Judeo-Christo-Muslim sources of messianism. Might we recycle that possibility more efficaciously? Do its justice, its renewal of the face of the Earth? Of course, we do not know what to do, how to do it, and whether it will matter. But amidst the uncertainty, it may be this amorously charged entanglement that turns the inescapable interlinkages of our history, our context, our mutual immanence, into places of planetary hope.

Hope? As I write this, harrowing elections come and go, leaving us to face the mounting improbability of addressing climate change. The transnational sovereignty of neoliberal capitalism holds the drivers of extinction in place. The crises of human injustice—race, class, immigration—distract democratic politics from the subtle degrees of global warming. The improbable threatens to harden into the impossible. We find ourselves, selves of any or no religion, circulating in the ambience of the biblical apocalypse. As time runs out, we have the option of taking ourselves out—by way of Christian right denialism or of honest progressive nihilism—and so by default, taking out the human future. Or? of breaking through the determinism of most apocalypticism, in a revelation, *apokalypsis*, disclosure, not closure, that signals inception.

Either way, we face the dense entanglements of a history from which there is no escape. There is no exit from our entangled becoming—in its radical immanence. And, in its immanence, opens the perspective of the *to come*. In its Derridean "messianicity" that *coming* becomes indubitably invested in a certain transcendence: not in escape, but in a hope without guarantee, the possibility of that other, unconditional condition or spectral *arrivant*, yet to come. If it comes from an eschatological exterior, it is tempted to view itself as the pure alterity of an absolute difference, divine or otherwise. Absolute difference perches at the opposite pole to absolute immanence, along the poststructuralist spectrum. The temptation to a transcendence of pure otherness might be Abrahamic or it might be deconstructive. Indeed, it might read out as an auto-deconstruction of theological transcendence.

However, as it comes unbound—even in the name of justice—from the flesh of the Earth, from intercarnational immanence, we might resist the temptation. A *coming* in its very transcending, its moving beyond, iterates Walter Benjamin's messianic moment as *ungeheueren Abbrevatur*: the uncanny recapitulation of an immense, monstrous, history.[3] It evokes not a moment out of history, but of history's deep *within*. It opens to a beyond which does not take place (if it does take *place*) outside of its relations, but only within new ones— relations to that material history, in its planetary diaspora, its world. Its recapitulation of its world, like the Deleuzian repetition with a difference, exceeds what has been, precisely by enfolding it otherwise.

From the impure perspective of that self-exceeding inside, one is no longer capable of the absolute, either of pure immanence or pure transcendence: the radicality of the *to come* springs from the *radix* of becoming. The radical is not *the pure*. It is the base matter of becoming. From it, beginnings repeatedly and differently take place. The beginning shoots up from a root, not as a vertical singularity, but from a root system, or a rhizome, horizontally spreading, deep or shallow, below the ground.

In other words, if the messianic *to come*, with its peculiarly attractive force of transcendence, means separation from the mottled, material practices of Earthly becoming, it will continue to betray its own chances. It might as well stick with the one-off *Christology, Inc.* And then the forcefield of Earthly disregard, initiated in otherworldly terms, can continue to waste the planet, and do so with the untouchable sovereignty of its secularizations. It would have the last word: the logos of the end of the world. The messianic option would have been (might still be) to bring its Christo-logos into resonance with all things Earthly. An intentional relation to the *logos*, the *principium*, as it already, everywhere, unfolds in all, *pan*, intensifies the creativity of our becomings. (John 1. 2 does not read that all things were created through the logos, but *ta panta egeneto*, through the logos "all *became*").

In contrast to an exceptionalist incarnation and return, the *to come* is read as the messianic calling of our immanent becoming, its transcendence is no longer that of a supersessionist descent or of a supernatural future, of an abstraction, or an extraction from the Earth. In its "movement beyond" (*transcendere*), it crosses historical, not supernatural, boundaries. Such transcending would precisely signify the transformative potentiality of immanence: the capacity of entangled bodily becomings to exceed what precedes them. In a language of political theology, they become sites, not of the sovereign exception, but of the possible inception.

An *Earth becoming*, that shoots up vegetally, can no longer be opposed to the *messianic coming* that in Walter Benjamin "flashes up." In that case, immanence has become hospitable to a transcendence that therefore ceases to evacuate materiality.[4] It is no accident that a great thinker of the new materialism, Karen Barad, works the Benjaminian opening of that flashing novum into a current constellation of science, politics, and Jewish mysticism. Benjamin himself, tapping Isaac Luria's notion of praxis and the revolutionary possibilities for redemption, read his kabbalah through a lens of Marxism. The repair of the world (*tikkun olam*) signifies an immanent praxis, shot through with, and made possible by, the flashing up of the infinite from within the finite. Here is the dramatic swerve Barad contributes, "The messianic—the flashing up of the infinite, an infinity of other times within this time—is written into the very structure of matter-time-being itself."[5]

Barad then wires this messianicity, conducting its political charge into contemporary crises of race, immigration, and climate, right into her own game-shifting reading of quantum entanglement. Quantum ". . . entanglements are not intertwinings of separate entities, but rather irreducible relations of responsibility. There is no fixed dividing line between 'self' and 'other,' 'past' and 'present' and 'future,' 'here' and 'now,' 'cause' and 'effect.'" Barad clearly states that no reduction of difference to a single or a universal One arises from the nonseparability of the quantum ones. "Entanglements are not a name for the interconnectedness of all being as one, but rather specific material relations of the ongoing differentiating of the world. Entanglements are relations of obligation—being bound to the other—enfolded traces of othering. Othering, the constitution of an 'Other,' entails an indebtedness to the 'Other,' who is irreducibly and materially bound to, threaded through, the 'self'—a diffraction/dispersion of identity." Here, Haraway's diffraction supports Derridean difference. "'Otherness' is an entangled relation of difference (*différance*). Ethicality entails noncoincidence with oneself."[6]

My book, *Cloud of the Impossible*, hosts a dense discussion of quantum entanglement, involving Barad, along with several other physicists who might count as "new materialists" in their push beyond the old materialism locked into a mechanical immanence. In *Cloud of the Impossible* I note that the quantum nonlocality hints at spatiotemporality of such radically interlaced becomings as to demand theological contemplation. What flashes up with messianic incipience exposes, in its mystical bottomlessness, Jewish and more.

In the tension of the becoming and the to come, I am creeping around to the question of panentheism: the all-in-God. Does it carry the radicality of immanence, or rather the authority of the Abrahamic West? As for the latter, speaking with sanction for "the sibling rivals of the family of Abraham" (A.O. Miller), or even one of them, whoever does? Panentheism does speak for a developing network, indeed a living rhizome among the minority discourses, often not Christian, directly, or indirectly related to Whiteheadian thought. As a word, panentheism, marked by its 'en,' conveys the immanent explicitly. In this it differs from pantheism, which communicates an identity more than an interiority.

One could say that panentheism is more radically, if less purely, immanent than pantheism, but it is precisely the need to oppose the two that can strip panentheism of its radicality. It may be a crucial, rhetorical move vis á vis the institutions of monotheism, however, as we learn from Mary-Jane Rubenstein, the theological habit of marking oneself off from pantheism continues a bizarre tradition of demonizing a barely extant constituency. Whatever else we are, we are *not that*, that pantheism, that confusion of God and world, or worse, God and pan, that goat, that monstrous stinky hybrid of divinanimality; and that heresy—as Spinoza's attackers shuddered—gives God a body.[7] Of course, panentheism cannot mark itself off from that heresy anymore, since such different thinkers as Charles Hartshorne, Sallie McFague, and Ivone Gebara have yoked their panentheism to the metaphor of the universe as the body of God. It is precisely demarcated from a simple identity of divinity and world, which would collapse the zone of immanence—its *in*—and thus disallows any relationship *between*.

Unless one's materialism dictates a simple identitism of mind and body—a reduction not at all implied by so-called new materialists—God's having, or indeed being body, does not reduce God to a finite or bound entity. Matter itself has morphed into a forcefield of materializations, of events of embodiment in "agential intra-action," admitting of no final boundary of space or time. The vibratory indiscernibility of matter and energy in twentieth century physics began, in Whitehead, to disclose its theological implications, effecting a universe of vibratory interrelation in which God serves, not as first cause, otherworldly exit, or final terminus, but as the eros that teases forth new becomings and takes them in—the inception as once beginning and as interiorization.

The mutual immanence of God and world reflects, indeed exemplifies the mutual immanence by which Whitehead defines all actualities as relations between becomings, and so as materializations:

> Every actual occasion exhibits itself as a process: it is a becomingness. In so disclosing itself, it places itself as one among a multiplicity of other occasions, without which it could not be itself.[8]

Whitehead's actual occasion thus transmutes the enduring substance—*res cogitans* or *extensa*—of separable individuals into the relationally constituted moments of becoming. This "becomingness" instigates a profound shift (not a transcendence) of ontology. Relations are no longer external. It is a matter, indeed a materialization, of "mutual immanence" or of "internal relations," constituting emergent subjects (superjects) rather than of attributes possessed by substances. For Whitehead, every subject—quantum, queen, queer— experiences and responds spontaneously to its world. Each process of becoming counts as an affective materialization of its world. In other words, relation no longer signifies an interaction between beings that existed before the interaction itself.

The resonance with Karen Barad's innovative language of "intra-active becomings" is striking. She has composed—drawing on a different, later, and largely continental philosophy—a full-fledged "relational ontology" as the basis for her "posthumanist account of material bodies." These bodies do not appear as classical agents, merely *inter*acting with their objects from the outside. "Rather, phenomena are the ontological inseparability/entanglement of intra-acting 'agencies.'"[9] The cogent, indeed urgent, applicability to matters of *climate weirding* is close at hand. For an ethics of responsibility is built from the cosmological bottom up, by way of an ability to respond, into human relations to the nonhuman. In *Meeting the Universe Halfway* Barad asserts that, ". . . we are always already responsible to the others with whom or which we are entangled, not through conscious intent but through the various ontological entanglements that materiality entails."[10]

Put differently, immanence would signify a mutuality of intra-action, not that of an interaction between preexistent entities, because the actual entity takes place in the time, the moment, of the becoming. This is different than immanence understood as an identity of all, or as a totality, or even as a single encompassing matrix. Its encompassing "matricial, khoric, spatiality" is itself being constituted moment by moment everywhere. Only abstractions of itself can form an immanence that would "contain" all becomings. No, there is nothing actual beyond this immanence. In this sense, an immanence of mutual entanglement is more radical—more elementally constitutive—than an immanence of unitary inclusion, which would always transcend its constituents. Or certainly than an immanence of total identity.

Nonetheless, the panentheist deity does not lack transcendence, hence Whitehead's chiasmus, ". . . it is as true to say that God transcends the World as that the World transcends God."[11] I hope it is clear that this is precisely not the take-out transcendence of metaphysical alterity, but the transcending that is an exceeding, a going beyond, not as supersession—for the past is carried forward within it—but as the effecting of "novel becomings" to which, then, this divine Eros has the pleasure and the pain of relating. A wide variety of process theological texts have made the case for the greater fidelity of this panentheos to the biblical legacy than can be found in the dualistic versions of classical theism. As process theology has developed over decades, it thinks in rigorous service to the ecology and economy of the Earth; as, then, the panentheism of God's body in McFague and Gebara becomes ecofeminism. The process proclivity to activism finds amplification in the important engagement of Whitehead by secular theorists associated with the new materialism; notably Jane Barad and Donna Haraway, and most explicitly, William Connolly.[12]

There is, however, an argument against the panentheistic deity that comes, not from the side of classical theisms or pantheisms, but rather from the discourse of the *to come*, of the Derridean sense of alterity as developed by John Caputo, "Panentheism says that God's existence is in ours and ours in God's. But in a theology of 'perhaps,' God does not exist; God insists, and it is our responsibility to bring about something that exists." Caputo has somehow missed the key process-theological dynamism of becoming, in which there is no flat and given existence, not of any actual entities, creaturely or divine. There is only the process of intertwined actualization in a moment. He then goes on to make exactly the sort of argument against the theodicy that panentheism has already made, rightly noting that the "hoary theological 'problem of evil' thus has nothing to do with all the choices that a sovereign, omnipotent, and omniscient God could have made but failed to make. . ." He does this as though it falls in a continuum with the critique of panentheism, without acknowledging the nearly century-long history of Whiteheadian deconstruction of precisely that sovereignty. And, still on the same page, announces that "the insistence of God is the chance that God can happen anywhere . . . the existence of God is liable to break out at any time. . ."[13] The breaking out suggests the messianic eventiveness of Barad's "flashing up." But how can one possibly read this proposition as complying to his announcement of God's simple nonexistence?

The existence of any bodies, of God's body, or indeed of God as the eros of possibility in Whitehead, exists to debunk any standard ontological or onto-theological existence. Indeed, he performed, nearly a century ago, the decon-struction of any substantialist, self-identical, transparently present, flat notion of existence—the notion that Caputo is presuming, wrongly projecting onto panentheism, but rightly rejecting. Admittedly, when process theology is in Sunday School mode it may not emphasize the becoming eventiveness of its divine existence, nor its interdependence with the world, for as Caputo says, "God depends upon us to exist." And, as Whitehead put it, God is ". . . creativ-ity's first creature." But here, the dependence is not just "upon us" and our human constructions. God emerges in relation, in the immanence of mutual participation in and with all that becomes. The aspect of God that precedes and exceeds the cosmos of bodies (the primordial as distinct from the conse-quent nature) does not, in Whitehead, exist but is pure possibility.

Panentheism remains in this way, in tension, at times quite creatively, with the messianicity of deconstruction, but closer, ironically, to a forthrightly god-less immanence—to a philosophy of radical immanence and becoming, such as that of Deleuze. This mutual immanence does, however, materialize some-thing like the "plane of immanence" that Deleuze lays out as thinking itself, "Concepts are like multiple waves, rising and falling, but the plane of imma-nence is the single wave that rolls them up and unrolls them."[14] It would in this, not accidentally, mimic the enfolding *complicans* of Cusa's infinite, and its unfolding of all, pan, "in and as the multiple," but only as all things are in all things, and each in each, as mediated by the universe. Cusa is my prime, pre-Whiteheadian ancestor for panentheistic immanence, naturally accused of pantheism in his day.[15]

The thinking of the plane of immanence does not restrict itself to any thought, mind, or indeed epistemic register: it is "movement that can be carried to infinity." It seems to signify, or to enliven, the relational dynamic of becom-ing. As Deleuze posits, "One does not think without becoming something else, something that does not think—an animal, a molecule, a particle—and that comes back to thought, and revives it. The plane of immanence is like a section of chaos and acts like a sieve."[16]

Deleuze makes clear that, ". . . wherever there is transcendence, vertical Being, imperial State, in the sky or on Earth, there is religion; and there is Philosophy whenever there is immanence," however Greek and agonistically, still among friends and as a "ground from which idols have been cleared."[17] This grounding immanence can be said to perform its own transcending of all sovereign transcendence—in a moving beyond, a becoming that carries beyond any *fin*, an infinite movement in and through the molecules of matter, the geophilosophy of Earth. In the newer materialisms, the Deleuzian influence remains ferocious. Many forms of it operate in and out of religion, with little friendship for theology, as in the ecologically energized work of Brian Massumi or Timothy Morton. Yet the agonism is fruitful, if it veers from antagonism, and all the more so when, with the help of such nontheists as William Connolly, Jane Bennett, and Karen Barad, alliances with theological traditions become more manifest on the plane of immanence. The chaosmic plane becomes constituted by the mutual participation of multiple religions and irreligions engaged in planetary projects of ecopolitical sanity. The mutual immanence of multiple spiritual traditions flashes up in the work of this very volume—inviting difficult crossings, at the same time, between multiple methods of religious practice; Indigenous, interreligious, secularized.[18] Great future vibrancies of the Earth in and as its vulnerable populations remain—whatever anthropocene proportions of planetary life and elemental integrity will not be healed—possible.

And yet, do we abandon what cannot be healed? Here, another dimension of "thinking within the planet" presses into our conversation. Reflecting on why the call for new materialism has become urgent, Mel Chen lands, ". . . well beyond rejecting either secularism or spirituality," instead wishing ". . . for an ethics of care and sensitivity that extends far from humans' (or the Human's) own borders. It is in queer of color and disability/crip circles, neither of which has enjoyed much immunity from the destructive consequences of contemporary biopolitics, that I have often found blossomings of this ethic . . . queerings of objects and affects accompanied by political revision, reworldings that challenge the order of things."[19] Chen's "interarticulate" reflection on her own, unhealable dis/ability—and usually what is termed disability is precisely not subject to remediation, incapable of being returned to the norm—has opened her to an interdependence, fluidity, and attention to the inanimate as the site of "mattering."

One might read Chen alongside Sharon Betcher's pneumatological "prosthesis," with its call for "intercorporeal generosity," an immanent ethics of the "obligation of Social Flesh." Under the banner of "the ruin of God," she writes, "By bringing persons close to flesh, one is reminded of the basic passage of a world of becoming, and we thus can begin to break through the Western effort to standardize the body, to value and offer status only to bodies so tightly normalized."[20] As the codes of the normalization of ability, sexuality, and coloniality break down, the great accompanying breakdowns become apparent. "As with earlier death of God or radical theologies, God here—in this disability theology—becomes ruined, emptied, that nothing-something, so that we are face-to-face with each other—with the sensual flush of sentience and its precarious vulnerability, its injurability."[21] Elsewhere, Betcher suggests that the metaphor and the practice of "disability" prepare us to deal with the growing urban and ecological damage of the Earth itself.[22] In other words, mindfulness of crip/tography may help us to prepare for the difficult adaptions required on a feverish planet, with its mounting multiplication of disabled ecosystems.

God and all the other ruins—all the dis/abled, denormalized, bodies of Earth—pose a fragile picture of the imminent future. An immense solidarity if we might work our mutual immanences, nonetheless. Or all the more so, the eschatological future-past of the new Heaven and Earth emerged only amidst unbearable injury. Its *to come* once carried the royal memory of Davidic anointing. If its sovereignty has gone to ruin, its "weak messianism" flashes up in new ways (even Deleuze invoked the "new Earth and new people"). When our plane of immanence moves toward such a future, a particular past comes enfolded. So, we keep thinking with the Hebrew *hashamayim*, the heavens, not of a supernatural beyond or metaphysical transcendence, not a separate plane, but a watery atmospheric plurality of the mysterious, darkly luminous expanse of the world—beginning with the first meaning, "sky."

We might call it "the new atmosphere and Earth"—requiring in its shifting, subtle materiality a new atmosphere of us—an atmosphere of impure immanence in radical ecodemocratic alliance. Among academics, that may entail conversations like ours, here, now. New crisis will remain imminent, requiring a new thinking that energizes, rather than depletes, the human planetary future; that situates it in an Earth and atmosphere severely wounded, multiply dis/abled, yet incapable of promising, animating, inceptions. Of the "multispecies becoming-with" Donna Haraway proposes *Terrapolis*, ". . . an open, worldly, indeterminate, and polytemporal" place. It plies not just any new materialism, "Terrapolis is rich in world, inoculated against posthumanism but rich in compost, inoculated against human exceptionalism but rich in humus, ripe for multispecies storytelling."[23] Responsible entanglements come to be as the multiple, embedded exceptionalisms come undone.

A political theology of the Earth will gather together the ecological forcefield of multiple philosophies and religions for the sake of their deglobalizingly, Earth-enfolded potentialities. Whitney Bauman's "polyamory of place" signals the disclosive mood and attractive force, beyond the fear of doom, of such immanent ingathering.[24]

I hope I have sketched, if in an expressionist manner, waves of a Western panentheistic reformulation of doctrines—God-creation-incarnation-eschaton—amorously offering their unfinished histories to whatever is to come. May it come becomingly.

Notes

1. Catherine Keller, *Intercarnations: Exercises in Theological Possibility* (New York: Fordham University Press, 2017).

2. Catherine Keller, *A Political Theology of the Earth: Our Planetary Emergency and the Struggle for a New Public* (New York: Columbia University Press, 2018).

3. Walter Benjamin, "Theses on the Philosophy of History," in *Illuminations: Essays and Reflections*, ed. and intro. Hannah Arendt (Boston: Mariner Books, 2019).

4. Ibid.

5. Karen Barad, "What Flashes Up: Theological-Political-Scientific Fragments," in *Entangled Worlds: Religion, Science, and New Materialisms*, ed. Catherine Keller and Mary-Jane Rubenstein (New York: Fordham University Press, 2017), 63.

6. Karen Barad, "Quantum Entanglements and Hauntological Relations of Inheritance: Dis/continuities, SpaceTime Enfoldings, and Justice-to-Come," *Derrida Today*, Vol. 3 (2) (2010): 265.

7. Mary-Jane Rubenstein, "The Matter with Pantheism: On Shepherds and Hybrids and Goat-Gods and Monsters," in *Entangled Worlds: Religion, Science, and New Materialisms*, ed. Catherine Keller and Mary-Jane Rubenstein (New York: Fordham University Press, 2017). See also her full-length study, *Pantheologies: Gods, Worlds, Monsters* (New York: Columbia University Press, 2018).

8. Alfred North Whitehead, *Science and the Modern World* (New York: The Free Press, 1997), 175–76.

9. Catherine Keller, "Tingles of Matter, Tangles of Theology: Bodies of the New(ish) Materialism," in *Intercarnations*, 60–82. (Also published in *Entangled Worlds*.) I also engage Barad in relation to the whole history of quantum entanglement in Chapter 4 of *Cloud of the Impossible: Negative Theology and Planetary Entanglement* (New York: Columbia University Press, 2014), 127–167.

10. Karen Barad, *Meeting the Universe Halfway: Quantum Physics and the Entanglement of Matter and Meaning* (Durham: Duke University Press, 2007), 393.

11. Alfred North Whitehead, *Process and Reality: Corrected Edition*, eds. David Ray Griffin and Donald W. Sherburne (New York: The Free Press, 1985), 348.

12. See especially the works of political philosopher William Connolly that explicitly engage Whiteheadian thought: *A World of Becoming* (Durham: Duke University Press, 2011); *Climate Machines, Fascist Drives, and Truth* (Durham: Duke University Press, 2019); and *The Fragility of Things: Self-organizing Processes, Neoliberal Fantasies, and Democratic Activism* (Durham: Duke University Press, 2013).

13. John D. Caputo, *The Insistence of God: A Theology of Perhaps* (Bloomington: Indiana University Press, 2013), 49.

14. Gilles Deleuze and Felix Guattari, *What Is Philosophy?* (New York: Columbia University Press, 1994), 36.

15. Barad, "Enfolding and Unfolding God: Cusan *Complicatio*," in *Cloud*, 87–123.

16. Deleuze and Guattari, *What Is Philosophy?*, 42.

17. Ibid., 43.

18. John Grim and Mary Evelyn Tucker, *Ecology and Religion* (Washington, D.C.: Island Press, 2014). See also Mary Evelyn Tucker and John Grim, eds., *Living Cosmology: Christian Responses to Journey of the Universe* (Maryknoll: Orbis, 2016).

19. Mel Y. Chen, *Animacies: Biopolitics, Racial Mattering, and Queer Affect* (Durham: Duke University Press, 2012), 237.

20. Sharon V. Betcher, *Spirit and the Obligation of Social Flesh: A Secular Theology for the Global City* (New York: Fordham University Press, 2014), 108.

21. Sharon V. Betcher, "Crypt/ography: Disability Theology in the Ruins of God," http://www.jcrt.org/archives/15.2/betcher.pdf, 113.

22. Sharon V. Betcher, "Of Disability and the Garden State," Spotlight on Theological Education. *Religious Studies News* (2013).

23. Donna J. Haraway, *Staying with the Trouble: Making Kin in the Chthulucene* (Durham: Duke University Press, 2016), 11.

24. Whitney Bauman, *Religion and Ecology: Developing a Planetary Ethic* (New York: Columbia University Press, 2014).

On the Matter of Hope: Weaving Threads of Jewish Wisdom for the Sake of the Planetary

O'neil Van Horn

Introducing the Impossible: "Hope" Amid the Anthropocene

The Anthropocene and its associated climate injustices challenge any viable concept of hope. Mass extinction, ocean acidification, desertification—these are just a few of the seemingly insurmountable matters of climate catastrophe. To tend to these concerns, which are only a fraction of what anthropogenic climate change is already realizing, it is critical to craft a robust notion of hope—distinct from any flat optimism—if any meaningful form of survival, much less flourishing, will be possible. This chapter will consider Judaism's contributions to this opaque hope. Central to this excursus is the following provocation: hope is not to be "had" so much as "made." This assertion is nothing new but is, in fact, composed of the very fibers of an ancient wisdom threading through Judaism: *tiqvah*, cord, hope. Entwined throughout the tradition, this term, *tiqvah*, suggests an *active, material* hope—not merely emotive but a catalyst for work, for weaving. It shirks any sense of "providence" or the likes, instead remaining closer to a form of immanental world-building in the face of, indeed in the wake of, cataclysm.

Using weaving as both conceptual metaphor and poietic praxis, I will trace some historical tangles of Jewish hope in order to gesture at an eco-ethic made possible by this vibrant wisdom. I will suggest that this concept may yet serve as fertile grounds for embodied way-making, without sliding into a naïve optimism, characteristic of many visions of hope. And yet, against all odds, this mattering hope might just animate *livable* modes of inhabitation, crafty forms of dwelling. If a future of planetary wholesomeness is possible, it will, by some earthly grace, be stitched together by weary hands whose faith knows no bounds.

(Re)Defining Hope: Preliminary Clarifications

"Hope," like "love," may well be one of the most bastardized words around. It is hard to miss—embroidered on everything from throw-pillows to tea towels. Hope is tossed around flippantly: sometimes connoting a general emotion of desire, other times aiming at some expectation yet-to-come, and *still yet* others sliding into the fixed confines of some projected future. Hope runs the gamut from shallow sentiment to unwavering certainty—at least in popular discourse.

Needless to say, some preliminary remarks about the nature of hope should come across as warranted. Hope, as used here, rests outside the binary of sheer doubt and unswerving certitude, perhaps only articulable through the vocabulary of *faith*. Hope is something of a cloud: tangible but elusive, present yet ungraspable. While it may be tempting to equate hope with "certainty," I will propose the opposite here: uncertainty is a necessary, constitutive catalyst provoking any possible spark of hope.

Uncertainty is that key facet of hope that prompts its movement, its energy, even at times its "felt" absence. Uncertainty keeps hope in-process, for hope without uncertainty easily becomes one of two general things: *either* some sort of static "knowledge" *or* some version of nihilistic apathy. Neither concrete knowledge nor pure doubt, hope is the frayed fringe of the in-between: neither sure plan nor utter resignation, hope is the marginal flicker of a divine creativity embodied in the present. Hope uncomfortably rests as a veritable paradox.

Ancient Threads Made Present

If nothing else, hope is a confluence of temporalities. While hope resides nowhere but the present, it emerges from that aporetic seam stitching together pasts for the sake of futures. If hope brings with it the possibility of *real* creativity, then it must inhabit the unknowable space between what-was and what might-yet-be. In as simple of theological terms as possible: hope can be *neither* pure transcendence *nor* pure immanence but, instead, abides in the misty possibilities between and among. Hope figures a certain, vibrantly material capacity that we might refer to as mystical—which is *not* to say "abstract."

When it comes to the Jewish inflection of hope that I seek here, something of a paradox reveals itself: *hope cannot exist without memory*. As David Hartman puts it, hope "has its roots in memory."[1] Memory serves an irreplaceable function within Judaism: *Remember! Do not forget!* These commandments reimagine futurities through the echo of past wisdoms. Hope materializes from the aporetic grounds of memories, even those "forgotten," as affective lures toward possibility.[2] This differential past instills wisdom that may yet be adaptive; it whispers a possibility that may yet be improvisational.

Here-now, I intend to explore Jewish pasts for the sake of uncovering a hope not only faithful to the material memories of Judaism *but also* sufficiently faithful in addressing the unprecedented trials of the Anthropocene.

While there are many words for "hope" across the writings of the Jewish tradition, I will use one as the contextual matrix from which I intend for the remainder of this work to emerge. That word, tacitly hidden among many, is *tiqvah*.

Tiqvah (also transliterated *tiqva*, *tiqwa*, and so on) means, literally, "cord." But, in its verbal sense, it means to "hope for," "await," "stay." Both senses ("cord" and "hope for") are important and should remain connected. At the risk of "projecting" my own theological desires onto this term, I will judiciously contend over the course of this piece that *tiqvah* can serve as a model for an *actively material* hope—one that might incite some sort of meaningful survival as tides rise—to be woven. *Tiqvah* stitches possibility in a resonant fashion with *tikkun*, the Jewish (and specifically Kabbalistic) notion of world-mending.

We find *tiqvah* across the prophets, psalms, and wisdom literature of the *Tanakh*; and given the many terms connoting "hope" in the Hebrew texts, *tiqvah* offers a bit of specificity as to *which particular phenomenon of hope* one might mean, appearing "in contexts where the uncertainty of human life and its fulfillment is sensed, leading to reflection on one's own prospects, hopes, and expectations."[3] Again, uncertainty remains a co-constitutive factor that incites hope. Similar to other related terms (including *seber* and *tohelet*), *tiqvah* "characterizes a human life that can be considered secure on the social level and ethically meaningful."[4] This iteration of hope remains utterly committed to the here-and-now, the fleshy, the radically material; accordingly, hope, at least in this instance, *does not* extend beyond the grave.[5] This caveat will make for some difficulty, especially when discussing the Shoah.

And yet, a vitally material hope like *tiqvah* does not necessitate its divorce from the glint of some divinely creative spirit, at least if one wants to take seriously its religious roots.[6] *Tiqvah* impossibly stitches possibility into the spiritual fabric of Jewish theo-philosophical framing, differentiating itself from any inflection of "secular humanism." The Rabbinic tradition faithfully inhabits this breach, carefully tacking the fabrics of the secular and sacred—never separate ontological categories so much as axiological principles in Judaism.[7]

The advent of Rabbinic Judaism is nothing less than the exemplification of *tiqvah*. The destruction of the Temple in 70 CE posed Jews with a seemingly insurmountable obstacle—out of which emerged Rabbinic Judaism. While it is debatable as to the extent to which the Temple "mattered" at all to first-century Jews living outside of Palestine,[8] transforming Judaism from a Temple-cult of sacrifice into a religion primarily of study and general piety was surely no small feat. The catastrophe was met with an inconceivable response. As Robert Goldenberg puts it, the building of "a Jewish way of life that made the Temple unnecessary in practice while it remained indispensable in theory."[9]

Rabbinic Judaism, in the wake of the Temple's destruction, needed to address massive theological quandaries: How would Jews achieve atonement without the Temple-cult? How would they explain the disaster of the Temple's destruction? How would they devise an entirely new way of life?[10] To study and pray became the creatively adaptive response of Jewish leaders, reimagining the tradition in order to recognize acts of lovingkindness as atoning,[11] to establish formal fixed communal liturgy, to reorganize common life around the Synagogue,[12] and so forth.

It would be something of a gross oversight to convey this knotted hope without making a brief reference to one of the primary animating forces of hope in Judaism: *messianism*.[13] Messianism is one of the critical catalysts undergirding Jewish hope, and yet it cannot be taken as synonymous to *tiqvah*; messianism can neither be separated from nor adhered to "hope," but the two twist and tangle. This chapter cannot tackle the messy problem of messianism in-depth, but elucidating one crucial fissure trending through the conversational arc will prove useful.

Primarily emerging from Talmudic texts (*Talmud Sanhedrin*, especially), Jewish messianism can be understood as fraying into two general strands: those who believe that the *mashiach* (messiah) arrives in response to human action and those who believe that the *mashiach*'s arrival is predestined, lacking any reference to human conduct. At base, the issue is this: is the *mashiach*'s arrival an inevitable event irrespective of human activity, or does human activity influence (either by accelerating *or* delaying) when the *mashiach* may come? Problems emerge in each: interpreting the *mashiach* as "predestined," on the one hand, arguably renders the need for the Jewish people's sanctification less urgent, thus making it less "material;" on the other hand, interpreting the *mashiach* as "influenced" arguably diminishes divine agency and the miracle of the divine "breaking into" history. The former can be accused of absolving Jews from the need to act in the face of injustice; the latter can be accused of too-fervently trying to strong-arm the *mashiach* into appearing. Those on either side of this split do tend to agree on one crucial matter: the *mashiach*'s arrival is not predictable, regardless of whether the *mashiach* is swayed by human activity.

Despite this one agreement, the general tension remains: when it comes to hope, what good is human effort, and yet what might the world become without that effort? To assume that one can *force* the "hoped for" is nothing short of a fallacy; indeed, one of the gravest errors in Judaism is to "press for the end."[14] But, Emmanuel Levinas also reminds us that, " . . . not to build the world is to destroy it."[15] What would become possible if we were to allow *tiqvah* to dwell in the misty borderlands between these poles? I suggest that *tiqvah* may yet stitch together these "incompossible" possibilities.

It is precisely because *tiqvah* materializes from these entangled knots—of the secular and sacred, of spirited materialism—that it can weave the aporetic; a loom, not to be too cheeky about it, can only function *because of* tension, after all. *Tiqvah* does not divorce creativity from contextuality, nor does it guarantee the former from the latter; it weaves tense openings and sacred knots, holding the two close in a complex web.

However, the shuttles of a loom can catch, snag, splinter—this hope *does not* guarantee that it will "all work out." This hope is no mere incremental "progress," and it tangles anything remotely resembling the linear. After all, one cannot go back to believing in secular "progress" after Auschwitz, nor can one return to the same God.[16] That the cords of hope can fray indiscriminately, can be torn asunder, cannot be taken for granted. The fallacy of steady progression toward some telos will not be proffered here; that violence can leave the threads of hope in tatters, possibly irreparable, *must* be considered.

In the Face of, In the Wake of, Cataclysm

The Shoah presents no small problem for any inflection of Jewish hope. What of hope post-Shoah? Must it die off, along with the sacred toward which it aims? Or might it persist—somehow?

As Levinas suggests, the question to ask post-Shoah *is not* "Does God exist?" *but rather* "Do we still believe in the excellence of Judaism?"[17] Not to dismiss the theodical questions that often overwhelm, it seems pertinent to heed Levinas's prompt to explore what good this web-like tapestry of hope might offer Jews here-now in the wake of the Shoah, or here-now facing the possibility of climate catastrophe.

From gas chambers to crematoria, hangings to ghettos, firing squads to dysentery, there is little room, if any, for offering any semblance of hope in any sort of meaningful way when it comes to the Shoah. Of course, many survivors of the unthinkable tragedies of the Shoah have reflected on the forms of hope that they were encouraged by other prisoners to preserve—if only as a survival

mechanism. One might think of Elie Wiesel being told upon entering Auschwitz, "We mustn't give up hope, even now as the sword hangs over our heads.
So taught our sages. . ."[18] It appears as if the cultivation of any hopeful inkling
of meaning or reason for survival seemed crucial—at least to some.

Of course, Wiesel was adamant that it was nothing short of chance or hap-
penstance that enabled his survival.[19] The utter evil of Nazism, while structural
and systematic in general, was amplified with the hope-destroying arbitrariness
with which many of the most gross injustices were carried out: the unpredict-
able few seconds in which prisoners were evaluated during "selections," the
capricious whim of "doctors" determining wellness, and so on. To take seriously
the unutterable perversities of the Shoah demands that one *refuse to connect
hope with survival*; hope, while amply material, cannot be fixed to any particular outcome with any degree of certainty. *One can live without hope in just
the same way that one can be killed with it.*

Even if hope never guarantees survival, it *must* remain vital to catalyzing
or inciting the world imagined-otherwise for it to have any meaning at all. But
make no mistake, hope cannot, on its own, *"bring about"* some end. The rub
is this: hope cannot guarantee anything, but it must still *do something* if it is
to remain a useful or viable concept.

The danger one faces in wrestling with hope in the wake of the Shoah is
this: to be naively "hopeful" is to be blind to the sufferings of more than six
million, but to reject hope outright is to deny the faithful persistence of many
and to forsake the dead as irredeemable.[20] Weaving hope is turbulent, always
somehow risking *either* the denial of the grave traumas of the Shoah *or* leaving
the dead to rot. Avoiding either chasm feels absolutely impossible and is, thus,
downright necessary.[21] This problem presents an irresolvable aporia. The point
of hope might then be simply to *resist*: to *refuse to resolve* such an aporia into
some neat or tidy conclusion. As Gregor of Wiesel's *The Gates of the Forest*
exclaims, "If their death has no meaning, then it's an insult, and if it does have
a meaning, it's even more so!"[22]

If hope inevitably remains a paradox by nature, then any attempt to explicate
it runs the risk of "explaining away" the mystical. This is all the more dangerous
for Jews post-Shoah, where an appeal to the mysterious could easily slide into
mystification-for-mystification's-sake, heightening a two-pronged denialism of
both Shoah and Anthropocene. One senses this risk in Wiesel's response to
the question he was asked in a 2005 interview, "Do you still have faith in God
as the ultimate redeemer?" He replied, "I would be within my rights to give
up faith in God, and I could invoke six million reasons to justify such a decision. But I am incapable of straying from the path charted by my forefathers,
who felt duty-bound to live for God. Without the faith of my ancestors, my

own faith in humanity would be diminished. *So my wounded faith endures.*"[23] A decision one way or the other betrays either those murdered in the Shoah or one's faithful ancestors; neither option will do! This hope is not an *aufhebung* in any sense but instead is an *abiding*, a *tending*—even in silence. The faithful, if we take Wiesel's notion of "wounded faith" seriously, must *refuse to resolve or "heal"* this aporia into some higher unity, instead opting to dwell in the ruins, to stay with the trouble.[24]

Tiqvah refuses resolution in that it is not ever an *end* or a desired "object," *nor* is it exactly the *means* by which that "end" or "object" is sought after. Hope is not just some "tool" used for fixing something; it is not some "thing" that is added to a situation in order to achieve an outcome, nor some disposable product that can be discarded after an end is obtained. Neither end nor means, hope persists elusively as that in-between drive, that spirit in the borderlands,[25] that messianic flashing-up;[26] it is *not* that which is desired, *nor* is it exactly that which "brings about" that which is desired. And it is certainly not some incremental rationalism. *Hope is the non-linear function of the in-between, the imagination of the impossible—without abandoning the material for some flight into the abstract.*

Levinas stresses this very point, reminding us of the religious spark integral to such a hope, ". . . to hope then is to hope for the reparation of the irreparable."[27] Eric Severson, interpreting Levinas on this matter, proffers, "To have hope is to settle neither for an irreparable past nor any anticipatable future . . . Hope refuses to resign the future to an extension of the past . . . " Severson continues, "Hope ushers the subject to the brink of salvation, but that salvation must come from beyond the subject, beyond the instant, beyond history."[28] Hope is inseparable from material action, *all the while remaining irreducible to it.* The characteristically Jewish paradox of the hidden, dispersed spark of the divine *Shekhina* becomes manifest here in such a way that refuses to allow hope to be definable purely in terms of "immanence" and "transcendence." There is a messianic alterity at the brink of the present, at the edge of the possible.

The Shoah remains an ever-unexplainable event, an incomprehensible rupture in *thought itself.* As Wiesel reminds us, "Auschwitz negates all systems, opposes all doctrines."[29] Hence, the proposed hope does not "solve" anything. It does not "answer," "justify," nor "explain" anything about the Shoah. The hope that emerges in the wake of the Shoah need not *"comprehend"* anything at all; but, despite this unavoidable failure to "comprehend," this resistance-hope must indeed *confront* the Shoah. This confrontation is nothing other than what it means to *bear witness*—to *remember* for the sake of never replicating the past's woes, to *not forget* for the sake of refusing to allow injustice to materialize again. This Jewish hope echoes the great Rabbi Tarfon's teachings, "It is not incumbent on you to complete the work [of perfecting the world]. *But you are not free to evade it.*"[30]

Here is where *tiqvah* may yet map onto the crises of the Anthropocene. This web-like hope need not "comprehend" the Anthropocene, nor must hope produce perfect solutions to the problems it poses; rather, this thread of hope need only, consistently and faithfully, confront the Anthropocene—undoing the knots that strangle without succumbing to the temptation to "explain away" the undue violence committed against communities, creatures, watersheds, and habitats. Put differently, this hope will not, and simply *cannot*, "make sense" of the impending, the already-here, calamities of climate injustice— though this does not make it any less revolutionary. Alternately, it steadily weaves responses without (fore)closing the undecidable aporias before us.

Weaving (for the Sake of) the Planetary

I would like to suggest that *tiqvah* may yet bear the capacity to prompt faithful living in the Anthropocene—faithful to efforts of sustainability, climate-related reparations, resource redistribution, and beyond. More specifically, the goal here is to briefly note the usefulness of the proposed *tiqvah* in embodying those principles of faithfulness to environmental justice.

Judaism, especially in Kabbalistic strands, has long held views amenable to new materialist concerns: from *kashrut* to exiled sparks of *Shekhina*, from *tikkun olam* to its view on Torah scrolls, Judaism has exhibited the curious capacity to constructively blur the material and the spirited, the secular and the sacred. Rabbinic Judaism, even in its earliest manifestations, has long been characterized by its affirmation of the material, the bodily, as complementary to the spiritual—rather than as conflictual or contradictory.[31] For some time it has been the case that, as Reuven Kimelman puts it, "The physical is not overcome, superseded, or consumed in the spiritual. Rather, the physical, the bodily, the carnal partake of the spiritual."[32] In this way, hope presents itself in the *absolutely ordinary* matters of Jewish tradition. As Eliezer Berkovits comments on Jewish ritual blessings, "The sacred *is not*, but has to be *brought into being* as the result of someone's action or behavior."[33] Accordingly, one must not fail to recognize a spark of hope that reveals itself in matters as simple as making the mundane *sacred*—blessing and breaking bread, resting, planting trees. The many blessings in the Jewish tradition of and for objects, activities, and so forth remind us not only that the material *matters* but that *the material is not simply inert or inactive*. In Kimelman's words, "The assumption of all these blessings is that *no aspect of the world is devoid of spiritual resonance*."[34] These blessings instill a glint of holiness in the material—recognizing the vibrancy of the ter-restrial and its deserved flourishing.

Tiqvah may yet serve as one possible way for engaging, and sustaining, this vibrantly material *olam*. I have argued that *tiqvah* limns possibilities for *living into*, for *confronting*, planetary catastrophe. This hope *does not* answer the questions of theodicy nor of extinction, but it *does* serve as a ground for stitch-ing together livable futures in the here-now. This hope provisionally mends and creatively adapts in concert with whatever resources may be at hand; that is to say, *tiqvah* remains on-the-way—never final, ever unfolding.

When it comes to the Anthropocene, elastic solidarities across all sorts of registers, locations, capacities, and matters will be, indeed *already are*, abso-lutely critical. Responses will need to be improvisational but resolute, innovative but steadfast. What is needed is a hope that catalyzes a faithfulness to *becoming*, to the ever-shifting nature of nature.

Judaism remains unwavering on this: one is not to force an "end," or "the world to come." Any attempt to do so will fail. Forcing a "solution" in the An-thropocene may prove to be not only naïve but fatal; "geo-engineering" solutions may prove to be as devastating as the very nearly successful "Final Solution" not many decades ago. *And yet, Judaism remains unswerving on the other end of things*: "doing nothing" is *not* an option. Daily practices of study, prayer, *mitzvoth*, habitual acts of lovingkindness, and *tikkun olam* are modes of sacred mending, transforming the abject or otherwise-objectified into spheres of vi-brancy, hopefully meriting their sustained flourishing; these deeds are faithful to a certain understanding of religiosity, but they are also faithful in other ways: not least, as modes of resisting the capitalist demands of "productivity," extri-cating oneself from webs of degradation and harm. The answer to the Anthro-pocene is obviously not "hermitry," as it were; but "doing something" in the Anthropocene may, paradoxically, look a lot like "doing less"—not in the sense of apathy so much as slower modes of living, smaller spheres of community, steadier measures of mending. This provocation, of course, is directed at those of us inhabiting industrialized nation-states, who bear the disproportionate responsibility for much of the activities instigating this climate crisis—differ-entially impacting communities who are already, undeservedly, experiencing the worst effects of rising tides, drying streams, and intensifying storms.

At the Fringe: A Postlude

There is a question that has remained unstated until now: why go to these specific sources in these specific times? The destruction of the Temple in Jerusalem—Rabbinic Judaism's shocking impetus—and the Shoah—

contemporary theology's haunting specter—are each a moment in which a community, in the wake of disaster, found some semblance of a creative elasticity sufficient for reimagining a whole tradition, an entire lifeway. I have sought to learn from these sources' tenacity to weave hope amid ruin, praying that these lessons might cultivate something in us, here-now, as we face an unknowable and undecidable future on this planet.

I make no claim about the *singularity* of this *tiqvah*-hope, and I make no defense for those who wield a perverted "hope" as a weapon, which includes the state of Israel in no small way. I have discovered *tiqvah*, then, *not just* in these texts: I have found it while walking the streets of the Aida refugee camp in Bethlehem, Palestine;[35] it is present in creative agricultural efforts taking place under the undue violence of Israeli occupation;[36] its spirit is evident in the artful *resistance to* and *on* the "separation wall."[37]

That is part of the paradox of hope: it cannot be contained, for it was never really ours, never really anyone's, to control, to delineate, to box in, or to wall out. So, while *tiqvah* does emerge, in part, from Jewish imagination, it is found also far beyond the tradition, sparking revolutionary desire in the disenfranchised or otherwise cast aside.

Hope parading as power over, or dominion understood as domination, is not hope: it is violence. Hope, embodied, can only mean to be among, with, and beside. It is my prayer that this might yet be possible: mending broken relationships with land and neighbor, undoing the tangles of unjust systems of colonialism, exploitation, and greed. Weaving a future of shalom—for *all*.

Notes

1. David Hartman, "Maimonides' Approach to Messianism and Its Contemporary Implications," *Daat: A Journal of Jewish Philosophy & Kabbalah*, Vol. 2 (3) (1978): 5. Other versions of research for this chapter have appeared in Rowe, *Of Modern Extraction: Experiments in Critical Petro-theology* (New York: T&T Clark, Bloomsbury, 2023) and "The Matter of Oil: Extraction Vitalisms and Enchantment," in *Religion, Materialism and Ecology*, ed. Sigurd Bergmann, Kate Rigby, and Peter Manley Scott (New York: Routledge, 2023).

2. Cf. Karmen MacKendrick, "Remember—When?" in *Sexual (Dis)Orientations: Queer Temporalities, Affects, Theologies*, eds. Kent L. Brintnall, Joseph A. Marchal, and Stephen D. Moore (New York: Fordham University Press, 2017), 277–291.

3. "Tiqwâ," in *Theological Dictionary of the Old Testament*, XV, eds. Johannes G. Botterweck, Helmer Ringgren, and Heinz-Josef Fabry (Grand Rapids: Eerdmans Publishing, 2006), 761. See also, William Wilson, "Hope," in *New Wilson's Old Testament Word Studies* (Grand Rapids: Kregel Publications, 1987), 222.

4. "Tiqwâ," 761.

5. "Tiqwâ," 761.

6. Walter Benjamin, "On the Concept of History," in *Selected Writings*, 4 Volumes, eds. Marcus Bollock and Michael W. Jennings (Cambridge: Harvard University Press, 1996–2003), 4, 390.

7. Eliezer Berkovits, *Faith after the Holocaust* (Newark: Ktav Publishing, 1973), 61.

8. Jacob Neusner, "Emergent Rabbinic Judaism in a Time of Crisis," in *Early Rabbinic Judaism: Historical Studies in Religion, Literature and Art* (Leiden: Brill, 1975), 34–49.

9. Robert Goldenberg, "The Destruction of the Jerusalem Temple: Its Meaning and Its Consequences," in *The Cambridge History of Judaism*, 8 Volumes, ed. W.D. Davies and Louis Finkelstein (Cambridge: Cambridge University Press, 1984–2018), 4, 202.

10. Neusner, "Emergent Rabbinic Judaism," 34–49.

11. Ibid., 46–48.

12. Reuven Kimelman, "Rabbinic Prayer in Late Antiquity," in *The Cambridge History of Judaism*, 8 Volumes, eds. W.D. Davies and Louis Finkelstein (Cambridge: Cambridge University Press, 1984–2018), 4, 573–611.

13. Of course, there are many who would argue that messianism is not an essentially Jewish notion, given that messianism is only barely "present" in the foundational texts of Judaism. See William Scott Green and Jed Silverstein, "Messiah," in *The Encyclopedia of Judaism*, 3 Volumes, eds. Alan J. Avery-Peck, Jacob Neusner, and William Scott Green (New York: Bloomsbury, 1999), 874–888; and Jacob Neusner, *Messiah in Context: Israel's History and Destiny in Formative Judaism* (Philadelphia: Fortress Press, 1984).

14. Cf. Gershom Scholem, *The Messianic Idea in Judaism: and other Essays on Jewish Spirituality* (New York: Schocken Books, 1971), 1–37.

15. Emmanuel Levinas, *Nine Talmudic Readings*, trans. Annette Aronowicz (Bloomington: Indiana University Press, 1990), 112.

16. Cf. Emil Fackenheim, *God's Presence in History: Jewish Affirmations and Philosophical Reflections* (New York: New York University Press, 1970), 67–98.

17. Emmanuel Levinas, *Difficult Freedom: Essays on Judaism*, trans. Seán Hand (Baltimore: Johns Hopkins University Press, 1990), 258.

18. Elie Wiesel, *Night* (New York: Hill and Wang, 2006), 31.

19. Ibid., viii.

20. Cf. Emil Fackenheim, *To Mend the World: Foundations of Post-Holocaust Jewish Thought* (Indianapolis: Indiana University Press, 1994), xxxi–xlix.

21. Ibid., 254.

22. Elie Wiesel, *The Gates of the Forest*, trans. Frances Frenaye (New York: Schocken Books, 1966), 197.

23 Aron Hirt-Manheimer, "On God, Indifference, and Hope: A Conversation with Elie Wiesel," *Reform Judaism*, 2005. https://reformjudaism.org/jewish-life/arts-culture/literature/god-indifference-and-hope-conversation-elie-wiesel; emphasis mine.

24. Donna J. Haraway, *Staying with the Trouble: Making Kin in the Chthulucene* (Durham: Duke University Press, 2016); and Anna Lowenhaupt Tsing, *The Mushroom at the End of the World: On the Possibility of Life in Capitalist Ruins* (Princeton: Princeton University Press, 2015).

25. Cf. Gloria E. Anzaldúa, *Borderlands/La Frontera: The New Mestiza* (San Francisco: Aunt Lute Books, 1987).

26. Benjamin, "On the Concept of History," 4, 397.

27. Emmanuel Levinas, *Existence and Existents*, trans. A. Lingis (Pittsburgh: Duquesne University Press, 1978), 93.

28. Eric R. Severson, *Levinas's Philosophy of Time: Gift, Responsibility, Diachrony, Hope* (Pittsburgh: Duquesne University Press, 2013), 62, 63; emphasis mine.

29. Elie Wiesel, "Art and Culture after the Holocaust," in *Auschwitz: The Beginning of an Era?*, ed. Eva Fleishner (New York: Ktav Publishing, 1977), 405.

30. *Pirkei Avot, Vol.* 2 (16); emphasis mine.

31. Reuven Kimelman, "The Rabbinic Theology of the Physical: Blessings, Body and Soul, Resurrection, and Covenant and Election," in *The Cambridge History of Judaism*, 8 Volumes., eds. W.D. Davies and Louis Finkelstein (Cambridge: Cambridge University Press, 1984–2018), 4, 946

32. Ibid., 946.

33. Berkovits, *Faith after the Holocaust*, 59.

34. Kimelman, "The Rabbinic Theology of the Physical," 951; emphasis mine.

35. *Lajee Center,* http://www.lajee.org/; "A Visit to Aida Refugee Camp," *The Israeli Committee Against House Demolition*, 2016, https://icahd.org/2016/06/09/a-visit-to-aida-refugee-camp/; "Who We Are: Alrowwad Cultural and Arts Society," *Alrowwad Cultural and Arts Society*, https://www.alrowwad.org/en/?page_id=9423; "Interactive Map," *B'Tselem: The Israeli Information Center for Human Rights in the Occupied Territories*, https://www.btselem.org/map.

36. "Palestine Institute for Biodiversity and Sustainability of Bethlehem University," https://www.palestinenature.org/; Tent of Nations Educational and Environmental Farm, "Tent of Nations: People Building Bridges," http://www.tentofnations.org/.

37. Khaled Diab, "The Art of Palestinian Resistance," *HuffPost*, January 16, 2013, https://www.huffpost.com/entry/the-art-of-palestinian-resistance_b_2121029; Amahl Bishara and Lajee youth, *The Boy and the Wall* (2005): https://www.shoppalestine.org/product-p/book_kids_boy_wall.htm.

Oily Animations: On Protestantism and Petroleum

Terra Schwerin Rowe

How does one live in a time when the unthinkable continually asserts itself? Empty streets in the city that never sleeps, insurrections at North American capitals, public misinformation, and preposterous disregard for democratic principles would be enough to face, enough to make one dizzy with incomprehension—but an end to human civilization within one's lifetime is simply unintelligible.[1]

Yet, as Amitav Ghosh explains, in climate change we are confronted with the unthinkable. As a novelist, Ghosh finds that although he is personally concerned with and occupied by climate change, he cannot write it, at least in the mode of the modern novel with its limits on the improbable. "Probability," he writes, "and the modern novel are in fact twins, born at about the same time, among the same people, under a shared star."[2] Echoing Max Weber, Ghosh explains how the modern novel helped convey a new rationalization of life through the banishment of enchantment, unpredictability, and the animation of the other-than-human world. By contrast, Ghosh reflects on surprise attacks of Bengal Tigers in the Sundarbans and the uncanny sense of being hunted rather than hunter—of being thought by the tiger—provoking him to query: "'Commonplace'? 'Moderate'? How did Nature ever come to be associated with words like these?"[3] According to Ghosh, more than a measure of how appalling or unprecedented climate change is, its unthinkability is due to modernity's explicit exclusion of its characteristics and symptoms from definitions of thought.[4] Calling to mind the Derridean *divinanimal*, within the rationalization and "regularity of bourgeois life," spread with efficiency by the accounting logic made so desirable by a Protestant work ethic, both the other-than-human and divinity are linked in their exclusion from the world of the probable.[5] As a result of these exclusions, climate change strikes moderns not merely as unexpected, unbelievable, or appalling—but *unthinkable*.

Given the modern disenchantment, death, or mechanization of the other-than-human world with key gender implications, many with feminist materialist concerns have promoted the reanimation of the material world.[6] Others address the desacralization of the material world more forcefully and emphasize the ethical importance of religious re-enchantments of nature. From an unexpected Christian perspective, for example, Mark Wallace reacts against what he sees as an antagonistic dualism constructed between Christianity and animism, proposing instead a "Christianimism" that might inspire more profound attachments to the world and thus a more resilient environmental ethic among the Christian faithful.[7]

While climate change does confront Moderns with the "force of things" (Bennett, *Vibrant Matter*) excluded from the definition of thought, this is not the first time industrialized humanity has had to face such inscrutability. While oil has introduced the very conditions of climactic unthinkability addressed by Ghosh, ironically, the first "discoveries" of oil in the U.S. elicited remarkably similar reactions. During the late nineteenth century, oil disrupted basic world views and expectations for nature. With its mysterious defiance of Newton's law of gravity and its uncanny disruption of Cartesian conclusions about consciousness and agency in the other-than-human world, oil introduced profound questions about the rationality and controllability of nature.

Intertwined with climate change, current petroculture theorists have similarly been confronted with the unthinkability of oil. A branch of the environmental humanities, intertwined with energy humanities that Ghosh also played a key role in establishing, petroculture refers to the sense that the influence of fossil fuels has not merely been operational or mechanistic, but affective, aesthetic, intellectual, religious, and ideological.[8] Analyzing such profound humanistic impacts of oil, petroculture scholars have questioned how modern society could have gone so long without really seeing oil—without being conscious of its formative conceptual and ideological power. The Canada-based After Oil collective, for example, concludes that, ". . . the power of oil is unconscious; we cannot grasp it and we don't perceive it."[9] Flying under the radar of conscious, critical reflection, oil has been the "magic that powers modernity" and, as such, has largely gone unthought.[10]

Oil challenges the well-worn and often well supported narrative of disenchanted matter in need of re-enchantment. Early accounts of oil discovery in America reinforce theses that call for the critical analysis of gendered nature and yet disrupt both the common critique of Christianity as promoting a view of the material world as inanimate, as well as hopes that reanimating nature with divinity necessarily provides environmentally friendly results. While predominant articulations of Christianity have indeed converged with a philosophy of mechanism in modernity to emphasize the inanimacy of the material world,[11] early accounts of oil discovery in the U.S. demonstrate that, when it comes to the matter of oil, Protestant Christianity has not universally promoted the disenchantment of matter. From this historical vantage point, between oil and modernity, enchantment is indeed *precisely* the issue, though whether curse or cure might depend on a host of other factors.[12] Consequently, the imperative role of new materialisms in the age of oil and climate change is not merely the resurrection of a dead world, but critical attention to its enchantments and disenchantments; the ways in which matter becomes gendered and racialized, profaned and sanctified, by whose logic and toward what ends.

More than resurrecting dead matter, religious new materialisms must disrupt a rigorously held binary between life and death—a distinction Christianity, and even some animisms, routinely defend. Perhaps, along with the aim of critically analyzing enchantments and disenchantments, religious new materialist interactions might also help remap such binaries to a more relational sense of responsive animation in a way that might allow for broader recognition and proper response both to climate change and the pervasive influence of oil.

Unthinkable oil

Reporters, returning from sites of new oil discovery in the late 1860s consistently expressed disbelief, shock, or an inability to comprehend or communicate what they had witnessed; "'astonished beyond measure'" at this "'mystery of the age,'"[13] they had "no language at [their] command by which to convey to the minds of [their] readers'"[14] such extraordinary scenes. They referred, in part, to human behavior in response to oil and this new race for wealth, but more than bizarre social acts, observers were particularly shocked by the *behavior* of *oil*—its defiance of the basic laws of nature, its intractability, and animacy.

The uncontrollability of oil was a particular source of both frustration and amazement. Environmental historian Paul Sabin has researched early oil narratives extensively and found that entrepreneurs in particular were "surprised and confused when oil refused to bend to their will."[15] Such behavior disrupted basic expectations about nature as controllable because it followed basic natural laws. Yet, more than the control of nature, oil disrupted the profound sense that other-than-human matter was inert or inanimate.[16] One reporter explained that upon hearing witnesses describe the way oil would start and stop at will and spout out from the ground "'One is almost constrained from his intuitive notion of the natural world to suspect such a story is a whop-per'" that the "'man who talks in this manner of oil flowing up, has been drinking poor whisky.'"[17] Less "intuitive" than Cartesian-Newtonian, this sense of the world as orderly, dead, and mechanistic had, by the later part of the nineteenth century, become so pervasive that it could easily be confused with good, old-fashioned, objective common sense.

Oil's disobedience to man's control, starting and stopping at will and thus causing "considerable consternation among investors and observers alike,"[18] was one thing, but its animacy could evoke a frenzy. In an attempt to impress the massive crowds of gawkers that an oil discovery could attract, oilmen would summon oil to "perform." As with a lion and its tamer, teasing and testing the edge of control, oilmen provoked their wells, getting them "stirred up" to en-tertain first time viewers.[19] Wells would often make rumbling, digestive noises as one journalist reports, "During the upheaving of the gas, it seems as if the very bowels of the earth were being all torn out and their sides must soon col-lapse."[20] Crude's strange behavior, suggesting liveness and animacy, disrupted what had once been clearly defined and rigorously reinforced conceptual di-vides between dead matter and animalistic vitality.

Closely connected to narratives of oil as uncontrolled, animalistic, and in need of taming were the portrayals of oil as a temperamental female "whom the male worker could not fully understand but still hoped to master."[21] Cor-responding with the rise of Cartesian-Newtonian views of nature, by the end of the nineteenth century, feminine imagery and metaphors for nature had receded. But in the 1860s, the inability of rational science to fully explain or predict oil had the effect of sparking a reemergence in gendered metaphors for nature. In poetry, songs, and witness and journalistic reports, oil "assumed the characteristics of a productive and temperamental female best managed by ingenious oil men."[22] Alternatively, reference to nurturing aspects were also common in imagery of oil as a nursing or birthing mother.[23]

More than other resources and industries like gold, cotton, and railroads, the mysteries of striking and then containing oil disrupted a Cartesian-Newtonian sense of an orderly, predictable, and controllable material world that could be regulated by a Protestant work ethic. In the following decades, this kind of control became more possible with advances in geological science and the imposition of control measures that became a hallmark of Rockefeller's Standard Oil style.[24] However, such regulations only came after key resonance between the unruly spirits of oil and less orderly aspects of religion and spiritu-ality. Such resonances amplified both the new oil endeavor as well as spiritualities that emphasized unconventional knowledge of a profoundly interconnected spiritual and material, physical and metaphysical world.

Divining Oil

While Sabin insightfully emphasizes the gendered implications of oil's unthinkability, he does not adequately account for similar seismic shifts in religious spheres. The popularity of divining rods and "other mystical practices" used to locate subterranean oil wells is briefly referenced, but the evident religious resonances are not explored further.[25] Examining the religious reverberations of oil more extensively, one finds that in the face of the inability of Cartesian-Newtonian knowledge systems to account for the animation of oil or accurately predict where to find it, such uncertainties also evoked new spiritual practices and religious beliefs where the metaphysical was employed to elicit empirical, quantifiable, physical results.

Such "practical spiritualism" was often personified in local figures whose advice soon became widely sought. One famous figure was Abraham James who, in 1867, near Pleasantville, PA became "forcefully possessed by his spirit-guides," fell unconscious, and woke to announce oil could be extracted from that very spot.[26] More Biblically based traditions similarly found the practical application of religion for oil extraction enticing. Historian Darren Dochuk has published the first comprehensive history of oil and religion in the U.S., focusing on characters like Baptist Patillo Higgins who employed prayer and Biblical interpretation to predict the location of the first major oil gusher west of Pennsylvania. Texas' Spindletop gusher produced more oil than was previously imaginable, defying common belief that there was no major oil to be found in the West, and is thus credited with ushering in the modern petroleum industry. For his ability to exegete land and scripture together, Higgins was honored with the title, the "prophet of Spindletop."[27]

As historian Rochelle Raineri Zuck notes in her research on James, Christian spiritualism, and oil, the connections between religion and the discovery of oil in the U.S. go deeper than "temporal and geographic proximity."[28] Each fulfilled the needs found in the other. When, in the early period of oil extraction, entrepreneurs needed to increase interest in the practical applications of oil, the spiritualist movement and religious organizations, with their established social connections, speaker circuits, and widely read publications, were able to efficiently spread the word and drum up interest. Reciprocally, spiritualist and religious people increasingly turned to the oil industry, not just because of its "lucrative financial opportunities," but because of its "potential to demonstrate the 'practical' applications of spiritualism."[29] Such connections appeased modern empirical desires for practical applicability and concrete verifiability, realms and aspirations modern religion generally could not satisfy. Zuck reports, for example, that a group of Christian spiritualists created the Chicago Rock

Oil Company, not just for material gains, but because they "hoped that oil itself could function as a kind of medium and persuade nonbelievers that spiritualism could make practical contributions to modern life."[30] Oil entrepreneurs, spiritual diviners, and religious faithful made certain assumptions about connections between the seen and unseen, material and immaterial worlds, believing that the unseen (under the earth or in the spirit realm) could animate or vivify the material world. Their shared views resonated, amplifying one another when it was found they could serve mutual aims and desires.[31]

Oil Theophanies

Clearly, the animacy of matter can be a lucrative endeavor—all the more so when it is both unexpected and accompanied by the possibility of transcendent meaning. Looking more closely at the entanglements of oil and religion than most petroculture scholars have been willing to do, we find not just animation and gender attributed to oil, but divine agency and manifestation as well. Indeed, early reports give a sense of oil as theophany, a visible manifestation of God. Sometimes in these narratives oil manifested itself as divine judgement, but more often an oil theophany surfaced as redemptive.

From its earliest "discovery" in Titusville, PA (1859), oil was consistently hailed as a savior, an agent of God's redeeming action in the world. As the first oil wells aligned with social division leading up to the Civil War (1861–65), many expressed hopes that petroleum would act as a savior for the country, providing a way out of war, replacing cotton to maintain economic stability, and freeing enslaved bodies from the bondage of physical labor.[32] As such, oil seemed to emerge as divine endorsement for the cause of the North—a sentiment later expanded as evidence of divine sanctification of the U.S. in general.[33]

Portrayals of oil as a redemptive agent only increased after the war. Resonating with the well-established trope of the "redeemer nation,"[34] as well as the deep-seated and long-standing Western philosophical and religious conceptions of energy and the rational, White male as fully developed human, early U.S. oil narratives often framed the story of oil in terms of salvation, redemption, theophany, and Christ-like figuration.[35] A prime example can be found in John J. McLaurin's 1896 *Sketches in Crude-Oil*. While the text has frequently been considered a significant source on early U.S. oil history, what seems to have passed under the radar of commentators is the way McLaurin's text functions as no mere work of history, but a redemption theology.

True to form for a traditional theology of redemption, McLaurin starts with the first moments of creation. He writes, "mineralogists think [petroleum] was quietly distilling 'underneath the ground' when the majestic fiat went forth: 'Let there be light!'"[36] And there was light—illuminating, transformative, redemptive light. In McLaurin's framing, oil follows the trajectory of the gospel message; just as the Good News starts in the East and finds special realization in Western Anglo society which positions it for a global spread, so too the story of oil begins in the East with ancient recognition by Old Testament authors and is fully realized in the West which poises it for global redemption. The text climaxes with a "Petroleum Idyl [sic]" serving as a hinge point between both East/West and oil/Christ: here the "star of the East" (both oil and Christ) is revealed in a conversion story of an Arab who jubilantly exclaims, "Hurrah for Jesus! . . . Hurrah for Petroleum!" Ultimately, McLaurin concludes that, "With John Wesley [petroleum] may exultingly exclaim: 'The whole earth is my parish.'"[37]

These early oil narratives reveal a sense in which the particular materiality of oil mysteriously elided animation, agency, and matter, thus taking on animation, feminization, and divinity. Aligned with such sacrificed foundations of modernity,[38] this is part of what continues to make oil culture so unthinkable and intractable today. Clearly, technological, industrial, and infrastructural investments in oil have been significant. More than this, spiritual/religious investments in oil have outlived their usefulness and must become thinkable.

Trickster Oil and the Sacred

Early oil histories clearly complicate both the modern narrative of Christianity as a monotheism monolithically opposed to animacy, as well as recent attempts to revitalize a kind of Christianimism with hopes of improved environmental affects and ethics. Some Christianimisms have clearly been functioning more consistently than is usually accounted for—and not all with laudable environmental effect.[39] Since oil has been able to emerge as the "magic that powers modernity,"[40] at least in part because it has been able to siphon off Christian articulations of divine vivification or redemption of the material world, adequately registering and responding to both oil and climate change might entail something other than the resurrection of dead matter.

Where nineteenth-century Christians and spiritualists attributed life-giving, animating, vivifying power of the material world to an ultimately divine source, as Stephanie LeMenager emphasizes, in a modern petroculture, "Liveness, as in seeming to be alive, now relies heavily upon oil."[41] Twenty-first century Western modes of identity and social construction now rely so heavily on the fossil fuel industry and petro-fueled media that LeMenager wonders whether, "the category of the human [can] persist without such forms indebted to fossil fuels."[42] An Enbridge Energy advertisement campaign asserts as much from a markedly less critical perspective: "E=life itself."[43] As scholars in the After Oil collective emphasize, this ad constructs a reality where, ". . . our social and personal lives . . . are only possible with the energy that Enbridge provides . . . Enbridge uses these expertly crafted images to tell us that happy and fulfilling lives depend on them."[44] Communicated here is not just a sense that energy is fundamental to life, but, that today, energy *is* petroleum, and that life remains unimaginable, if not impossible, outside a petroculture. By its omnipresence, petroleum is granted omnipotence.

How does one proceed if, with feminist materialists and ecotheologians like Wallace, one sees the ecological and gender justice significance of challenging the Cartesian-Newtonian, mechanist view of the material world as dead; and yet with LeMenager, and informed by late nineteenth century petro-Christianimisms, one also wants to challenge the pervasive way petroleum has come to define what it means to be alive? How do we both challenge a view of the other-than-human world as inert and yet emphasize that, in some sense, oil has remained a bit too lively?

While challenging the way a petroculture reduces life and energy to petro-leum, LeMenager also reminds us how the materiality of oil itself deconstructs any binary between life and death: ". . . oil challenges liveliness from . . . [an] ontological perspective as a substance that was, once, live matter and that acts with a force suggestive of a form of life."[45] As former lives, capturing the sun's energy in the form of carbon, recycled now as "dead matter" that can animate machines, let alone modern industrial society, petroleum troubles any solid boundaries between dead and alive, animate and inanimate, active agent and passive receiver.

As life recycled, this oily trickster resists pure demythologization that would merely render enchanted oil as dead, quantifiable matter—as well as common petro-critical aesthetics of oil as bad, dirty, and disgusting. In Mel Y. Chen's analysis of animacy they remind us that the boundary between life and death is often policed with key gender, race, and dis/ability implications by associating uncleanliness, disgust, and blackness with death. In *Animacies*, Chen explains, "matter that is considered insensate, immobile, deathly, or otherwise 'wrong' animates cultured life in important ways." From this perspective, oil remains unthinkable not just through its alliance with the animal, feminine, and sacred, but through association with death, disgust, blackness, and immorality as well. Images of oil-soaked birds and humans frolicking in oil gushers—important as they may have been for mobilizing public outcry that led to stronger regulations of offshore oil drilling—are intended to evoke disgust, to draw on a gut-level association with oil as dirty, ugly, and inherently immoral. As a resident of Santa Barbara, who has studied the oil spill of 1969, one could sympathize with

LeMenager if her book's cover image of a jubilant human positively dripping with petro seemed aimed at evoking disgust, or if she repeatedly, throughout the book, asked and responded to the question, "why is oil so bad?"[47] Yet does the persistent association of oil, purely with immorality and disgust, not also reinforce a sense of oil as dead matter rather than reincarnated lives? And does an affective association of oil as dirty, bad, and thus dead, not allow the pervasive ways oil animates modern industrial life to remain obscured—unthinkable? This trickster-like oily slippage between death and life, animate and inanimate, good and bad is profoundly unsettling for the basic binaries of Western philosophy, religion, and their justice traditions. Perhaps in this sense, oil has become the Achilles heel of this worldview: it lubricates a slide between life and death, good and bad, and resists categorization as either salvation or damnation.

Chen's proposed approach to the modern Western problem of matter rendered dead seems insightful here. Rather than ". . . reinvest[ing] certain materialities *with* life," Chen suggests, "remap[ping] live and dead zones away from those very terms leveraging animacy toward a consideration of affect and its queered and raced formations."[48] Chen also posits moving away from a binary consideration of death and life to a continuum of animacy. Rather than attending to innate or inherent characteristics of liveness and deadness, Chen suggests attending to evoked responses—to that which animates. Here, animacy emerges not as an inherent, stable quality or category, but as phenomena that emerge in responsiveness when bodies meet. With the aim of attending to the animacies that emerge when bodies encounter one another rather than the resurrection of dead matter, we are better able to see oil in all its queer, ontological slippages and recognize when it animates spiritualities, theologies, politics, industries, societies, and climactic shifts.

Life, Death, and Response in a Time of Climate Change

A profound irony emerges at the heart of an analysis of the death of nature in a time of climate change. At the moment, the more immediate—and yet profoundly interrelated—problem seems to be that *humanity* remains dead, lifeless, inanimate, and inert, frozen by a globalized economic system and international and domestic political standoffs. From this perspective, an animated material world will not provide meaningful change unless humans can somehow also become animated by—responsive to—the liveliness of sea level rise, ocean acidification, and glacial retreat.

Rendering oil thinkable, while avoiding a full demythologizing move to mere dead matter, we begin to see and think of oil in ways inextricably linked with the ability to finally think/feel/respond to climate change. We might recall that Derrida, in identifying the structure of sacrifice of the divinanimal at the heart of the *ego cogito*, suggests that thought does not start with reflection on the self-same, but in the encounter with the other who is wholly other.[49] This encounter of difference—materialized for him in the face of the wholly otherness of a cat who sees/perceives/thinks/feels him—animates his ability for thought. We might further reflect, along with Donna Haraway, that such encounters do not just initiate thought but animate thinking, feeling, and responding as well.[50] In emphasizing animations rather than inherent characteristics of life or death, attention is drawn to exchanges, meeting places of bodies—of bodies within bodies within bodies. As Chen suggests, attending to animacies not only changes patterns of thought, but in calling one to respond and drawing attention to responses of bodies, new ways of relating are also opened.

Such attention to animacies also makes available new ways of envisioning the sacred. Rather than relating to and envisioning the sacred as inhering in one thing or another, divinity emerges in exchanges, the liminal, in-between spaces and intra-actions where possibilities for the continuation of diverse vivification are not shut down, but continually find breathing space. This is, after all, what has happened for millennia: animates come together in the face of the unknown, commune by exchanging bodies, breath, and all that sustains their animation—and in those exchanges they discover they have "brushed up against the experience of the sacred."[51]

Notes

1. David Spratt and Ian Dunlop, "Existential Climate-Related Security Risk: A Scenario Approach," *Breakthrough*, National Centre for Climate Restoration, Melbourne, Australia (May 2019): 1–10. Accessed August 13, 2020.

2. Amitav Ghosh, *The Great Derangement: Climate Change and the Unthinkable* (Chicago: University of Chicago Press, 2017), 16.

3. *Ibid.*, 21.

4. *Ibid.*, 30.

5. On divinanimality, see Jacques Derrida, *The Animal That Therefore I Am*, trans. David Wills, ed. Marie-Louise Mallet (New York: Fordham University Press, 2008), 132. Ghosh, *The Great Derangement*, 19, citing Franco Moretti.

6. Val Plumwood, *Feminism and the Mastery of Nature* (New York: Routledge, 1993) and Carolyn Merchant, *The Death of Nature: Women, Ecology, and the Scientific Revolution* (New York: HarperCollins, 1980). On feminism and new materialism, see Stacy Alaimo and Susan Hekman, eds., *Material Feminisms* (Bloomington: Indiana University Press, 2008).

7. Mark I. Wallace, *When God Was a Bird: Christianity, Animism, and the Re-Enchantment of the World* (New York: Fordham University Press, 2019).

8. Imre Szeman and Petroculture Research Group, "Introduction," in *After Oil* (Edmonton, Alberta: Petrocultures Research Group, 2016), 9. See also, Sheena Wilson, Adam Carlson, and Imre Szeman, eds., *Petrocultures: Oil, Politics, Culture* (Canada: McGill-Queen's University Press, 2017) and Imre Szeman and Dominic Boyer, *Energy Humanities: An Anthology* (Baltimore: Johns Hopkins University Press, 2017). Amitav Ghosh's 1992 essay, "Petrofiction" is consistently cited as a key initiating text of petroculture studies, "Petrofiction," in *The New Republic* (1992): 29–34.

9. Szeman et al., *After Oil*, 48

10. *Ibid.*

11. Edward B. Davis, "Myth 13: That Isaac Newton's Mechanistic Cosmology Eliminated the Need for God," in *Galileo Goes to Jail and Other Myths About Science and Religion*, ed. Ronald L. Numbers (Cambridge: Harvard University Press, 2009).

12. Similarly, Karen Bray, in this volume, wonders whether new materialists have "too quickly forgotten, or too lightly engaged, or too willfully ignored, or too hastily redeemed the very fact that affect, material entanglement, and planetary relationality are not simply salvific, or not merely the way out of this climate colonial mess." Karen Bray, "Gut Theology: The Peril and Promise of Political Affect," in *Earthly Things: Immanence, New Materialisms, and Planetary Thinking*, eds. Karen Bray, Heather Eaton, and Whitney Bauman (New York: Fordham University Press, 2023).

13. Paul Sabin, "'A Dive Into Nature's Great Grab-bag': Nature, Gender and Capitalism in the Early Pennsylvania Oil Industry," *Pennsylvania History*, Vol. 66 (1999): 477.

14. *Ibid.*, quoting *Crawford Journal*, Sept 4, 1860.

15. *Ibid.*: 475.

16. *Ibid.*: 473.

17. *Ibid.*: 477.

18. *Ibid.*: 475.

19. *Ibid.*: 483, citing *The Warren Mail*, August 18, 1860.

20. *Ibid.*: 481, quoting *Crawford Journal*, Sept 4, 1860.

21. *Ibid.*: 479.

22. *Ibid.*: 473.

23. *Ibid.*: 481.

24. See Sabin's comparison of oil to other industries and resources. See also Darren Dochuk, *Anointed with Oil: How Christianity and Crude Made Modern America* (New York: Basic Books, 2019), 82.

25. The *Oil City Register*, for example, reports that in Pennsylvania in 1865 the use of divining rods was "generally practiced." Sabin, "Dive," 476.

26. Rochelle Raineri Zuck, "The Wizard of Oil: Abraham James, the Harmonial Wells, and the Psychometric History of the Oil Industry," in *Oil Culture*, eds. Ross Barrett and Daniel Worden (Minneapolis: University of Minnesota Press, 2014), 19.

27. Higgins was by no means alone in the practice of intertwining geology and Biblical exegesis. Several other leading religious figures, particularly in the South, took up this dual vocation of pastor-geologist. See Dochuk, *Anointed with Oil.*

28. Zuck, "Wizard," 20.

29. *Ibid.*, 21.

30. *Ibid.*, 27.

31. See also connections between electricity and metaphysical belief which influenced the connections between oil and the spiritual in Darryl Caterine, "The Haunted Grid: Nature, Electricity, and Indian Spirits in the American Metaphysical Tradition," *Journal of the American Academy of Religion*, Vol. 82 (2014): 371–397.

32. Dochuk, *Anointed with Oil*, 8.

33. Andrew Nikiforuk, *The Energy of Slaves: Oil and the New Servitude* (Vancouver, CA: Greystone Books, 2012).

34. Ernest Lee Tuveson, *Redeemer Nation: The Idea of America's Millennial Role* (Chicago: University of Chicago Press, 1968).

35. Terra Schwerin Rowe, *Of Modern Extraction: Experiments in Critical Petrotheology.* (New York: Bloomsbury/T&T Clark, 2022).

36. John J. McLaurin, *Sketches in Crude Oil: Some Accidents and Incidents of the Petroleum Development in All Parts of the Globe* (Harrisburg: Published by the author, 1896), 61.

37. *Ibid.*, 46.

38. Derrida, *The Animal That Therefore I Am*, 132.

39. For more on nineteenth century evangelical Christianimisms, see Brett Malcom Grainger, *Church in the Wild: Evangelicals in Antebellum America* (Cambridge: Harvard University Press, 2019). Although Grainger, like Wallace, remains hopeful about the environmental implications of Evangelical Christianimisms, I would emphasize—as Grainger's chapter on electric theology suggests—that the history of spiritualism and religion in nineteenth century oil extraction demonstrates their environmental ambiguity.

40. Szeman et al., *After Oil*, 48.

41. Stephanie LeMenager, *Living Oil: Petroleum Culture in the American Century* (New York: Oxford University Press, 2014), 6.

42. *Ibid.*, 6.

43. *Ibid.*, 47.

44. Szeman et al., *After Oil*, 47.

45. LeMenager, *Living Oil*, 6.

46. Mel Y. Chen, *Animacies: Biopolitics, Racial Mattering, and Queer Affect* (Durham: Duke University Press, 2012), 2.

47. LeMenager, *Living Oil*, 68, 70, 80, 92, 100.

48. Chen, *Animacies*, 13.

49. Derrida, *The Animal That Therefore I Am*, 65.

50. See Haraway's admiration and critique of Derrida's reflections: "He did not fall into the trap of making the subaltern speak . . . Yet he did not seriously consider

an alternative from of engagement either, one that risked knowing something more about cats and how to look back, perhaps even scientifically, biologically, and therefore also philosophically and intimately. He came right to the edge of respect . . . but he was sidetracked by his textual canon of Western philosophy and literature and by his own linked worries about being naked in front of his cat." Donna J. Haraway, *When Species Meet* (Minneapolis: University of Minnesota Press, 2008), 20.

51. "If it has sometimes brushed up against the experience of the sacred, I guess it's because that is just what happens when people come together in the face of the unknown, finding a sense of communion that is often lacking elsewhere, and making room for the strange kinds of words that point towards the wordless" (Dark Mountain Project, "The Devil's Door." Accessed August 13, 2020, https://dark-mountain.net/the -devils-door-a-call-for-contributions-to-issue-12/). My thanks to Jacob Erikson for introducing me to the Dark Mountain Project.

Interreligious Approaches to Sustainability Without a Future: Two New Materialist Proposals for Religion and Ecology

Kevin Minister

"You can't study sustainability anymore because I am not sustainable." This declaration came from my spouse shortly after being diagnosed with Stage IV colon cancer. My reflections on the relationship of sustainability and the future began in this moment, as many theoretical insights do, as a way of making sense of events in my own life. On the one hand, it seemed obvious to me in the moment that the death of an individual and even the injection of toxic, polluting, chemicals into the body to extend the individual's time until death were fully reconcilable with the concept of sustainability. On the other hand, I was struck by the dissonance I felt between the concepts of sustainability and the imagination of a future in this moment.

Since that encounter, I have tried to take that feeling of dissonance *too seriously* as an impetus to rethink the relationship of sustainability and time and how that fits with my ongoing work in the field of interreligious studies. This chapter represents my attempt to articulate the reflections emerging from this inquiry as part of a larger conversation with all the authors in this volume regarding the insights that come from thinking together about new materialisms, ecology, and religious studies. I make two proposals for the significance of new materialisms in the study of religion and ecology. The first relates, more directly, to the understanding of sustainable ecology and the second to theories of religion. First, in opposition to appeals for a future as a redemptive space for ecological well-being, I suggest that new materialisms orient the study of religion to a present sustainability without hope for a future as we know it. Second, in contrast to the historic importance of the world religions model to the study of religion and ecology, I suggest that new materialisms shift the study of religion toward the lived terrain of interreligious encounters.

Sustainability without a Future

Much of the discourse about sustainability is oriented around hope for the future. Both fearful projections of the horrors of climate change and calls for preserving hope for the possibility of life as we know it make sustainability conceptually subordinate to futurity. Melinda Harm Benson and Robin Kundis Craig have argued that the language of sustainability is no longer culturally useful because it relies so heavily on an imagined future that will either be stably controlled or apocalyptically chaotic.[1] Because the process of climate change is a complex, self-organizing, adaptive system that makes continual changes to our new normal, imagining a stably controlled future as we know it renders sustainability an illusion, primarily serving the maintenance of the status quo. This sort of inequitable, future-oriented concept of sustainability manifests most overtly when climate change policy conversations focus on limiting the potential future harms of carbon emissions from the significant portion of the world that currently has a small carbon footprint because of limited access to energy resources, while countries like the United States, with the highest per capita emissions, regularly miss targets on their promises for future emission reductions. In contrast to future-oriented sustainability, I posit here how sustainability, without a future, orients religion and ecology towards attuning to ways of life that support equitable resilience in the present without appeals to the future.

Discourses about sustainability tend to operate in a combination of S. Lochlann Jain's "prognostic time" and Alison Kafer's "curative time." For Jain, prognostic time attempts to make sense of the present by projecting a future rendered by a diagnosis of the conditions that define our past.[2] Through this projection, we grieve the loss of life as we know it. The planet's prognosis is an unsustainable future of rising temperatures, melting ice caps, increasingly powerful storms, droughts, and flooding, all while mosquito-born diseases spread. This prognosis demands sacrifices in the present in hopes of preserving life as we know it in the future. The demands to sacrifice and stage interventions in the present, in order to improve the outcomes of the future, move the grief and mourning inherent to prognostic time into what Alison Kafer calls "curative time," a sense of time that requires interventions in order to obtain a cure for the present ills.[3] These interventions preserve hope for a future with life as we know it. Jain and Kafer's notions of time play on both fear of an anticipated future and a need for a future that preserves life as we know it.[4] Thinking in terms of prognostic time offers our best way of projecting what will happen in a future based on our past, while thinking in terms of curative time suggests what present interventions will make the most significant difference in maintaining life as we know it.

While these experiences of prognosis and intervention are inescapable in the present, grieving the loss of our current, unsustainable life as we know it and calling for sacrifices in order to preserve a future with a life as we know it, demand substantial negative emotional resources that leave us with little energy for generating alternative way of life in the present. Fear of the future, like all threats to well-being, sends people into fight, flight, or freeze states that shift us away from the safe and social states in which we can engage in the complex, collaborative, and empathic problem-solving processes necessary to generate sustainable ways of life in the present. As anyone who has worked in a university or organization that is struggling to survive knows, fear is a poor driver of sustainable innovation. Hope for a future that preserves life as we know it—already an unsustainable way of life—bases present interventions on desires for an unsustainable way of being in the world. As Catherine Keller and John J. Thatamanil reflect, in response to the apocalyptic threat of the COVID-19 pandemic, perceiving the impending end of the world as we know it can reveal the fissures, precarity, and inequity of the world we have built, allowing us to forge a new world in light of that revelation, rather than trying to reinforce an unsustainable situation.[5] Neither the fear of the future, nor the need for a future as we know it, are sustainable orientations to life.

We must give up hope in the future and reorient life towards sustainable orientations in the present. I suggest that this is a central part of what Christopher Key Chapple is getting at in his chapter in this volume (see, "Immanence in Hinduism and Jainism: New Planetary Thinking?") when he argues that practices of immanence in Hindu and Jain traditions provide internal stabilization and a mode of ethics oriented to how we live in the present that breaks through the Big History of human progress.[6] Likewise, Chris Ives points out in his essay, "Mountains Preach the Dharma: Immanence in Mahāyāna Buddhism," that, "insofar as nirvāṇa denotes an awakened way of living, it is very much here in this world and hence 'immanent.'"[7] These arguments gesture toward religious ways of being that move toward sustainability by living in the present, without an attachment to a future as we know it.

While sustainability without a future may sound hopeless, Tommy Lynch proposes an apocalyptic divestment from this world as a way of finding hope at the end of the world. Through engagement with the work of Lee Edelman and Frank Wilderson, Lynch argues that the inescapable, violent, material structures of race, class, and heteronormativity organize the material world and that there is nothing outside of this world in which to hope.[8] While dynamism and a diversity of orientations to life exist within this world, there are no outsides in which to place our hope for a future not structured by race, class, and heteronormativity. Thus, the only time in which to place our hope is the end of this world: "Hope for the end is still a kind of hope—the hope of apoc-alypticism. It is the hope, not in the possibilities of this world (no matter how remote), but in the possibility of something new becoming possible through its annihilation."[9] There is no sustainability to be found in the future of this world as we know it, so we need to stop fearing the end of this world and start hoping for it. Placing our hope in the end of the world as we know it shifts the orientation of sustainability from the future to the present. As Lynch suggests, "Living apocalyptically is the constant investigation of what it means to engage in this refusing, of cultivating habits of refusal and of developing the capacity to sustain this refusal as a mode of negatively being in the world. It is a pure negativity, not interested in refusing in favour of a determinate alternative."[10] Sustainable ways of being are rooted in a present refusal of the demands to create a future in a racist, classist, and heteronormative world.

The present refusal of the demand to create a future as we know it taps into the desires and energies frequently quashed by the ascetic demand to forgo the present in the name of the future. This is one of the deepest insights I find in Kathi Weeks's anti-work theory. Her anti-work theory is not about rejecting the value of work, but about rejecting the unsustainable demand to work for a living.[11] There is no sustainable future in which the value of life is subordinated to an ascetic, White supremacist, heteronormative work ethic. The refusal to work for a living is about "getting a life" in the present, oriented around low impact ways of meeting the needs and desires foreclosed by the demand to work for a living, allowing new ways of being together to emerge without pre-determination. Refusing the demand to work for a living frees up the energies of a multitude of repressed desires to empower the pursuit of a sustainable way of living together in the present without determination of what sort of future that will bring. Whereas the ascetic demand to forgo the present in the name of a future as we know it pursues sustainability by quashing desires in the pres-ent, cultivating habits of refusal pursues sustainability by releasing desires.

In opposition to mainstream environmentalism's call for austerity in hopes of a future as we know it, I find inspiration in Nicole Seymour's, "The Queer-ness of Environmental Affect," as a way of feeling out the spaces of the present that cultivate desires for sustainable ways of being in the present. Seymour's analysis examines how desires for sustainability can be attached to present and past queer or non-normative spaces (rather than the future) by mining queer theory's work on attachment to places, like the gay bar. Seymour recognizes that attachment to "nature" or "the ecological" might be considered inherently queer insofar as it expands the scope of the social, but Seymour invites us to move beyond the pristine images of wilderness as we think about sustainability, and onto the darkened, foggy, dancefloors of queer nightlife, particularly those inhabited by queers of color.[12] Examining artistic representations of these queer environments, Seymour argues that attachment to these spaces, both past and present, refuses postapocalyptic narratives in which human needs are reduced to essential commodities, but is, rather, inseparable from aesthetic, cultural, and environmental resilience.[13] Seymour proposes "'inappropriateness', 'over-investment,' and 'excessiveness,' not as traits to be disavowed, but rather as rallying points for environmental activists."[14] Exploring sustainability in the wee hours of the morning, amidst a queer, dancing community that is moving together in a rumbling warehouse, one can learn and practice forms of sustainability driven by attachments and desires, foreclosed by heteronormative, White supremacist, capitalism. I affirm Seymour's assertion that these spaces hold all sorts of "bad" (meaning both negative and non-normative) affects that can become an alternative energy for cultivating sustainable ways of being both in and beyond these liminal spaces. In her essay "Dancing Immanence: A Philosophy of Bodily Becoming," Kimerer LaMothe names this ethical way of being in the present, "We do so not (just) by thinking our way through, per se, but by cultivating a *sensory awareness* of the bodily movements we are making and of what those movements are creating in us and around us in our relations with others."[15] This sensory awareness in present movements is a source for cultivating empathy and compassion, dancing new ways of being into life at the end of the world.

Embracing sustainability without a future uncouples ecological movements from the unsustainable desires for a heteronormative, White supremacist, capitalist future of working for a living and empowers us to live and dance in the present, at the end of the world. Sustainability without a future opens up the present for the critique, cultivation, and pursuit of desires foreclosed by either the need for a future as we know it or the fear of an indeterminate future. We cannot escape living in the prognostic time of climate change and, undoubtedly, we need technologies of mourning to help us cope with the lost promise of a future as we know it. But the power of cultivating sustainable ways of living together will come from freeing up the energies of unfulfilled desires to be pursued in the present by refusing the ascetic demand to create a future as we know it.

Interreligious Approaches to Sustainability

Engaging new materialisms not only shifts approaches to sustainability, it also shifts how we approach the study of religion. Like much of the broader study of religion, the "world religions model" has been the dominant paradigm in the study of religion and ecology. In the world religions model, the ecological dimensions of religion are approached through the lens of a singular religious tradition: Hinduism and Ecology, Buddhism and Ecology, Islam and Ecology, Christianity and Ecology, Indigenous Religion and Ecology, etc. Even when analyzed in modes of comparative religions, comparative theology, or religions of the world, these approaches tend to be built on the world religions model; bringing distinct traditions into conversation primarily at a theoretical level, rather than first at an already interconnected ecological level. In this section, I will argue that engaging new materialisms shifts the study of religion, and religion and ecology, toward the lived terrain of interreligious encounters.

When I use the phrase "interreligious approaches," I mean to suggest that religious identities and ways of being are constituted in dynamic relationship with their environment. The "inter" in "interreligious" refers to the interactive, intersectional, and interpersonal understanding of religious ways of being.[16] New materialisms compel scholars to take seriously the multiplicity of religious ways of being, perceiving religion as being materially constituted by the interactions of social difference (including race/ethnicity, gender/sexuality, and class/labor), placed in global context, and religious difference. My proposal that engaging with new materialisms shifts the study of religion and ecology toward interreligious approaches builds on my work on religious experience and new materialisms. In my previous work, I have argued that new materialisms help us articulate religious experience as a mode of conversion that organizes bodies. In conversation with Sara Ahmed, I suggested that religious ways of being are, "constituted through conversions—patterns of turning in space that give the sense that the community is there."[17] Religion is interactive and intersectional at a material level, formed through environmental interactions that, "divide bodies and space, set out ways of inhabiting the spaces they demarcate, and direct bodies in certain directions."[18] Religious ways of being generate flows of energy as (social) power in the production of material relationships. By moving towards an interreligious approach, I am going beyond my previous work in two ways; the interreligious approach illuminates the ways in which religion is ecologically constituted and, the manner in which religious ways of being are already constituted in interaction with other, different ways of being.

The ecological constitution of religious ways of being can be seen in several chapters featured in this volume. Elana Jefferson-Tatum's analysis of African and Africana religious traditions (see, "Africana Sacred Matters: Religious Materialities in Africa, the Caribbean, and the Americas") demonstrates how metaphysics, material, and natural are all interrelated within these traditions as gods, spirits, persons, and things that are mutually bound in a common cosmos, "Creation itself is, therefore, testimony to the fundamental interrelatedness and interdependence of the spiritual and the material, of the Earthly and the terrestrial."[19] African and Africana traditions show how interdependent religious traditions are with their ecological contexts, as part of their ecological context. Likewise, John Grim's articulation of, "Indigenous humanism [as] not centered on the human-as-separate but rather on a hominizing perspective of person-in-the-world seeing and being seen as human,"[20] shows that persons, communities, and ways of being are ecologically constituted in Indigenous traditions (see, "Indigenous Cosmovisions and a Humanist Perspective on Materialism"). Both of these examples reveal the shortcomings of the world religions model that so rarely addresses these traditions or, if it does, tends to homogenize and abstract them from their ecological contexts as "Indigenous religion." An interreligious approach that highlights the interdependence of religious traditions with their ecological contexts, brings these traditions back into focus as the distinct traditions that they are, emphasizing their importance to the study of religion.

An interreligious approach to the study of religion and ecology also helps us relocate the so-called major world religions within their ecological contexts as constituted interactively with their environment, and as changing with changing ecological contexts. I see several examples of this in other chapters featured in this volume. Chris Ives' (see, "Mountains Preach the Dharma: Immanence in Mahāyāna Buddhism") engagement with the role of nature in the understanding of emptiness—whether in ecologically affirming modes like Thich Nhat Hanh's practice of being the forest, or modes that identify nature with suffering to be escaped—demonstrates both that Mahāyāna ways of being are constituted materially in relationship to their environment, and that Mahāyāna ways of being are differentiated from one another by their relationship with their environment. We must also account for the agency of the biosphere in the constitution of religious ways of being, as Philip Clayton's ethics of matter suggest (see, "Matter Values: Ethics and Politics for a Planet in Crisis").[21] Religious ways of being are constituted as an ecological way of being in the interaction of persons with their environments and other ways of being in that environment.

Interreligious approaches prioritize "lived religion" as the dynamic encounters of religious subjects with their environment, including other ways of being. Shifting the study of religion toward the terrain of lived encounters through engagement with new materialisms, perceives that religious ways of being are always constituted in dynamic interaction with other, different ways of being rather than as pure, whole forms.[22] Both the emphasis on the purity of lineage and the derivative concept of multiple religious belonging[23] are colonial constructions. The shift towards interreligious approaches, which focus on the terrain of lived encounters in which religious ways of being are always already constituted in dynamic interaction with other ways of being, can function to decolonize the study of religion.

The commitment to the encounter with lived religion, closely reflects the humanistic method for the study of religion advanced by Tyler Roberts in conversation with Robert Orsi and Saba Mahmood, as well as with the emergence theory of religion offered by Kevin Schilbrack in this volume (see, "Emergence Theory and the New Materialisms"). In opposition to historicist reductions of religion to social formation or ideology, Roberts seeks to take seriously the agency of religious subjects in how they live, and the transformative responses that encounters with such religious subjects can elicit from scholars.[24] Schilbrack demonstrates that the materiality of religion in the frame of social emergence rejects reductionist approaches which treat religion as only a property of individuals or which treat religion as purely a social construction.[25] Schilbrack argues that, "An emergence theory of religion holds that there are social structures that exist and operate whether or not one is conscious of them and whether or not one has named them."[26] Religious ways of being are material realities that emerge in the interactions of persons, groups, and environments in ways that cannot be reduced to the sum of their parts or abstracted from the ecology in which they emerge.

Interreligious approaches perceive religion as a dynamically constituted part of an ecosystem, requiring attention to the vibrant and fragile nature of material ways of being. As scholars of religion, this means that our study and scholarship are always interactive in the religious ecology in which we live and work. This is the attentiveness to which Karen Bray's essay, "Gut Theology: The Peril and Promise of Political Affect," calls us when she invites us to sit with the guilt for the harm our work does and our inability to trust our feelings as guides.[27] As Kimerer LaMothe articulated in "Dancing Immanence," "The process of coming to believe in a relational world view entails learning to *participate as consciously as possible* in a nexus of bodily becoming that is always already generating human persons, in and through their own bodily movements."[28] The scholar of religion, just like religious identities and communities, is constituted in an interactive, intersectional ecology of being and must think, write, act, move, teach, and love in perpetual self-reflection on the conditions of their own emergence.

Conclusion

Thinking with new materialisms, I have sought to reorient ecological thinking towards a present sustainability without hope for a future as we know it, as well as the study of religion towards an ecological understanding of religion in which religious ways of being are always already constituted by their inter-actions with their environment and other ways of being. Giving up hope for a future as we know it reorients practices of sustainability towards the present pursuit of desires that have been foreclosed by the demand for heteronormative, White supremacist, capitalist futures. This proposal is practical, ethical, and political as it taps into repressed social energies to empower resistance to un-sustainable social orders, and to create spaces for the emergence of alternative ways of being. This means, on the one hand, reorienting discourses of sustain-ability around present spaces that cultivate sensory awareness of the diversity of ways of being and moving together and, on the other hand, the recognition that the pursuit of environmentally sustainable ways of being together in the world is inextricably tied to resisting heteronormativity, White supremacy, and the demand to work for a living.

I also seek a theoretical shift in religious studies, moving the focus of ap-proaches to religion and ecology away from a world religions model and toward the lived terrain of interreligious encounters—because religion itself is ecolog-ical in nature. This shift calls for the study of religion and ecology to employ intersectional approaches in order to attend to the ways in which environmental sustainability is inescapably tied to gender/sexuality, race/ethnicity, and class. There are already excellent exemplars of this in the field, including Carol Wayne White's *Black Lives and Sacred Humanity* and Whitney Bauman's ed-ited volume *Meaningful Flesh*. But we are still some distance from intersectional approaches being normative in the field of religion and ecology. This theoretical proposal also suggests that religion and ecology have an essential role to play in understanding and articulating the nature of religion as a dynamic, material way of being, constituted through the interrelation of persons and their envi-ronments, and different ways of being in those environments. Approaching religion in this way treats it as a vibrant, material, ecological way of life.

Making these two proposals together invites the field of religion and ecology to focus on spaces that create openings for exploring lived practices not bound by the demand for heteronormative, White supremacist, capitalist futures. These spaces bear the possibility of practices of religiosity pursing sustainable desires, reflecting Whitney Bauman's articulation of religiosity as practices of attunement in his essay, "Developing a Critical Romantic Religiosity for a Planetary Community."[29] Together, these proposals shift the focus of religion and ecology away from hoping in the possibility of world religions to persuade the masses to live sustainably in the name of a future as we know it and towards the way localized spaces of practice attune persons to both the diversity of life, and how to move together in ways that affirm this diversity. While there is no doubt that this can happen in spaces that have been traditionally identified as religious, this also opens up the study of religion and ecology to spaces not traditionally thought of as religious, like a queer warehouse dance party where practitioners can develop their proprioception, in attunement with the movement of the diversity of life flowing together. Whether in traditional religious communities or nontraditional spaces like the dance floor, these practices of attunement must be perceived as ways of being in the present that are defining themselves against other ways of being in the present, most specifically against heteronormativity, White supremacy, and the capitalist demand to work for a living. For, it is precisely in such communities of practice which attune us to the diversity of life, that we shall learn to dance a new world into being in the present.

Notes

1. Melinda Harm Benson and Robin Kundis Craig, *The End of Sustainability: Resilience and the Future of Environmental Governance in the Anthropocene* (Lawrence: The University Press of Kansas, 2017), 2–3, 9–11. While I appreciate their analysis of resilience as a more useful category for environmental governance, I choose to continue using the language of sustainability here precisely because I am rejecting the use of it as a construct for preserving a future. As Benson and Craig are aware, switching to the language of resilience leaves open the question of "resilience for whom?" (which they encourage governmental bodies to ask) that threatens to subordinate the present to a future as we know it. Benson and Craig, *The End of Sustainability*, 136.

2. Jain analyzes how the sense of time of persons who are diagnosed with cancer is affected by living into their prognosis, the statistical probabilities of how long they have to live, and their likelihood of disease-free survival. According to Jain, prognostic time lives in the presence of impending death without welcoming it, and thereby becomes a "technology of mourning, holding together the future and the past." S. Lochlann Jain, "Living in Prognosis: Toward an Elegiac Politics," *Representations*, Vol. 98 (1) (2007): 90.

3. Kafer contrasts "curative time" with "crip time." Curative time presumes that a cure can always be found to correct illness and disabilities to a "normal" state and expects that interventions in the name of a cure should always be pursued with hope. Crip time challenges curative time by living in light of the reality that limitations and illness are inevitably part of life, meaning that the decisions we make in the present, and the futures we imagine, must always include limitations and illness as part of life, without giving sway to an illusion of a stable norm that doesn't include these things. Alison Kafer, *Feminist, Queer, Crip* (Bloomington: Indiana University Press, 2013), 27.

4. Consider for example, the public presentation of governmental and scientific findings from the fifth Intergovernmental Panel on Climate Change report that focuses both on prognostic predictions for the likely range of climate change impacts in the coming century and advocating for urgent interventions to cure

climate change before its effects become irreversible (even as it acknowledges that is impossible). Working Groups of the Intergovernmental Panel on Climate Change, "Climate Change 2014 Synthesis Report Summary for Policy Makers", last accessed October 18, 2019, https://www.ipcc.ch/site/assets/uploads/2018/02/AR5_SYR_FINAL _SPM.pdf.

5. Catherine Keller and John J. Thatamanil, "Is This an Apocalypse? We Certainly Hope So—You Should Too," *Religion & Ethics*, last updated April 20, 2020, https://www.abc.net.au/religion/catherine-keller-and-john-thatamanil-why -we-hope-this-is-an-apo/12151922?fbclid=IwAR0Yu6vmZqTVzdgwU4zHv83pB-A -9EFlpqrg15BtxuVroBu8Ir4SAlsL7MU.

6. Christopher Key Chapple, "Immanence in Hinduism and Jainism: New Planetary Thinking?" in *Earthly Things: Immanence, New Materialisms, and Planetary Thinking*, eds. Karen Bray, Heather Eaton, and Whitney Bauman (New York, Fordham University Press, 2023).

7. Christopher Ives, "Mountains Preach the Dharma: Immanence in Mahāyāna Buddhism," in *Earthly Things: Immanence, New Materialisms, and Planetary Thinking*, eds. Karen Bray, Heather Eaton, and Whitney Bauman (New York, Fordham University Press, 2023).

8. Tommy Lynch, "A Political Theology for the End of the World" (paper presented at the *Political Theology Network: Inaugural Conference*, Atlanta, Georgia, February 15, 2018). Lynch's recently released book on *Apocalyptic Political Theology* offers an important contribution on religion and apocalyptic thinking that further develops the analysis in this presentation. Thomas Lynch, *Apocalyptic Political Theology: Hegel, Taubes and Malabou* (London: Bloomsbury, 2019), Chapter 5.

9. Ibid.

10. Ibid.

11. Kathi Weeks, *The Problem with Work: Feminism, Marxism, Antiwork Politics, and Postwork Imaginaries* (Durham: Duke University Press, 2011), 46.

12. Nicole Seymour, "The Queerness of Environmental Affect," in *Affective Ecocriticism: Emotion, Embodiment, Environment*, eds. Kyle Bladow and Jennifer Ladino (Lincoln: University of Nebraska Press, 2018), 240–241.

13. Ibid., 246.

14. Ibid., 251.

15. Kimerer LaMothe, "Dancing Immanence: A Philosophy of Bodily Becoming," in *Earthly Things: Immanence, New Materialisms, and Planetary Thinking*, eds. Karen Bray, Heather Eaton, and Whitney Bauman (New York: Fordham University Press, 2023).

16. For further details on my definition of interreligious studies as interactive, intersectional, and interpersonal, see Kevin Minister, "Decolonizing Interreligious Studies," in *Interreligious Studies: Dispatches from an Emerging Field*, ed. Hans Gustafson (Waco: Baylor University Press, 2020).

17. Kevin Minister, "Organizing Bodies," in *Religious Experience and New Materialism: Movement Matters*, eds. Joerg Rieger and Edward Waggoner (New York: Palgrave Macmillan, 2016), 59.

18. Ibid., 60.

19. Elana Jefferson-Tatum, "Africana Sacred Matters: Religious Materialities in Africa, the Caribbean, and the Americas," in *Earthly Things: Immanence, New Materialisms, and Planetary Thinking*, eds. Karen Bray, Heather Eaton, and Whitney Bauman (New York: Fordham University Press, 2023).

20. John Grim, "Indigenous Cosmovisions and a Humanist Perspective on Materialism," in *Earthly Things: Immanence, New Materialisms, and Planetary Thinking*, eds. Karen Bray, Heather Eaton, and Whitney Bauman (New York: Fordham University Press, 2023).

21. Philip Clayton, "Matter Values: Ethics and Politics for a Planet in Crisis," in *Earthly Things: Immanence, New Materialisms, and Planetary Thinking*, eds. Karen Bray, Heather Eaton, and Whitney Bauman (New York: Fordham University Press, 2023).

22. The critique of the purity of religious lineage has been most impactfully presented in Jonathan Z. Smith, *Relating Religion: Essays in the Study of Religion* (Chicago: University of Chicago Press, 2004), 171.

23. See for example, Paul Hedges, "Multiple Religious Belonging after Religion: Theorising Strategic Religious Participation in a Shared Religious Landscape as a Chinese Mode," *Open Theology*, Vol. 3 (2017): 48–72; Devaka Premawardhana, "The Unremarkable Hybrid: Aloysius Pieris and the Redundancy of Multiple Religious Belonging," *Journal of Ecumenical Studies*, Vol. 46 (1) (2011): 76–101; Jeannine Hill Fletcher, "We Are All Hybrids," in *Monopoly on Salvation?: A Feminist Approach to Religious Pluralism* (New York: Continuum, 2005); and Michelle Voss Roberts, "Religious Belonging and the Multiple," *Journal of Feminist Studies in Religion*, Vol. 26 (1) (2010): 43–62.

24. Tyler Roberts, *Encountering Religion: Responsibility and Criticism After Secularism* (New York: Columbia University Press, 2013), 114–118.

25. Kevin Schilbrack, "Emergence Theory and the New Materialisms," in *Earthly Things: Immanence, New Materialisms, and Planetary Thinking*, eds. Karen Bray, Heather Eaton, and Whitney Bauman (New York: Fordham University Press, 2023).

26. Ibid.

27. Karen Bray, "Gut Theology: The Peril and Promise of Political Affect," in *Earthly Things: Immanence, New Materialisms, and Planetary Thinking*, eds. Karen Bray, Heather Eaton, and Whitney Bauman (New York: Fordham University Press, 2023).

28. LaMothe, "Dancing Immanence."

29. Whitney Bauman, "Developing a Critical Romantic Religiosity for a Planetary Community," in *Earthly Things: Immanence, New Materialisms, and Planetary Thinking*, eds. Karen Bray, Heather Eaton, and Whitney Bauman (New York: Fordham University Press, 2023).

Which Materialism, Whose Planetary Thinking?

Joerg Rieger

Introduction

Ever since the publication of Lynn White's classic article, "The Historical Roots of Our Ecologic Crisis" in 1967, Western Christianity has been suspected of being a major cause of the rampant environmental destructions that mark our age. According to White, Western Christianity is, "the most anthropocentric religion the world has seen,"[1] playing humanity and the natural world against each other and leading to the devaluation and destruction of the latter on a grand scale. White argued that this attitude has shaped Western culture to the core and influenced modern science. He also noted that even post-Christian Western culture continues to be shaped by a disregard for nature.

Another way to frame White's suspicion would be to question the impact that religions (and in particular Western Christianity), emphasizing the non-material, ethereal, and transcendent, have on the environment. This is one of the common threads of the chapters in this volume, and there seems to be broad agreement among scholars of religion that the most appropriate response is to give more prominent voice to religious traditions that emphasize the material, non-ethereal, and immanent. This is where materialism and the new materialism enter the conversation.

The materialisms to be engaged here are, of course, not of the reductionist or crude determinist kind that rule the day in certain discourses in the natural sciences.[2] The materialist traditions that interest us are conversant with dialectical traditions that have emerged out of particular historical struggles—traditions that are still often neglected in the field of religious studies and even in some strands of the new materialism itself. My expectation is that these traditions can help engage religious discourses in ways that are congenial to some of their fundamental concerns, starting with the Abrahamic traditions. Moreover, engaging these particular materialist traditions is one way to address the ethical implications of religious discourse.[3]

Materialism—What's New?

New materialist approaches, broadly conceived, are useful because they can help reclaim aspects of the long-suppressed dialectical legacy of certain forms of materialism (and in some cases certain kinds of Hegelian idealism) for the study of religion. At stake is not the abortive discussion of whether "material" or "ideal" factors are all-determinative, but how these factors influence and shape each other. New materialisms can contribute to a constructive rethinking of the category of religion, and how it intersects with and is part of material reality. In addition, new materialisms invite analyses of the existence of alternative religious practices that do not conform to the dominant powers, providing deeper understandings of their nature and promise.

New materialist religion scholars Clayton Crockett and Jeffrey Robbins, for instance, reclaim Ludwig Feuerbach's materialist critique of religion with a positive twist: that human concerns play a major role in the formation of religion, they note, is not reason for rejecting religion, but for reclaiming it.[4] Religion, in the account of these new materialists, can become a force for empowerment and social change. New materialist scholars of religion invite the field of religious studies to take seriously material and physical realities and to reconceive the roles of economics, ecology, and energy in the production of religious experience.

This broadening of older, dialectical materialist traditions is promising for several reasons. First, it includes a wealth of new insights produced in the natural sciences, from quantum physics and genetics to neurobiology, some of which are explored in other chapters in this volume. The natural sciences have come a long way from the days of Newtonian physics, where cause and effect, subject and object, were easily distinguished and every question had a straightforward answer. This does not mean, however, that the natural sciences need to be followed uncritically. As the sciences are taken more seriously in the humanities and religious studies, we need to keep in mind that "sciences (and technologies) and their societies co-constitute each other," as Sandra Harding has pointed out from a feminist and postcolonial perspective.[5]

Second, new materialisms reshape and broaden our understanding of agency, as agency does not need to be rooted in ideas and good intentions. Realizing that there is a "mismatch between actions, intentions, and consequences," new materialists Diana Coole and Samantha Frost advocate an open systems approach to the interactions between socioeconomic and environmental conditions, combining biological, physiological, and physical processes. By the same token, matter needs to be considered as having agency in its own right, as new materialists emphasize "the productivity and resilience of matter."[6] Matter is always in a process of becoming rather than merely being. One of the great advantages of these approaches is that the weight of transformation does not rest on the shoulders of well-meaning individuals.

New materialism offers some inspiration, although we need to gain further clarity about who the agents are in these models and how such agency can be realized under the conditions of capitalism, which seeks to harness every form of agency, human as well as nonhuman, large and small, for its own purposes. While it is commendable that, as Coole and Frost point out, in the new materialisms "the capitalist system is not understood in any narrowly economistic way but rather is treated as a detotalized totality that includes a multitude of interconnected phenomena and processes,"[7] we need continued investigations of how capitalism affects us and our world, including religion. After all, historical and dialectical materialisms emerged amidst the tensions of capitalism, and this history needs to be considered as we engage new materialisms.[8] This brings us back to the overarching question explored by the contributors to this book, even though it is not always posed explicitly—"What are we up against?"

One approach to this question might be Coole and Frost's emphasis on the "immense and immediate material hardship for real individuals" who lost their savings, their pensions, their houses, and their jobs in the meltdown of the economy after 2007.[9] Even a decade and a half later, many have never really recovered from the Great Recession, as most of the benefits of the economic recovery went straight to the top. A majority of the jobs that were created in the wake of the recession were not of the same quality as the jobs that were lost, with fewer benefits, lower compensation, and less influence and power. The COVID-19 pandemic only added to this situation, even though some forms of labor, mostly lower-paid, are now classified as "essential." Within this landscape, the structures of neoliberal capitalism require another look, both by those seeking to reclaim materialism and by those investigating matters of ecology and religion and what is called the "planetary."

Moving the Conversation Forward

Referring to neoliberal capitalism in an effort to answer the question "What are we up against?" is a starting point that does not seem to need much justification, as this form of capitalism shapes not only large structural realities around the globe, but also personal relationships and human subjectivity at the micro level. Of course, many of these dynamics are still awaiting investigation from a religious studies perspective.[10] Nevertheless, linking the topics of materialism and planetary thinking to the developments of neoliberal capitalism seems appropriate, as all the topics we are discussing are thoroughly shaped by it: not only the material realities of the environment, and what we call nature but also the material realities of agency and affect, as well as all the ethereal realities one can possibly think of, including ideas and angels. Keep in mind that even in classical theist traditions, aseity is only claimed for God— and even this claim is relativized by most of the practiced religious traditions, for instance, when people pray to God, thus robbing the divine of its splendid isolation.[11]

The surprise is not that the topic of capitalism would come up, but that this topic is not reflected in a more central fashion even in discussions regarding the new materialism and religion, and the growing number of efforts to address links between ecology/environment and religion. For comparison's sake, if the conversations that are taking place in the chapters of this book were located in Germany between 1933 and 1945, how would later observers feel about a lack of references to fascism?

While new materialists tend to operate with a stronger awareness of capitalism than scholars of religion (including religion scholars dealing with new materialism), a more substantial assessment of neoliberal capitalist reality from the perspective of working people and laborers might make a difference in the development of the study of religion. This is where remembering the historical analyses of older dialectical materialisms can be beneficial, for instance in various Marxist traditions. These analyses were tied both to capitalism in general and to the tensions that arise in the world of labor, as this is where capitalism is rooted, from where it takes off, and where many of its greatest tensions manifest themselves. This is true even in the days of contemporary financial capitalism—why else would the pushback against working people and their associations (unions, etc.) continue to be so severe? Moreover, this is also where concerns for agency, affect, and even the ability to think and believe differently, have deep roots. We are talking about matters highly relevant to the study of religion and theology.

It is surprising, then, that new materialism and its manifestations in the study of religion and theology hardly mention labor or work. When labor is mentioned at all (let alone in one collection of essays that represents the spectrum of new materialisms) the claim is that the new materialisms should go beyond the focus on labor that has been characteristic of materialisms in the past. Rosi Braidotti, a leading materialist feminist, develops her proposal as if labor did not exist, promoting instead a "biocentered egalitarianism" that "breaks the expectation of mutual reciprocity," concluding that we have to give up ideas of retaliation and compensation.[12] While retaliation and "tit-for-tat" may indeed not be the most productive ways of relating to others, giving up notions of compensation and reciprocity altogether is not really an option for people who have to work for a living—the proverbial "99 percent." In the Abrahamic traditions, regulations of debt and the forgiveness of debt correspond to these concerns.[13]

To move forward, we need to take a step back. Where Crockett and Robbins "posit earth as subject"[14]—a factor that is indeed too often overlooked in religious discourse and definitely deserves greater attention—what if we were to add working people as subjects—including both productive and reproductive labor—as these are the agents doing most of the work that sustains humanity and whose contributions to subjectivity are often overlooked? Historical and dialectical materialisms cannot afford to do without those whose subjectivity is neglected in dominant discourse, stretching from politics to economics and religion (Caesar built Rome, Henry Ford produced automobiles, and Paul expanded the church's mission). At present, not only are both the productive and reproductive contributions of the Earth and of working people casually overlooked, they are systematically repressed—and this is true, not only for the world of labor and the production of goods and services, but also for the production and reproduction of ideas, affects, and agency.

While in the current climate it may be easier to talk about the Earth as subject rather than working people as subjects—even tense topics of ecology seem to be easier to discuss than labor in many communities—we may not be able to do one effectively without the other. That which devalues the agency of working people (by cutting salaries, benefits, hours at work, pushing employment at will, the gig economy, and exploiting unpaid reproductive work) also devalues the agency of the planetary.

Examples of efforts to downplay the role of working people (who constitute the majority of humanity) as subjects of religion are legion, ranging from the backlashes against Latin American liberation theology, feminist theologies, Black theologies, and any other theologies from the margins that keep demanding more than mere inclusion within the dominant system. At the same time, the study of popular religion, which has become a hot currency in recent years, seems to be perceived as less of a threat, as it does not experience similar backlashes when popular religious experience is studied in itself, in isolation from dominant religious experience. In Christian theology, for example, Jesus preaching good news to the poor and uplifting them, is perceived as less of a threat than Jesus proclaiming woe to the rich, although the two belong together (Luke 6:20, 24).

Recalling Karl Marx's critique of Ludwig Feuerbach's materialism is instructive here. Going beyond Feuerbach, Marx observes that material objects and matter itself are not mere givens but are produced by labor and commerce so that materialism needs to take into account the produced nature of matter. Material reality is never a static entity, as it is constantly produced and reproduced. Today, this is truer than ever, even for the so-called "natural" world itself, a fact which did not escape Marx even though the human control of nature was much less developed in his time. Human agency is part of the picture here, specifically in the agency of working people, as both nature and workers are constantly engaged in "changing the form of matter." This is where capitalism is rooted, as wealth is generated from the interplay of labor and nature.[15] While the new materialisms have done an excellent job of deepening our understanding of the productive capacities and the agency of the "natural" world, far beyond what Marx and his contemporaries could have known, we still need a deeper analysis of the fundamental—and therefore potentially revolutionary—contributions of human labor (both productive and reproductive) in the current economic situation.

Material Practices Revisited

Following these materialist intuitions, the task of religion and theology is to study material relations that are produced in the context of the particular relations of power rather than material relations in universal terms—this is the mistake that many theists and atheists (and some nontheists) make. In order to identify and connect with alternative forms of religion, the materialist study of religion needs to take into account the history of how power is shaped and reshaped in particular social relations and in relation to social movements, as well as pay attention to who benefits and who does not in a given system (including both people and the Earth). That is what is missing in Feuerbach, who mistakes dominant religion for religion in general, as well as in some of his contemporary admirers.

Jason Edwards makes a start in his conclusion to an important collection of essays on the new materialisms, when he argues for a return "to a kind of historical materialism that focuses on the reproduction of capitalist societies and the system of states, both in everyday practices of production and consumption and in the ideological and coercive power of states and the international system."[16] Still, we need to raise the question of how such a focus on "material practices" can benefit from a closer connection to the realities of labor; and how the study of religion and theology, as well as concerns for ecology and the environment, might benefit.

Taking a closer look at productive and reproductive labor will help us become more aware of the produced nature of virtually everything that surrounds us, nature and religious practice included. This does not imply a negative judgment, as being produced is not a drawback; but it reminds us of the fact that nothing ever just "fell from the sky" and, equally important, that there may be alternative mechanisms of production we can explore and harness.

In addition, there is an odd sort of transcendence that occurs when produced objects are commodified. This is part of the problem that we are up against. Marx uses the example of a table. In terms of its "use value," there is nothing mysterious about a table. Wood, produced by nature, is altered by human labor to produce a common thing used for particular purposes, a dining room table, a desk, a kitchen table, etc. Capitalist transcendence enters the equation in terms of the exchange value of the table. In economic exchanges, the table becomes a commodity and what matters is no longer the labor, the materials, or the use value, but the profit made when it is sold. Because profit is usually thought of as a relationship between things, what is concealed is that it is actually produced in a relationship between people.[17] In Marx's words, "A commodity is therefore a mysterious thing, simply because in it the social character of men's labour appears to them as an objective character stamped upon the product of that labour; because the relation of the producers to the sum total of their own labour is presented to them as a social relation, existing not between themselves, but between the products of their labour."[18] Marx compares this to religious ideas, where "the productions of the human brain appear as independent beings endowed with life."[19]

While the significance of labor and of human relationships is covered up with this process of commodification (this then leads to "commodity fetishism"), labor and human relationships do not disappear. Sara Ahmed, another new materialist, adds that the example of the table also reminds us of other divisions of labor, manifest for instance in the division between who usually works at a desk and who usually works at a kitchen table. In this example, the kitchen table represents the racial and class-based divisions of labor, as desk work is often supported by the domestic labor of Black and working-class women.[20] Observing these relationships matters, and ecological and environmental relationships become clearer from this point of view.

These reminders of specific material relations in the world of labor help rethink the options for the study of religion and theology. What Marx calls religion in the above example is how philosophical idealists and many theologians view religion, as "independent beings endowed with life." But there is no reason religion cannot also be viewed from a more materialist perspective that takes into account the relation of life to material realities. The same is true for the notion of transcendence: transcendence does not have to be an idealist concept, but neither does it have to be determined by capitalist relations. There are materialist ways to conceive of transcendence, for instance when it is defined, not in opposition to immanence, but as transcending one kind of immanence in favor of another. This is in sync with the Jewish traditions, where the notion of salvation has to do with the flourishing of life, rather than with escape to another world. In Christianity, this is one way to understand the incarnation of Christ, where the Roman Empire is transcended by God's solidarity with peasant movements rather than by an escape into the ethereal.[21]

While capitalism needs to cover-up the contributions of labor, the study of religion can resist this cover-up and benefit from taking labor into account.[22] The rethinking of working people as subjects might throw new light on notions of the Earth as subject as well—Whitney Bauman's (following Gordon Kaufman's) notion of "biohistories" might prove to be helpful here.[23] What are the contributions of the planet, how might its agency be conceived?

It is not necessary to demand that material practices be limited to activities that are directly related to processes of production. They may include, as Edwards suggests, "all those practices involving material bodies—organic and nonorganic—that . . . can be seen as a totality of practices that reproduce the relations of production over time."[24] At a time when the nature of labor is shifting and more and more people are being pushed into the informal sectors where they might hold down casual jobs in the gig economy, temp jobs, or no formal jobs at all, this is an important reminder. It also reminds us of the importance of the kind of work that is done without compensation, like housework or volunteer work, and other productive and reproductive activities that people are doing off the clock. In the bigger picture, this includes the work of nonhuman agents and ecosystems. What Edwards calls "the constitution of experience through the manifold forms of material practice outside the immediate space of production,"[25] is also valuable, especially when thinking about religion. This is akin to what *Mujerista* theologian Ada María Isasi-Díaz called "*lo cotidiano*," the "everyday," which is at the heart of life for the majority of Latinas (as Isasi-Díaz emphasized throughout her work), as well as for most people who have to work for a living.[26] Some scholars of religion and theology are currently rediscovering the fact that many of the Abrahamic traditions are more concerned with the material and the everyday than with the ethereal.[27]

Still, in the current cultural climate in the United States there is very little reflection on how subjectivity, including religious subjectivity, is constituted through the regular processes of production at work.[28] There have been hardly any studies in recent decades on how class, as a relational category, shapes subjectivity and religion.[29] Without relating back to these basic processes that occupy the bulk of the waking hours of most people—the 99 percent is defined as people having to work for a living—we may not be able to develop a clear enough understanding of the importance of material practices more broadly conceived. Merely talking about the material practices of "everyday life" can also be misleading if it is not acknowledged that everyday life, invariably, is shaped by labor relations, both formal and informal; the same is true for the growing discussions and engagements of affects.[30]

At stake is not merely analyzing the impact of material practices on religion, but also the identification of alternative ways of life and religious experiences that grow out of these material practices, and the potential that reshaping processes of production and reproduction might have for reshaping religious experience and practice. This is not a romantic dream about life outside the dominant system—and it has even less to do with conventional understandings of transcendence. As Edwards argues, "the material practices constitutive of modern life are the only grounds from which we could hope and expect to bring about important political and social transformations."[31] While material practices under the conditions of capitalist exploitation of labor can, and do, make us compliant to the status quo, they also harbor the potential for resistance and for producing alternatives.

The challenge is to figure out which material practices are currently producing the most fertile ground for emerging alternatives. In Jesus' time, the practices of the peasants seem to have provided this ground; in Marx's time it would have been industrial labor—the proletariat. Today, that question is more complex—some would point to the so-called precariat[32] and even to the agency of the Earth. Whatever the way forward, working people and labor movements (formal and informal) need to be part of the conversation, especially because many religions (including the Abrahamic ones) often started with working people, or were shaped by them in significant ways.[33]

Alternative religious subjectivities and practices are, therefore, closely linked to material practices and alternative ways of life. Once again, this is where labor is important, as it has a long track record of producing movements that have not only kept some of the worst abuses of capitalism in check, but also built communities at the local, national, and international levels that have made positive contributions to the world. Labor is where progress in the fight against sexism (according to the traditions of socialist feminism) and racism (according to Martin Luther King and W.E.B. DuBois among many others) has been made, and here is where progress in the fight against ecological destruction is finding increasing support as well.

Scholars of religion and theology would do well to investigate more thoroughly how alternative religious subjectivities and practices shape up in the history of particular movements of exploited working people, in solidarity with the agency of the exploited Earth, and what difference they might be making in the context of dominant religion. These scholars cannot produce such alternative religious subjectivities and practices synthetically and studying them effectively will entail some involvement in the various resistance movements of our time. Recent movements such as Occupy Wall Street, Black Lives Matter, Reproductive Justice, and Ecojustice have implications for religion, including its practices, its doctrines and beliefs, and its ways of life. In short, the confluence of materialism and religion needs to develop in terms of social movements.

Conclusions

Historical-dialectical thinking in the materialist traditions asks questions of power: What are we up against, where are the contradictions, and what are the alternatives? Among the most basic contradictions in neoliberal capitalism are still the exploitative relations of the ruling class and the working class, despite the developments of financial capitalism that currently fascinate many scholars of religion and theology.[34] These exploitative relationships are mirrored in capitalism's extractive relationships with the Earth. Knowing that we cannot escape these tensions—tensions that challenge the last pretentions of scholarly objectivity—scholars of religion and theology need to decide what to do with them and how to rethink them in the present. To do that, the question of class becomes relevant once again, after a long silence on the topic. Even when class is referenced by new materialists like Edwards, who is one of the few exceptions, it is only in order to note that class is not everything.[35] But how many voices are left today, whether in new materialism or the study of religion and theology, that would claim that class is everything?

Meanwhile, there are some explorations of how the movements of working people shape and reshape ecological concerns, from community gardens to worker cooperatives.[36] Rethinking class and ecology is a major battleground, as the fathers of neoliberal economics, such as Friedrich von Hayek, also draw on ecological models, providing the dominant paradigms of current material, planetary, and even religious thinking. In the case of neoliberal economics, a specific interpretation of biological evolution provides the basis for emphasizing competition as natural and necessary for flourishing. Challenging this particular material and planetary logic, what can scholars of religion and theology say about the agency of people and the Earth that is different and new, and from where would they draw inspiration if not from observations of embodied alternatives?

Notes

1. Lynn White, Jr, "The Historical Roots of Our Ecologic Crisis," *Science, Vol.* 155 (3767) (March 1967): 1205.

2. For a brief engagement with these materialisms see my chapter "Why Movements Matter Most: Rethinking the New Materialism for Religion and Theology," in *Religious Experience and New Materialism: Movement Matters*, eds. Joerg Rieger and Edward Waggoner (New York: Palgrave Macmillan, 2015), 135–156.

3. For an extended argument of topics developed in "Why Movements Matter Most," see Joerg Rieger, *Theology in the Capitalocene: Ecology, Identity, Class, and Solidarity* (Minneapolis: Fortress Press, 2022).

4. Clayton Crockett and Jeffrey Robbins, *Religion, Politics, and the Earth: The New Materialism* (New York: Palgrave Macmillan, 2012).

5. Sandra Harding, "Beyond Postcolonial Theory: Two Undertheorized Perspectives on Science and Technology," in *The Postcolonial Science and Technology Studies Reader*, ed. Sandra Harding (Durham: Duke University Press, 2011), 21.

6. Diana Coole and Samantha Frost, "Introducing the New Materialisms," in *New Materialisms: Ontology, Agency, Politics*, eds. Diana Coole and Samantha Frost (Durham: Duke University Press, 2010), 7.

7. Ibid., 29.

8. As such, historical and dialectical materialisms are part of what is called "modernity" but always pushing back against basic tenets of modernity as defined by capitalism.

9. Coole and Frost, "Introducing the New Materialisms," 31. I talk about this in terms of the "logic of downturn." See Joerg Rieger, *No Rising Tide: Theology, Economics, and the Future* (Minneapolis: Fortress Press, 2009).

10. One of the few examples of such investigations is Bruce Rogers-Vaughn, *Caring for Souls in a Neoliberal Age: New Approaches to Religion and Power* (New York: Palgrave Macmillan, 2016).

11. It is often overlooked especially in Christian theology that even a simple traditional act such as praying to God can call divine aseity into question.

12. Rosi Braidotti, "The Politics of 'Life Itself' and New Ways of Dying," in *New Materialisms: Ontology, Agency, Politics*, eds. Diana Coole and Samantha Frost (Durham: Duke University Press, 2010), 214.

13. Even in Christianity, debt, for the most part, refers to the economic problem of indebtedness of people and communities under the auspices of the Roman Empire, rather than some abstract religious phenomena. This is how materialist scholars of the New Testament read the petition for the forgiveness of debts that is part of the Lord's Prayer (commonly mistranslated as "trespasses").

14. Crockett and Robbins, *Religion, Politics, and the Earth*, xx.

15. "[Humans] can work only as Nature does, that is by changing the form of matter. Nay more, in this work of changing the form he [or she] is constantly helped by natural forces." Karl Marx, *Capital: A Critique of Political Economy*, Volume 1, Book 1, ed. Frederick Engels, trans. Samuel Moore and Edward Aveling (Moscow:

Progress Publishers, First English Edition 1887); https://www.marxists.org/archive/marx/works/download/pdf/Capital-Volume-I.pdf.

16. Jason Edwards, "The Materialism of Historical Materialism," in *New Materialisms: Ontology, Agency, Politics*, eds. Diana Coole and Samantha Frost (Durham: Duke University Press, 2010), 283.

17. Marx talks about these issues in *Capital*, 46–47, but since he does not mention the terms exchange value and profit in this section, it is difficult to follow.

18. Marx, *Capital*, 46–47.

19. Marx, *Capital*, 47.

20. Sara Ahmed, "Orientations Matter," in *New Materialisms: Ontology, Agency, Politics*, eds. Diana Coole and Samantha Frost (Durham: Duke University Press, 2010), 248–254.

21. See the comments on transcendence in Joerg Rieger and Kwok Pui-Lan, *Occupy Religion: Theology of the Multitude, Religion in the Modern World* (Lanham: Rowman & Littlefield, 2012), 71–76.

22. One of the fundamental shifts that I am suggesting in my book *No Rising Tide* is the move from the debate of economics from a focus on redistribution to production.

23. Whitney A. Bauman, *Religion and Ecology: Developing a Planetary Ethic* (New York: Columbia Press, 2014), 165.

24. Edwards, "The Materialism of Historical Materialism," 283.

25. Edwards, "The Materialism of Historical Materialism," 288.

26. Ada María Isasi-Díaz, *Mujerista Theology: A Theology for the Twenty-First Century* (Maryknoll: Orbis Books, 1996).

27. See the argument in Joerg Rieger, *Jesus vs. Caesar: For People Tired of Serving the Wrong God* (Nashville: Abingdon, 2018), chapter 3.

28. The so-called "New Working Class Studies" are among the exceptions.

29. To my knowledge, the notion of class as relationship has only been picked up lately in religious studies. See *"Religion, Theology, and Class: Fresh Conversations after Long Silence,"* in *New Approaches to Religion and Power*, ed. Joerg Rieger (New York: Palgrave Macmillan, 2013); and *Faith, Class, and Labor: Intersectional Approaches in a Global Context*, eds. Jin Young Choi and Joerg Rieger (Eugene: Pickwick, 2020). Class, when discussed at all, is often described in terms of stratification rather than in terms of relationship.

30. Even leisure time is not "off the hook," because it is designed for the reproduction of productive labor capacities and shaped by the trends and interests of the capitalist economy. Capitalist economy rests on the foundations of labor, a fact that it constantly seeks to cover-up.

31. Edwards, "The Materialism of Historical Materialism," 292.

32. Guy Standing, *The Precariat: The New Dangerous Class* (London: Bloomsbury Publishing, 2011).

33. This is argued in Joerg Rieger and Rosemarie Henkel-Rieger, *Unified We Are a Force: How Faith and Labor Can Overcome America's Inequalities* (St. Louis: Chalice, 2016).

34. Katherine Tanner, *Christianity and the New Spirit of Capitalism* (New Haven: Yale University Press, 2019).

35. Edwards, "The Materialism of Historical Materialism," 296.

36. See, for instance, *Sustainable Lifestyles and the Quest for Plenitude: Case Studies of the New Economy*, eds. Juliet B. Schor and Craig J. Thompson (New Haven: Yale University Press, 2014) and Naomi Klein, *This Changes Everything: Capitalism vs. the Climate* (New York: Simon & Schuster, 2014). See also, Annika Rieger and Joerg Rieger, "Working with Environmental Economists," in *T&T Clark Handbook of Christian Theology and Climate Change*, eds. Ernst Conradie and Hilda Koster (London: Bloomsbury / T&T Clark, 2020), 53–64.

Rewilding Religion for a Primeval Future

Sarah M. Pike

This essay is not so much about planetary *thinking*, but is concerned with *practices*, especially ritualized practices, which express and constitute immanent worldviews. There has been little attention to ritual practice in literature on the new materialism. Because I am an ethnographer who studies ritual practices, among other things, I focus on what people are doing with other species and other forms of matter and how they talk about their experiences with the more-than-human world. In this essay, I understand ritual as a distinctive way of enacting and constituting relationships among humans, as well as between humans and nonhuman animals, plants, rocks, and so on.[1] I am interested in how planetary and materialist thinking *emerges from and is expressed through* ritualized relationships with the more-than-human world. These cases are from my twenty plus years of research on contemporary North Americans' *new practices of immanence that reimagine older practices*. They are examples of rewilding religion/spirituality in contemporary America and of new animisms lived and expressed through ritualization.

I will also pick up on some of the themes explored in the other chapters of this volume that resonate with the rewilding movements and worldviews I discuss, such as the body's training and porosity, and relationships with consecrated objects. If many of the other chapters in this volume ponder the questions, "What are we to do?" and "How should we consciously engage with the material?" then my interlocutors would answer, "Do rituals to save the Earth," because that is exactly what many of them are engaged in: ritual practices that express, shape, contest, and constitute our relationships with other beings and landscapes on this planet with the explicit goal of responding to environmental change and crisis.

Three of the communities I have conducted fieldwork with over the past two decades—contemporary Pagans (Witches, Druids, and others recreating pre-Christian European traditions), radical eco-activists, and ancestral skills practitioners—are characterized by immanent worldviews. They adopt what Jane Bennett describes as "strategic animism" and look to the past for alternate ways to relate to the more-than-human world.[2] These movements took shape during a postmodern, post-industrial historical era in the U.S. (the 1970s and 1980s) when environmental problems were being identified and explicitly addressed. They created new, embodied, ritualized practices that draw on and reimagine older practices in order to shape close relationships with plants, rocks, tools, other animals, and landscapes. For them, what is sacred and meaningful is on the Earth and in the company of Earth's other denizens; they foreground immanence and reject transcendence.

These communities might be seen as outliers in the North American religious landscape and most of them are not indigenous to North America (my cases are in North America, but their counterparts exist in many other geographical regions). In many ways, these movements and communities are not really outliers; they are where some of our most pressing issues are getting worked on and reimagined and where responses to environmental crisis are being created and put into practice. Yet there are perils to experimenting with new ritual forms that animate, and make vibrant, our relationships with the things and beings around us. Rituals to save the Earth in these communities are complex and contested. This essay is an invitation to explore both the promise and peril of these possibilities for living on the planet.

Spiral Dances

In 2018, I received an email from Starhawk, a well-known feminist Witch, writer, activist, and permaculture teacher, about a Samhain ritual. (Samhain, on October 31, is the new year for many contemporary Pagans). In her email, Starhawk explained that her commitments are "teaching practical, spiritual, and strategic tools for challenging and regenerating both natural and social ecosystems." Her message read as follows:

> **Samhain, a Time of Power: Calling Upon the Ancestors for Guidance . . .**
> Autumn is here, the leaves are turning, and I can feel the veils thinning as we move toward Samhain—Halloween—and Dia de los Muertos. It's a time of power, when the ancestors offer their guidance and the descendants to come clamor for us to make a viable world for them to live in.

There has never been a more important time to gather all of the power we collectively carry and use it to further the values of nurturing, caring, and justice!

The veils she mentions are between our everyday world and the world of the dead. The email included information about a Spiral Dance in the San Francisco Bay Area called " A Ritual of Death and Rebirth for These Times," an event "Where land meets water, together with our beloved dead, [and] we stir a cauldron of tears and outrage to brew an elixir of radical peace and justice."[3]

Starhawk's Samhain events blend religion, politics, and environmentalism and are based on the assumption that ritual practices have an impact on political and environmental issues. Starhawk is one of the most influential contemporary Pagan writers and is at the convergence of the movements and communities I study: she has been an active leader in eco-activist and anti-capitalist protests since the 1970s and is currently a permaculture teacher who has participated in ancestral skills gatherings. Central to Starhawk's work is the practice of ritual as magical politics that can change participants' consciousness as well as the world around them.

The Spiral Dance, which Starhawk and other San Francisco Bay Area Witches developed in the late 1970s, is a ritual that contemporary Pagans have performed in a variety of contexts, including during environmental protests. Central to the organizers' initial purpose for the first Spiral Dance, held in 1979 in San Francisco, was an "attempt to integrate a political vision and a spiritual vision" that came in part from Bay Area Pagans' involvement in anti-nuclear politics.[4] A key component of the annual Spiral Dance, and of much Pagan ritual practice, is the creation of altars. For most Pagans, altars express this religion of immanence and include objects related to the natural world, especially the elements of earth, air, fire, water, and spirit.[5]

An American Pagan community, Reclaiming, turned to the Spiral Dance during protests against the Diablo Canyon nuclear power plant in the 1980s as well as anti-logging protests in the 1990s.[6] At one anti-logging protest in northern California, Starhawk, one of the founders of Reclaiming, invoked a spiral. In her account of the protest, she describes it as follows:

We gather in the woods to claim this forest as sacred space, to charge our letters, our petition, our phone calls, with magic, that extra something that may shift the structures just a bit, create an opening for something new. We sing, we chant, we make offerings, we claim this land as sacred space. . .we intend to conjure back the salmon, the ancient groves, the community of those indigenous to this place. We draw spirals in the dirt."[7]

The spiral image and embodied movement connect these ritual practitioners to the land they want to protect and at the same time inscribe their protective intentions on the land.

The Spiral Dance is found in contexts around the world, both within and outside the contemporary Pagan community. Across the ocean from Reclaiming's Spiral Dance, Pagans and activists in Ireland turned to the Dance for their environmental protests. In 2003, a group of contemporary Druids gathered at the Hill of Tara in County Meath, Ireland, to protest plans to build a highway through the prehistoric site. They linked hands and spun around each other in a Spiral Dance to protect Tara with a magical barrier.[8] These environmental protection rituals work to articulate Pagans' and eco-activists' connections to the Earth and construct the Earth and all its inhabitants as sacred and of value, a hallmark of immanent traditions like Paganism. Starhawk's reference to "conjuring back" is both metaphoric and practical. It involves symbolic actions, like the drawing of the spiral, as well as putting one's heart and hands to work in material practices that facilitate the restoration of other species.

Becoming Feral: Rewilding Self and World

Eco-activist Pagans like Starhawk are spiritual counterparts of a diverse movement that complements the rewilding visions of writers and activists like Dave Foreman, founder of the Rewilding Institute, and George Monbiot, author of "A Manifesto for Rewilding the World," who promote mass restoration of lost wild food chains and wildlife corridors.[9] Eco-activists also want to rewild humans by establishing a deep connection to other species through teaching and learning skills and practices of attunement to the more-than-human world.

From the 1970s on, in the U.S. and U.K., many eco-activists were influenced by both Paganism and anarchism, especially green anarchism.[10] According to historian Barbara Epstein, who studied anti-nuclear protests in the 1970s and 1980s, "the polytheism of Paganism has been attractive to anarchists and others around the direct action movement as an alternative to cultural imperialism that tends to be associated with Christianity or other monotheistic world religions."[11] By the end of the 1990s, while Paganism was the dominant form of religious identity, anarchism had become the most influential political view among radical eco-activists who linked environmental and nonhuman animal issues with social justice concerns and critiques of capitalism and the state. The green anarchy movement was one of many "frequent crossovers" between radical eco-activist groups like Earth First! and the anti-capitalist movement best known for the 1999 Seattle WTO protest and Occupy Wall Street.[12]

In part because of the influence of green anarchism, many eco-activists emphasize humans' animality and the need for us to "rewild" and "uncivilize" *our* lives, to become more feral than civilized by creating deeper connections with other species. "Nature" and "the wild," tend to mean places and states of being that are less domesticated and set in opposition to industrialized "civilization." Rewilding of the human self in a community of other species is a reaction against domestication, which eco-activists see as a social ill linked to capitalism.

As a protest against civilization, many activists imagine cities decaying and forests thriving. They endow plants and other animals with rewilding agency. Since most eco-activists see humans as a destructive, invasive species, they celebrate other species' invasion of the human-made world. In the zine, *Feral: a journal towards wildness*, the narrator of the poem, "Ned Ludd was Right," dreams of "ugly monstrosities of steel and glass and concrete" in ruins, "being eaten by a forest."[13] Eco-activist visions of a future after industrial collapse is one in which humans are no longer dominant, but one species among many others. Activists' music, poetry, and artwork clearly express these kinds of future visions. In the poem, "A Handful of Leaves," Sean Swain imagines the future as a return to the "Stone Age":

> A prayer for the children of the next Neolithic,
> That we leave to them
> A field of lilies where a Walmart once stood,
> Salmon upstream from the ruins of a dam,
> Kudzu vines embracing skeletons of skyscrapers,
> Cracked and overgrown ribbons of nameless super-Highways.[14]

Rewilding and re-enchantment are twin tools for reclaiming a "new Neolithic" that looks to the past for possibilities of a different kind of future. Christina Wulf, a participant in the anarchist, eco-activist gathering, Wild Roots, Feral Futures, explains that her activism is, ". . . a reconnection. Tracing our human bloodlines back a brief genetic distance, a handful of centuries, we can re-learn how to live on this earth in balance with breathing forests and all their inhabitants."[15] For eco-activists like Wulf, reconnecting to an ancestral past is the best way forward. Activists invoke a past that is both genetic (Wulf's reference to "bloodlines") and cultural, and urge "relearning" how to live on this Earth.[16] Eco-activist commitments are expressed through particular kinds of embodied practices, such as ritualized protests, which try to bring rewilding visions into being. They constitute and express relationships with the more-than-human world by invoking childhood and the historical past, and by reskilling and cultivating bodily senses to become more porous and vulnerable.

The lure of the past—one's childhood past and past eras—is powerful in these movements. Along with pre-industrial cultures, childhood experience is an important resource for Pagans and eco-activists. Eco-activists and Pagans both embrace childhood animism and draw on childhood experiences as resources. They reverse the developmental model that assumes children will grow out of talking with trees and animals, feeding fairies, and other childhood acts that activists see as formative in their childhood and desirable in their adult worlds.[17] By celebrating and nurturing "immanentist" views, Pagans and eco-activists practice what religious studies scholar David L. Haberman calls "deliberate anthropomorphism."[18]

Tree-sits as Ritualized Practices of Radical Porosity

Deliberate anthropomorphism that draws on childhood experience is at work during eco-activist protests. Specifically, these protests include techniques of cultivating attention to other animals, trees, rocks, and landscapes. In Richard Powers' 2018 novel, *The Overstory*, two activists who occupy a tree-sit in a redwood named Mimas attune themselves to flying squirrels, weather, and communications between trees. They come to understand that "There are no individuals in a forest, no separable events. The bird and branch it sits on are a joint thing . . . Forests mend and shape themselves through subterranean synapses. And, in shaping themselves, they shape, too, the tens of thousands of other, linked creatures that form it from within."[19]

Tree-sits have been one of the most common forms of eco-activist protests since 1985. They involve specific, ritualized practices and have been used to prevent logging and delay roadbuilding and pipeline construction. They are characterized by commitment and sacrifice: putting one's life in danger for the lives of other species. Tree-sits also transform tree-sitters, making their relationships to other species more intimate. In eco-activist Ron Huber's view, ". . . to occupy a wild canopy community at risk, to unite one's survival to it, is a powerful act . . . Unsurprisingly, given *Homo Sapiens* arboreal ancestry, living for a time as part of a canopy community seems to trigger a rewilding reflex that forever changes a person's relationship with Nature."[20] During tree-sits, forest species work physical and moral changes on activists in a rewilding process.

When activists live on platforms high in tree canopies for weeks or months, they experience their bodies differently. They acquire what philosopher David Abram describes as a special kind of perception, characterized by reciprocity with other species, as they touch the forest and are touched by it.[21] After living in Luna, an ancient redwood, for a year and a half, eco-activist Julia Butterfly Hill described the changes in her body, ". . . the tree had become part of me, or I her. I had grown a thick new muscle on the outer sides of my feet from gripping as I climbed. . . My fingers were stained brown from the bark and green from the lichen. Bits of Luna had been ground underneath my fingernails, while sap, with its embedded bits of bark and duff, speckled my arms and hands and feet. People even said that I smelled sweet, like a redwood."[22]

The blurring of boundaries between activists and trees unsettles and challenges our taken-for-granted categories. Eco-activists' tree-sits disrupt and vary the usual hierarchical classification of things and beings. Their vulnerable bodies positioned high up in treetops tell us that trees and other forest species are worth sacrificing for. Their bodies identify with other vulnerable tree and nonhuman animal bodies threatened by logging through ritualized interactions that constitute relationships of identification between human self and nonhuman other.

The deliberate anthropomorphism of activists who refer to "Mother Earth," describe trees as gods and goddesses, and imagine trees as having human personalities, increases their identification with other species and confirms their commitment to eco-activism. However, these practices of intimacy and identification have the drawback of reducing the worlds of nonhuman species to human experience and vocabulary. Is the tree rewilding the human or the human domesticating the tree? Some green anarchists critique Pagans and eco-activists who deify the nonhuman world. They argue for a different kind of immanence, as in these comments by anarchist Autumn Leaves Cascade:

> . . . place your hands in moss and soil. Feel a river's flow. Watch the dance of dragonflies. Behold the wonder of thunder and lightning. Speak with birds, hear their song. Conjure fire by friction. Feel the movement of wind. Practice tracking and botany. Chant and sing and dance together around campfires! Forage! . . . Whatever you do, do not adapt to the cage of the city with rituals of alienation. To rust metallic gods means to resist that which eradicates wildness and vitality. Free your feral heart, and find kinship among the bonfires. For ruins, not runes.[23]

Cascade notes that they were inspired by anarchist Comrade Black's essay, "Neo-Paganism is not the Answer—Climb a Fucking Tree," in which Black points to a more ancient, *pre-Pagan* past—a nontheistic nature religion. If you want to worship nature, argues Black, ". . . you don't need a sunwheel, pentacle, or a goddess to do so—go out and climb a fucking tree, sit in its branches, learn ecology, listen to the wind rustling the leaves through the branches, watch the squirrels . . . Then do whatever it takes to stop those fuckers who wanna cut that tree cause all they see is dollar signs."[24] For Black, to conceptualize trees as deities is to project human views and categories onto nonhuman nature. In activist communities, theistic Pagan perspectives and nontheistic anarchist perspectives are often in tension.

Fire-making and Focal Things

One antidote to excessive anthropomorphizing is to train the senses to attune to the more-than-human world on its own terms. Potawatomi scholar Robin Wall Kimmerer describes this kind of attunement as a "grammar of animacy," which characterizes my third example of an immanent worldview in practice: the ancestral skills movement.[25] This movement overlaps historically with both Paganism and eco-activism.[26] Ancestral skills practitioners learn "to speak botany," to borrow from Kimmerer's grammar of animacy, and practice other forms of attending to the more-than-human world around them. This does not mean observing "nature" from a distance. Classes at the gatherings I attended included: hunting and preparing wild animals to eat with simple tools; tanning their hides for clothing and bags; making baskets from willow gathered along a nearby river; foraging for edible and medicinal plants and fungi; creating cordage (twisting plant fibers into a rope or cord); making fire with two pieces of wood; and knapping rocks (shaping rocks by hand to make stone tools).[27]

Around a community circle at Falling Leaves Earth Skills Rendezvous in rural Georgia, participants sang the Pagan song "Angels Singing," that captures ancestral skills practitioners' orientation towards the world: "River, sea, redwood tree/Spirit of the wind will set us free/Angels singing, angels singing in my soul/wood, stone, feather and bone/Spirit of the Earth will call us home." During the numerous ancestral skills gatherings I have attended, ritual activities that created relationships among participants, as well as between people and plants, nonhuman animals, clay, rocks, and fire, helped bolster commitment to a common way of life, not identified with a particular religious tradition, but imbued with spirit and meaning.

Rocks are important in the ancestral skills movement and can be approached through the same kind of attunement process that Kimmerer takes with plants. Ancestral skills teacher Lynx Vilden observes that rocks, like plants, will teach us if we ". . . attune ourselves to them . . . The rocks have their voices and they're talking to us." She takes students to a river where they "listen to the rocks" and she instructs them to use all their senses in selecting stones for various purposes: "I suggest the students tap on the rocks and get a feel for their density, their hardness, their ability to strike a spark. I ask them to taste them, smell them—experience everything they can about them."[28] Experiencing everything through the senses requires being open to what rocks can communicate about their usefulness as tools, what they "want" to become in and with human hands. In this way, ancestral skills practitioners treat rocks as subjects with their own concerns and agency. An ancestral skills practitioners' worldview is composed of the relationships that emerge from practices like fire-making and with things like knives and stones.

The ancestral skills movement expresses a kind of paleonostalgia, a longing for a simpler life that can be retrieved from the past and made relevant to the present. Participants in the movement access a past way of life by learning skills that more directly connect them to the natural world. On the website for "Buckeye Gathering: Ancestral Arts and Technology" in northern California, the organizers observe that we still carry this past within us that can serve us well today: "We are at a unique juncture in history, with an increased awareness in the possibility of our own extinction from not living responsibly and in harmony. Relearning to tend the land will take patient, imperfect steps. . . Many of us long for the clans and tribes that we know in our bones." Most ancestral skills practitioners believe that we can bring back an ancient correspondence between humans and the rest of the world through sensual and embodied ways of interacting with the matter around us in meaningful ways.[29]

Objects like stone tools carry multiple meanings for ancestral skills practitioners and acquire what Jane Bennett calls "vibrancy" or "thing-power" as they are formed and used.[30] Ancestral skills practitioners consecrate the tools they use in similar ways. And in these practices of consecration, tool users both define, and are defined by, the objects they consecrate for the skills they need. In his "Elegy to a Fire Drill," Forager, a participant on PaleoPlanet, a message board for ancestral skills practitioners, describes his relationship with his fire-making kit, "It is with a note of sadness that I finally retire the best hand drill I've ever employed. Little could I imagine that this opportune Goldenrod stalk from a floodplain would rise above the Mugwort, Mullein, and Yucca . . . in terms of its dependably consistent performance. This stick became my 'go-to' rod of fire whenever I truly needed to spin out an ember."[31] Forager's fire drill was directly connected to specific plants and "a flood plain," adding to its "vibrancy." That the drill would be seen as having agency makes perfect sense, given that fire-making is one of the primary rituals of the ancestral skills movement and a direct connection to past peoples' fire-making work.[32]

Tools like the fire drill take on meaning because they directly connect practitioners to other species (mullein, mugwort), places (the goldenrod from a "flood plain"), and the past. They orient practitioners to a world of vibrant material meaning. Maron, who taught a hand drill fire-making class, described how he cultivated a particular kind of attention, so that when he walked in the woods or drove along a road, he was always noticing plants and fungi that might be particularly useful in a tinder bundle (a small bundle into which coals produced by the fire drill are placed), or as a spindle for the fire drill. Ancestral skills practitioners also learn how to harvest materials—coppicing willow for example, an ancient technique of shaping the willow in ways that help it thrive and make it easier to harvest. In these ways, in the ancestral skills movement, fire-making becomes a ritualized, learned process that expresses and constitutes many kinds of relationships with other species and surrounding landscapes full of meaning and power.

Being Eaten

In my final example, Pagans, eco-activists, and ancestral skills practitioners focus on rewilding death rituals to save the Earth. Green burial and other practices emphasize the porosity of our disintegrating bodies as they are eaten by earth and animals. In this example, immanent worldviews express their difference from transcendent religions: the body and all its potential energies are given to the Earth. Circle Cemetery in Wisconsin, established in 1995 by Circle Sanctuary, is one of the oldest Pagan nature sanctuaries in the U.S. It was also one of the first green cemeteries in the U.S., and combines the preservation of green space (including a restored prairie) with burial of cremains and non-embalmed bodies.[33] Peter Michael Bauer, a self-described "urban rewilder" and ancestral skills teacher, argues that death rituals, too, need to be rewilded: "When I die, I would like my body buried in the ground . . . I want a tree to be planted in my honor . . . If that is not possible, I would like a sky burial. I would like my body to be placed high atop a mountain and eaten by vultures."[34] Even in the context of death, Pagans, eco-activists, and ancestral skills practitioners want to return to past ways of ritualizing relationships to the more-than-human world. In these ritualizations they fold their immanent beliefs, not into an orientation towards the afterlife (though they may believe in one), but into the eternal cycle of decay and creation that characterizes life on this planet.

Notes

1. I draw on Michael J. Houseman, "Relationality," in *Theorizing Rituals, Volume I: Issues, Topics, Approaches, Concepts*, eds. Jens Kreinath, Jan Snoek, and Michael Stausberg (London: Brill, 2006).

2. Jane Bennett, *Vibrant Matter: A Political Ecology of Things* (Durham and London: Duke University Press, 2010), xvii and 18.

3. "Reclaiming Spiral Dance," accessed November 2, 2018, https://www.reclaiming spiraldance.org/.

4. Participants hold hands and move in a counterclockwise motion. As the leader comes near the end of the circle, they spiral around and begin moving clockwise while facing the rest of the dancers.

5. For an overview of contemporary Paganism in the U.S., see Sarah M. Pike, *New Age and Neopagan Religions in America* (New York: Columbia University Press, 2004).

6. Reclaiming emerged from Goddess-based feminist spirituality and grassroots activism in California. See Barbara Epstein, *Political Protest and Cultural Revolution: Nonviolent Direct Action in the 1970s and 1980s* (Berkeley: University of California Press, 1991).

7. Pike, *New Age*, 159–160.

8. Jenny Butler, "Druidry in Contemporary Ireland," in *Modern Paganism in World Cultures: Comparative Perspectives*, ed. Michael F. Strmiska (Santa Barbara: ABC-Clio, 2005), 95.

9. The idea and scientific approach of *rewilding* was developed by Michael Soulé in the mid-1990s.

10. The largest radical environmentalist movement in the U.S., since at least the 1980s, has been Earth First! which exemplifies the convergence of Paganism, anarchism, and anti-capitalism.

11. Epstein, *Political Protest*, 191.

12. Green anarchism is related to anarcho-primitivism, which aims to develop a blend of "primal" and contemporary anarchy, a synthesis of the ecologically focused, anti-authoritarian aspects of earlier cultures with various forms of anarchist critiques of power relations ("A Primitivist Primer," accessed August 10, 2020, http://www.eco -action.org/dt/primer.html).

13. N.d, n.p., San Francisco, CA.

14. Sean Swaine, "Stone Age," *Earth First! Journal* (2013): 72.

15. Christina Wulf, "Conserving Wild Nature in Virginia," *Earth First! Journal* (1999): 11.

16. The language of "blood" and "land" is contested in these communities, leading to an emphasis on confronting racism and cultural appropriation, which I discuss in *For the Wild: Ritual and Commitment in Radical Eco-Activism* (Oakland: University of California Press, 2017), 162–195.

17. For a discussion of how childhood experience shapes the developing commitments of eco-activists, see Pike, *For the Wild*, 71–103.

18. David L. Haberman, *People Trees: Worship of Trees in Northern India* (New York: Oxford University Press, 2013), 24.

19. Richard Powers, *The Overstory* (New York: W.W. Norton, 2018), 218.

20. Ron Huber, "Back to the Trees," *Earth First! Journal* (2005): 33.

21. David Abram, "Magic and the Machine: Notes on Technology and Animism in an Era of Ecological Wipe-Out," keynote address at "Wonder and the Natural World" conference at Indiana University, June 22, 2016.

22. Julia Butterfly Hill, *The Legacy of Luna: The Story of a Tree, a Woman, and the Struggle to Save the Redwoods* (San Francisco: Harper San Francisco, 2000), 227.

23. "To Rust Metallic Gods: An Anarcho-Primitivist Critique of Paganism," accessed March 26, 2015, https://hastenthedownfall.wordpress.com/2015/03/26/to-rust-metallic-gods-an-anarcho-primitivist-critique-of-paganism/.

24. "Neo-Paganism Is Not the Answer," accessed August 5, 2020, https://profan existence.com/2014/07/06/neo-paganism-is-not-the-answer-climb-a-fucking-tree/.

25. "Learning the Grammar of Animacy," accessed August 10, 2020, http://moonmagazine.org/robin-wall-kimmerer-learning-grammar-animacy-2015-01-04/.

26. Sarah M. Pike, "Rewilding Hearts and Habits in the Ancestral Skills Movement," *Religion*, *Vol.* 9 (10) (2018), https://www.mdpi.com/2077-1444/9/10/300.

27. There are hundreds of gatherings, workshops, and classes held throughout the U.S. One of the oldest gatherings is Rabbitstick Primitive Skills Gathering: https://www.rabbitstick.com/.

28. "Living Wild," accessed July 1, 2018, http://www.lynxvilden.com.

29. Within the ancestral skills movement there is an ongoing debate about cultural appropriation and a tendency to reify the Stone Age skills of "primitive" people.

30. Bennett, *Vibrant Matter*, 6.

31. "PaleoPlanet," accessed July 16, 2020, https://www.tapatalk.com/groups/paleo planet69529/.

32. Friction fires are an essential skill in the ancestral skills movement. For more information, see https://www.youtube.com/watch?v=CF9GiK_T4PA.

33. "Circle Sanctuary," accessed October 27, 2018, https://www.circlesanctuary .org/index.php/cemetery/circle-cemetery.

34. "Thoughts on Death," accessed August 12, 2020, http://www.petermichael bauer.com/thoughts-on-death/.

Planetary Thinking, Agency, and Relationality: Religious Naturalism's Plea

Carol Wayne White

Introduction

As the twenty-first century continues to unfold, citizens around the globe are increasingly confronted by scientific evidence concerning the dire state of the planet we inhabit. Current rates of carbon released into the atmosphere are unprecedented in the past sixty-six million years, and if the rapid rise of sea levels continues it is likely to displace millions of people. Other crises loom, as well, ranging from forced migration, mass species extinction, water scarcity, and ocean acidification.[1] As with other ecologically-minded folks, I do not view these as isolated, arbitrary events, but rather as a set of interrelated problems generated by the accelerated effects of human activity on the Earth's systems. As the scientific consensus indicates, humanity has become a global, geophysical force affecting the planet, comparable to other natural processes.[2] Among religious naturalists like myself, there is also the sobering realization that human animals are not only the first species that has become a planet-scale influence, we are also *aware of that reality*.[3]

This chapter offers critical reflection on this crucial point, with an eye toward assessing the value of New Materialism (also considered critical materialism in this paper) for planetary thinking. I posit religious naturalism as one specific orientation—and certainly one of several important vectors—for exploring the rich, theoretical potential of materialist discourses for planetary thinking. As a capacious, ecological worldview, religious naturalism reassesses who we humans think we are in the grand scheme of things, shifting humans' attention back to ourselves as natural processes, inextricably connected to other life forms and material processes. As a critical intervention in modern humanistic thought in the West, this religious worldview invites us to conceive and enact new forms of relationality with each other and with the more-than-human worlds that are an integral part of our existence. Given the immensity of the multifaceted problems identified above, my explication of religious naturalism will not be universal in scope or applicability. Rather, I share the humble view that new materialism is, for the moment, a "place from which to understand, criticize, and reconstruct—in ways that more traditional religion-and-science discourse hasn't yet accomplished."[4]

This chapter is divided into three sections. In the first part, I introduce religious naturalism as a new materialist discourse, outlining its basic tenets and then using it to formulate a naturalistic conception of humanity. These steps provide a rich point of departure for discussing the theoretical significance of critical materialist thinking. In the second section, I include religious naturalism among the cluster of narratives associated with the Anthropocene concept. This is a rich, multivalent term that has brought renewed focus to how we conceptualize agency, understand human-nature relations, and attempt to provide plausible responses to the challenges of climate change.[5] Lastly, I argue that religious naturalism points us in the right direction, offering imaginative visions of ethical import and hope within planetary thinking.

Unsettling Grounds: Religious Naturalism as Materialist Ontology

In *The Routledge Handbook of Religious Naturalism*, Donald Crosby and Jerome Stone articulate an important theme that I will advance in this chapter. They suggest that "thinking deeply about nature and our place as human beings in nature is an urgent and salutary activity for each of us and for the institutions of our societies, no matter what our personal religious or secular outlooks may be in this time of rampant species endangerment, global climate change, and looming ecological crisis."[6] As a type of critical materialist discourse, religious naturalism champions a communal ontology grounded in the observational conviction that nature is ultimate.[7] Additionally, the qualifier "religious" in religious naturalism affirms the natural world as the center of humans' most significant experiences and understandings. Religious naturalism does not posit any ontologically distinct and superior realm (God, soul, Heaven) to ground, explain, or give meaning to this world. Rather, attention is focused on the events and processes of this world to provide what degree of explanation and meaning are possible to this life. As suggested by Wesley Wildman, a shared conviction among religious naturalists is the ceaseless, explicit focus on myriad nature in "its beauty, terror, scale, stochasticity, emergent complexity, and evolutionary development."[8]

For my purposes, religious naturalism's theoretical appeal is its fundamental conception of humans as natural processes, intrinsically connected to other natural processes. The advances of the sciences, through both physics and biology, have served to demonstrate not only how closely linked human animals are with nature, but that we are simply one branch of a seemingly endless natural cosmos. Big Bang cosmology shows the world evolving naturally, based on the interconnection and interaction of all its fundamental components. I share Loyal Rue's contention that humans are "ultimately the manifestations of many interlocking systems—atomic, molecular, biochemical, anatomical, ecological—apart from which human existence is incomprehensible."[9] As by-products of other natural processes, and intimate participants with them, humans are material beings through and through. Consider, for example, Michael W. Fox's astute observation that ". . . our bodies contain the mineral elements of primordial rocks; our very cells share the same historically evolved components as those of grasses and trees; our brains contain the basic neural core of reptile, bird, and fellow mammal."[10] We are also structured by relationality. In *The Sacred Depths of Nature*, Ursula Goodenough also offers a lucid account of humans as relational, natural organisms, providing sound scientific data that supports our fundamental interconnectedness with other living beings. As she puts it, ". . . and now we realize that we are connected to all creatures. Not just in food chains or ecological equilibria. We share a common ancestor. We share genes for receptors and cell cycles and signal-transduction cascades. We share evolutionary constraints and possibilities. We are connected all the way down."[11] Goodenough's claims reinforce my view that humans are, by our very constitution, relational beings, and our wholeness occurs within a matrix of complex interconnectedness—in ways of conjoining with others that transform us.

In this context, I offer a naturalistic view of humans as complex social organisms, capable of loving, connecting deeply with others, and symbolizing our environment (or engaging in world formation) through values and language. We are multi-level psychosomatic unities—both biological organisms and responsible selves. Another important insight here is humans' heightened awareness of our ability, self-consciously, to make decisions, act on those decisions, and take responsibility for them. This point converges with Loyal Rue's descriptive account of human beings as star-born, Earth-formed creatures endowed by evolutionary processes to seek reproductive fitness under the guidance of biological, psychological, and cultural systems that have been selected for their utility in mediating adaptive behaviors.[12] In positing human animals as emergent life forms, however, I resist a particular reading of this claim that concludes human beings are the triumphant summit of natural development. Rather, informed by ecological studies, I affirm that "organisms of various types, including human beings, are inextricably bound together in a web of mutual interdependence for their continual flourishing and survival as they make common if varied use of the energy of the sun."[13] All members of an ecosystem are equally important, comprising it as a functional whole. As the later work of Aldo Leopold emphasized, we honor the radical interdependence of plants and animals in their natural environments, including the observation that human organisms are intimate participants in ecological relations and belong to a wider biotic community.[14]

In a religious context, the notion of humans seeking, finding, and experiencing community with others—an essential aspect of our humanity—is important. Thus, exploring the tenets of religious naturalism in conjunction with values discourse, I consider humans' embodied awareness and appreciation of our connection to "all that is," an expression of what we perceive and value as ultimately important. (Our embodied awareness is what I later describe as our constitutive relationality.) Value, in this sense, refers to an organism's facility to sense whether events in its environment are more or less desirable.[15] As Holmes Rolston suggests, minimally, this facility evokes the notion of adaptive value, which is the basic matrix of Darwinian theory. Within a larger ecological framework, however, this truth takes on fuller meaning: "An organism is the loci of values defended; life is otherwise unthinkable. Such organismic values are individually defended; but, as ecologists insists, organisms occupy niches and are networked into biotic communities."[16]

Binary Constructions and Configurations

My model of religious naturalism invites humans to be daringly positive *about* our materiality and to embrace materiality in its diverse manifestations. In doing so, it encourages us to resist binary thinking that demarcates certain spheres of life as superior and others as inferior, justifying the exploitative practices of the former.[17] This is an important insight that Mary-Jane Rubenstein addresses in her chapter, "The Animist, Almost Feminist, Quite Nearly Pantheist Old Materialism of Giordano Bruno," when discussing an influential set of binary differentiations intimately associated with the complex legacy of thinking materially in the West. I share Rubenstein's astute observation that attending to an influential binary structure also means attending to "the formation of its racially, environmentally, sexually, and theologically toxic

companions."[18] Centuries removed from what Rubenstein identifies as its paradigmatic articulation in Aristotle, the toxicity of this structural ordering has
remained in the popular imagination. One historical example is the influential
Euro-American colonial narrative of "civilization" overcoming wilderness, as
reflected in the diaries, personal writings, and memorials of White Puritans,
who measured progress in terms of how far a people could distance themselves
from Nature.[19] Intermingled in their accounts are cultural, religious, and social
forces that essentially made wilderness synonymous with darkness and sinister
forces. As Roderick Nash asserts, "The pioneers shared the long Western tradition of imagining wild country as a moral vacuum, a cursed and chaotic
wasteland. As a consequence, frontiersmen acutely sensed that they battled
wild country not only for personal survival but in the name of nation, race,
and God. Civilizing the New World meant enlightening darkness, ordering
chaos, and changing evil into good."[20] These perspectives fueled the legacy
of White supremacy in the United States, both justifying Black slavery and
the exploitation of the more-than-human natural worlds. Paul Outka aptly
describes aspects of this colonizing legacy as Whites viewing dark-skinned
peoples as part of the natural world, and then proceeding to treat them with
the same mixture of contempt and exploitation that also marks American
environmental history.[21]

Religious naturalism can support efforts intent on moving beyond these
problematic historical perspectives. It belongs to an American tradition of immanental thinking that rejects "the ontological and epistemic exceptionalisms
that have set humans over and against other forms of animal life and human
culture, society, and economy over and against the ecological systems upon
which they depend."[22] Specifically, religious naturalism challenges a colonizing
legacy that depends on the dominant cultural fantasy of human exceptionalism—a premise that assumes the human alone is not a spatial and temporal
web of interspecies dependencies. Emphatically rejecting this phantasm that
has lent theoretical support to popular myths of the self-made individual in the
U.S., religious naturalism encourages us to join with Donna Haraway in appreciating our intricate entanglement with other material processes:

> I love the fact that human genomes can be found in only about 10
> percent of all the cells that occupy the mundane space I call my body;
> the other 90 percent of the cells are filled with the genomes of bacte-
> ria, fungi, protists, and such, some of which play in a symphony neces-
> sary to my being alive at all, and some of which are hitching a ride
> and doing the rest of me, of us, no harm. I am vastly outnumbered by
> my tiny companions; better put, I become an adult human being in

company with these tiny messmates. To be one is always to become with many.[23]

From a religious naturalism standpoint, all human endeavors arise from the critical awareness that we are part of an inextricable network of natural processes that make the very category of the human itself intelligible. What I am suggesting here is that humans are relationally constituted, or that a fuller understanding of ourselves is gained through accepting our constitutive relationality. Our embeddedness with myriad nature invigorates a fuller sense of our expansive humanity as already, always entangled becoming—a point that shares theoretical resonance with Catherine Keller's materialist theological orientations in, "Amorous Entanglements: The Matter of Christian Panentheism," where she attests: ". . . there is no exit from our entangled becoming."[24]

Furthermore, religious naturalism exposes the deficiencies adhering to a generic, universal construction of "man" that has justified the devalued status of fleshy, material "women" and other embodied subjects relegated to minority status. Enlightenment configurations of this normative human have been associated with a coherent, White, propertied, and rational subjectivity.[25] Rather than assume that gender, race, class, abled-bodiedness, and other socially derived markers provide the basis of our humanity, we should recognize them as highly complex categories constructed in contested discourses and other social practices. When these constructions are used to support racism, speciesism, sexism, and other forms of cultural superiority, they become forced impositions on the wholeness of natural interrelatedness and the deep genetic homology that evolution has wrought.

Religious Naturalism, Anthropocene Narratives, and Ethical Concerns

The concept of the relational, material human featured in this chapter underscores why human histories—and not only those of the West—are intimately connected to the history of the Earth. Accordingly, we affirm inseparable ethical connections between humanity's relationality with other natural processes on the planet and humans' activities with each other. It is not an either/or situation. These linkages help situate religious naturalism among a cluster of Anthropocene narratives, described by Rolf Lidscot and Claire Waterto as sharing these fundamental convictions: (i) that Earth itself is a single system within which the biosphere is an essential component; (ii) that human impact is global and accelerating, now threatening the fundamental life processes of

Earth; (iii) that this change is traceable geologically, possibly implying a new geological epoch, "the Anthropocene" and; (iv) that there is a need to radically change current human activities in order to avoid this threat.[26]

Anthropocene narratives often invoke human agency in attempts to legitimize decisions and motivate actions in response to the problem of anthropogenic climate change and its myriad effects. Within the context of moral theory, these narratives also raise thorny questions involving the challenging task of moving from the descriptive to the normative.[27] In other words, when asking about humans' purposeful activity *in response* to the Anthropocene, these narratives draw attention to the perennial conundrum of how to get from the "is" to the "ought" in moral theory. As ethicist Maria Antonaccio observes, for a number of thinkers, the Anthropocene concept "cannot, by itself, support any conclusion for how we ought to behave."[28] While it is beyond the scope of this paper to flesh out a full position, I suggest that in positing humans as material processes embedded in nature, religious naturalism offers potential inroads for us to consider. Specifically, it provides one plausible rationale when advancing ethical concerns about the state of the planet, or when responding to the question, "What ought we to do?"

An implicit, normative, ethical stance of religious naturalism is found in its conception of humans as relational, material organisms that are evolutionarily equipped to ask "why" our activities matter within the universe. As value-laden, natural processes, humans are capable of being concerned about the effects of our activities as we enact our constitutive relationality in an appreciable universe. To articulate this contention better, I appeal to Crosby's concept of metaphysical perspectivism, which contends that the world we inhabit has a plurality of entities, each with its own individuality, particularity of expression, and distinctive perspective on everything else. As Crosby writes, "All the elemental particles, atoms, molecules, compounds, inorganic and organic entities and combinations of those entities, including human beings and their histories, cultures, and societies, and all of the actions, reactions, functions, qualities, and traits of these particular things and their relations are included. No two perspectives or systems of them are exactly alike."[29] Metaphysical perspectivism reinforces the fact that humans' perspectives are *included* with and inflected by the perspectives of other existents in the universe. As noted by Kevin Schilbrack, " . . . the world is deeply and richly structured even before humans arrive on the scene with their particular senses, ways of thinking, and discourses."[30] Humans reside within an appreciable universe. Human valuing thus becomes part of what Philip Clayton describes as a vital sphere of activity throughout "the biosphere, understood as a single interconnected system of value and valuing."[31] Clayton's insights on value amplify my earlier point about human-ity's mode of valuing within a matrix of processual, natural interactions. He observes, "The biosphere was packed with living, interpreting systems well before human beings came onto the stage, agents who already possessed many of the perspectives that are manifested in higher organisms. Every organism, every living thing is an agent composed of communities of living parts. They sense or perceive their environment, process data, and make appropriate re-sponses. The lifeworld is agent-centered; it is an ontology of living agents."[32]

An important question arises for me at this juncture: Does positing a distinct human perspective or mode of agency essentially make one susceptible to the charge of residual anthropocentricism? Perhaps. I am also not sure how to avoid that possible charge when describing humanity's capacity to reflect on our constitutive relationality. What I fully affirm, however, is the richness of per-spectival agencies. Rather than definitively and naively assuming a monolithic, ontology of agency, we might instead point to the functionary aims of different agents. For example, it is useful to feature human agency as possessive of a distinct valuing quality that other agential entities might not possess, or even advance with the same intensity. Consider Clayton's sense that "humans know that the primary agents in their lives—spouse, children, friends, extended family—are of value, which means that these others have a claim on us to be treated as valuable."[33] Humans' perspectival status necessarily involves value-laden interactions with other processes, not least because we experience value ourselves among ourselves.

Given that humans are not at the center of this vibrant universe, highlighting functional differences does not confer ontological superiority. Rather, we hum-bly and respectfully acknowledge belonging to and becoming with "more-than-human vitality and creativity," or existing in "a world in which agency and value are distributed through the whole of nature's patterns, processes and precarities rather than concentrated within or monopolized by a single species, ethno-racial group, nation, political economy, or concept of God."[34] It is im-possible to conceive of human agential activity in isolation from other natural processes that both constitute our being here and remain central to our iden-tities. This insight underscores my premise that there is ethical import in humans asking "why" we should be concerned with enacting our constitutive relationality. In the final analysis, we cannot escape asking the question of what we ought to do. I share Keller's conviction that humans are "always already responsible to the others with whom or which we are entangled, not through conscious intent but through the various ontological entanglements that ma-teriality entails."[35]

Religious Naturalism, Ethical Aspirations, and Hope

Given our embeddedness in materiality, humans are not independent, indifferent actors outside of the matrix of natural processes, nor triumphant agents in the world's perpetual unfolding. Consequently, I imagine humans' ethical activity in an appreciable universe as aspirational, as we remain aware of the multiple conundrums constitutive of human embeddedness in materiality. What I am suggesting is that our ethical activities be seen as valiant efforts to minimize violence, domination, and exploitation in our engagements with myriad nature. The presumption that any one of us can occupy a pristine, ethical position outside of such embeddedness is a type of egoic delusion. A helpful analogue is found in Loren Gruen and Robert Jones's discussion of the moral dilemmas facing vegans in, "Veganism as an Aspiration:"

> The belief that a rejection of industrialized livestock products allows one to avoid complicity in harming other animals is too simplistic and ignores the complex dynamics involved in the production of consumer goods of all kinds, global entanglements we engage with each time we purchase and consume food of all sorts. Vegan diets have "welfare footprints" in the form of widespread indirect harms to animals, harms often overlooked or obscured by advocates of V1. Industrialized agriculture harms and kills a large number of sentient field animals in the production of fruits, vegetables, and grains produced for human (not livestock) consumption.[36]

From the vantage point of veganism, Gruen and Jones argue that a naive ethical positioning ignores the complex dynamics involved in the production of consumer goods of all kinds, and the global entanglements we engage in each time we purchase and consume food of all sorts. Despite wanting it to be otherwise, vegan, or not, we cannot live and avoid killing. As a matter of fact, they assert, "All aspects of consumption in late capitalism involves harming others, human and nonhuman."[37]

This sobering recognition, along with many others, leads me to ask: What, then, should we do at all? What is ethically possible within an appreciable universe? There are no easy answers. Minimally, however, I think we ought to remember why we are even asking about ethical action in the first place. As suggested above, set within the epistemological context of metaphysical perspectivism, we cannot escape asking the question of what we ought to do. Given this fact, I believe human agents ought to acknowledge the violent structures and mechanisms we have created—both within our relations with each other and with the wider, more-than-human connections that constitute our being here. We must not fail to recognize the taxonomies of power and the hierarchies of worthiness operating in the logics of dualism, and in the variant expressions of human exceptionalism that have led to the colonization of myriad nature. As Donna Haraway suggests, ". . . different atrocities deserve their own languages."[38] In the current configuration of the world, widespread ecological devastation and danger, continued exploitation of species, including specific human populations and bodies, remain constant. Accentuating our mutual interdependence, human agents should continue to postulate ethical principles that provide, as fully as possible, inclusive, and global analyses of intersectional oppressions.

We need to problematize or rethink traditional modernist concepts (e.g., freedom) that have supported the illusion of human autonomy. In an appreciable universe, for example, it is untenable to retain "conceptions of unique human agency and the presumption of progressive norms, such as liberty, that the planet is capacious enough for individual acts to be thought of as disconnected from the peoples, species and processes once rendered as 'others.'"[39] Humans are always in relationships and that implies certain constraints on what we are able to do. Our individual freedoms are constrained by many things (e.g., age, race, gender, socio-economic status, hierarchies, and our participation in more-than-human relational ontologies, etc.). We are constantly making decisions and sacrificing our freedoms at every turn for the ones we acknowledge, love, and aim to value. Recognizing this principle on familial, societal, and planetary levels is an essential part of ethical reasoning.

Next, we need to recognize that climate change is sure to increase certain forms of injustice already operating in our various sociological and cultural settings. As Colin Polsky and Hallie Eakin have observed, recent "differential social outcomes associated with climate stress may have as much (or more) to do with historical inequities and disparities in the social and institutional contexts of human activity than with differential exposure to climate shocks."[40] Around the globe, eco-justice advocates have observed that the most vulnerable are the disadvantaged (often poor and ethnic groups), who disproportionally live in neighborhoods with much higher environmental risks. The moral imagination I am trying to sketch here sustains itself through our willing participation in "movements of scientific inquiry, movements of cultural expression, movements for global distributive justice, movements to eliminate needless suffering, and movements to preserve the ecology of our home planet."[41]

Finally, in addressing the sense of "overwhelm," religious naturalism helps resuscitate the concept of hope. As I have argued elsewhere, hope underscores the actions of relational, desirous, value-driven beings who reject dominant notions of humans as outside of nature.[42] This conception of hope shows human agents *not giving up on* our evolutionary capacities as biological organisms who can express love by creating alternative systems of interaction, or who can consistently question our values, behaviors, and resource uses. This hope is seen in our local forms of activism when protesting mountaintop mining for coal, drilling for oil/petroleum, and fracking for gas and oil; or when deciding on the food we eat, determining how it is produced and transported, and considering ways of decreasing food waste in the U.S. and elsewhere. Certain possibilities *may* occur when human organisms begin to align our actions with the deeper mystery that we are not at the center of all that is. And, as Wesley Wildman suggests, we reject manipulative or unreflective supernatural authorization of moral claims by any one individual or group.[43] With acute awareness of our entangled becoming within an appreciable universe, human agents continually revise, correct, or even forfeit older perspectives as newer forms of knowledge become available.

Notes

1. Richard E. Zeebe, Andy Ridgewell, and James C. Zachos, "Anthropogenic Carbon Release Rate Unprecedented during the Past 66 Million Years," *Nature Geoscience*, Vol. 9 (2016): 325–329. http:/doi:10.1038/NGEO2681. See also Mathew E. Hauer, Jason M. Evans, and Deepak R. Mishra, "Millions Projected to Be at Risk from Sea-Level Rise in the Continental United States," *Nature Climate Change*, Vol. 6 (2016): 691–695.

2. Rolf Lidskog and Claire Waterton, "Anthropocene—A Cautious Welcome from Environmental Sociology?" *Environmental Sociology*, Vol. 2 (4) (2016): 395.

3. Jeremy J. Schmidt, Peter G. Brown, and Christopher Orr, "Ethics in the Anthropocene: A Research Agenda," *The Anthropocene Review*, Vol. 3 (3) (2016): 188.

4. Philip Clayton, "Matter Values: Ethics and Politics for a Planet in Crisis," in *Earthly Things: Immanence, New Materialisms, and Planetary Thinking*, eds. Karen Bray, Heather Eaton, and Whitney Bauman (New York: Fordham University Press, 2023).

5. For discussion on why the "Anthropocene" concept is problematic, see Donna Haraway, *Staying with the Trouble: Making Kin in the Chthulucene* (Durham: Duke University Press, 2016).

6. Jerome Stone and Donald Crosby, "Introduction," in *The Routledge Handbook of Religious Naturalism*, eds. Jerome Stone and Donald Crosby (New York: Routledge, 2018), 2. For a sampling of religious naturalism studies, see Ursula Goodenough, *The Sacred Depths of Nature* (New York: Oxford University Press, 2000); Loyal Rue, *Religion Is Not about God: How Spiritual Traditions Nurture Our Biological Nature and What to Expect When They Fail* (New Brunswick: Rutgers University Press, 2006); Chet Raymo, *When God Is Gone, Everything Is Holy* (Notre Dame: Sorin

Books, 2008); Jerome Stone, *Religious Naturalism Today: The Rebirth of a Forgotten Alternative* (New York: SUNY Press, 2008); Donald Crosby, *The Thou of Nature* (New York: SUNY Press, 2013); Michael Hogue, *The Promises of Religious Naturalism* (Lanham: Rowman & Littlefield, 2010); and Carol Wayne White, *Black Lives and Sacred Humanity: Toward an African American Religious Naturalism* (New York: Fordham University Press, 2016).

7. I am following Donald Crosby's materialist metaphysics, or a view of reality that regards all existence as diverse forms and functions of matter. Rather than being reductionist, it is emergentist or expansionist in character. In "Matter, Mind, and Meaning," Crosby writes: "Approaches to a proper understanding of matter require the resources of all fields of thought, from physics, to chemistry, to biology, to psychology, to sociology, to philosophy, to art, to religion, and to the experiences of everyday life." David Crosby, "Matter, Mind, and Meaning," in *The Routledge Handbook of Religious Naturalism*, eds. Jerome Stone and David Crosby (New York: Routledge, 2018), 118–128.

8. Wesley Wildman, "Religious Naturalism: What It Can Be, and What It Need Not Be," *Philosophy, Theology and the Sciences, Vol. 1* (1) (2014): 41.

9. Rue, *Religion Is Not about God*, 25.

10. Michael W. Fox, "What Future for Man and Earth? Toward a Biospiritual Ethic," in *On the Fifth Day: Animal Rights and Human Ethics*, eds. Richard Knowles Morris and Michael W. Fox (Washington, D.C.: Acropolis Books, 1978), 227.

11. Goodenough, *Sacred Depths*, 73.

12. Rue, *Religion is Not About God*, 75.

13. Crosby, *The Thou of Nature*, 16.

14. Aldo Leopold, *Sand County Almanac: And Sketches Here and There* (New York: Oxford University Press, 1987), 204.

15. R. J. Dolan, "Emotion, Cognition, and Behavior," *Science*, Vol. 298 (2002): 1191.

16. Holmes Rolston III, "Environmental Ethics and Religion/Science" in *The Oxford Handbook of Religion and Science*, eds. Philip Clayton and Zachary Simpson (New York: Oxford University Press, 2006), 911.

17. White, *Black Lives and Sacred Humanity*, 34ff.

18. Mary-Jane Rubenstein, "The Animist, Almost Feminist, Quite Nearly Pantheist Old Materialism of Giordano Bruno," in *Earthly Things: Immanence, New Materialisms, and Planetary Thinking*, eds. Karen Bray, Heather Eaton, and Whitney Bauman (New York: Fordham University Press, 2023).

19. Marjorie Spiegel, *The Dreaded Comparison: Human and Animal Slavery* (New York: Mirror Books, 1997), 16.

20. Roderick Nash, *The Wilderness and the American Mind*, Fifth edition (New Haven: Yale University Press, 2014), 24, 27.

21. Paul Outka, *Race and Nature: From Transcendence to the Harlem Renaissance* (New York: Palgrave Macmillan, 2008), 3.

22. Michael Hogue, *American Immanence: Democracy for an Uncertain World* (New York: Columbia University Press, 2018), 12, 106–07.

23. Donna Haraway, *When Species Meet* (Minneapolis: University of Minnesota Press, 2008), 3–4.

24. Catherine Keller, "Amorous Entanglements: The Matter of Christian Panentheism," in *Earthly Things: Immanence, New Materialisms, and Planetary Thinking,* eds. Karen Bray, Heather Eaton, and Whitney Bauman (New York: Fordham University Press, 2023).

25. Donna Haraway, "Ecco Homo, Ain't (Ar'n't) I A Woman, and Inappropriate/d Others: The Human in a Post-Humanist Landscape" in *The Haraway Reader* (New York: Routledge, 2004), 48.

26. Lidskog and Waterton, "Anthropocene," 397.

27. Maria Antonaccio, "De-moralizing and Re-moralizing the Anthropocene," in *Religion in the Anthropocene,* eds. Celia Deane-Drummond, Sigurd Bergmann, and Markus Vogt (Eugene: Cascade Books, 2017), 122.

28. Ibid., 122.

29. Donald Crosby, *Living with Ambiguity* (Albany: SUNY Press, 2008), 67–68.

30. Kevin Schilbrack, "Emergence Theory and the New Materialisms," in *Earthly Things: Immanence, New Materialisms, and Planetary Thinking,* eds. Karen Bray, Heather Eaton, and Whitney Bauman (New York: Fordham University Press, 2023).

31. Philip Clayton, "Matter Values: Ethics and Politics for a Planet in Crisis," in *Earthly Things: Immanence, New Materialisms, and Planetary Thinking,* eds. Karen Bray, Heather Eaton, and Whitney Bauman (New York: Fordham University Press, 2023).

32. Ibid.

33. Ibid.

34. Hogue, *American Immanence,* 3.

35. Keller, "Amorous Entanglements."

36. Lori Gruen and Robert Jones, "Veganism as an Aspiration," in *The Moral Complexities of Eating Meat,* eds. Ben Bramble and Bob Fischer (New York: Oxford University Press, 2015), 157.

37. Ibid., 157.

38. Haraway, *When Species Meet,* 336, Note 23.

39. Schmidt, Brown, and Orr, "Ethics in the Anthropocene," 188.

40. Colin Polsky and Hallie Eakin, "Global Change Vulnerability Assessments: Definitions, Challenges, and Opportunities," in *The Oxford Handbook of Climate Change and Society,* eds. John S. Dryzek, Richard B. Norgaard, and David Schlosberg (New York: Oxford University Press, 2011), 207.

41. Wildman, "Religious Naturalism," 54.

42. Carol Wayne White, "Re-envisioning Hope: Anthropogenic Climate Change, Learned Ignorance, and Religious Naturalism," *Zygon: Journal of Religion and Science,* Vol. 52 (2) (June 2018): 579.

43. Wildman, "Religious Naturalism," 54.

Dancing Immanence:
A Philosophy of Bodily Becoming

Kimerer L. LaMothe

Those of us contributing to this volume hold out hope that worldviews, featuring some kind of divine *immanence*, may impel strategic interventions in the global rush to ecological destruction by countering the notions of divine *transcendence* that can and have been used to justify unfettered use of the natural world. In this chapter, I offer resources for affirming dance, not simply as an object of study, but as a critical category for illuminating the constitutive role played by rhythmic bodily movement in generating and maintaining worldviews that reject conceptual bifurcations between the spiritual and material, or human and nature, and instead endorse notions of divinity in which the creative source of all that is true, and real, and good, is itself present in or manifest as the "natural world."

Dance artists, as well as scholars in dance history, anthropology, and religious studies now admit that humans throughout time have relied on practices of dance to obtain knowledge—not only about what humans, nature, and the divine *are* in relation to one another—but about how humans should *act* amidst these relationships. Developing *dance* as a *critical* category can shed light on how ideas concerning divine immanence take root, and become real, for the humans who believe in them.

First, I introduce a *philosophy of bodily becoming* that is informed by and accountable to the experience of dancing. I then offer three examples of religious traditions in which the practice of dancing plays a critical role in educating the senses of those who dance, guiding them to perceive their own bodily movements as participating in the ongoing life of a spiritual reality which includes both humans and the natural world. Finally, looking at cases in this volume through the lens of dance suggests that *ideas* of divine immanence are not enough to promote Earth-friendly action. Rather, such ideas gain traction when grounded in *practices* of bodily movement that offer people an opportunity to know these ideas as true in relation to their own bodily selves.

A Philosophy of Bodily Becoming

Over the past twenty years, I have developed a philosophy of bodily becoming, drawing inspiration from Friedrich Nietzsche and the American modern dancers, Isadora Duncan, Martha Graham, and Ruth St. Denis; as well as from the latest research in evolutionary biology, neuroscience, developmental psychology, philosophy, and anthropology.[1] In this philosophical approach, movement—not matter or spirit—is primary. Accordingly, a human body is not a material object that moves, it is itself *movement*. A human body exists as a *potential to move* that is constantly becoming who it is by virtue of the movements it makes. A human being *is* a *rhythm of bodily becoming*; and dancing is a practice in which humans cultivate a sensory awareness that guides their participation in this rhythm of bodily becoming that they are.

I begin by acknowledging that "dance," as far as can be known, is a human universal. Every human infant, before walking or talking, will spontaneously respond to rhythms and visual cues by matching these patterns and making novel movements in response.[2] This *kinetic creativity* is vital for infant survival. Compared to other primates, humans are born hopelessly dependent on caregivers; to a notable degree, humans grow their brains after they are born in what is called "experience-dependent development."[3] This "experience" is mediated primarily through bodily movement. Every pattern of movement that infants make in reaching, holding, twisting, throwing, lifting, and touching builds their physical, emotional, and intellectual capabilities in particular directions; they create and become a capacity for sensing and responding that they in the next moment—*are*. These movement patterns form the perceptual templates that growing humans exercise, not only in learning about themselves and the world, but in cultivating life-enabling relationships with sources of sustenance, beginning with their own caregivers.

Infants must be able to *notice, recreate,* and *play with* the patterns of movement that appear to them; as they do, they *become* who they are in the world.[4] Regardless of whether or not people think of themselves as dancers, this set of movement-making skills—the kinetic creativity that forms the basis of every human's thinking and feeling, knowing and growing—is the set of skills that the act of dancing exercises and develops.

Several points are worth noting here. For one, this rhythm of bodily becoming, that every human is, is open-ended; it is not predetermined by genetic code nor oriented toward a specific telos. It does not depend upon a normative or abled "body." Every human alive is engaging in bodily movement—however minimal—that is the agent of their becoming. And no movement can be repeated in exactly the same way. Every movement a human *is* courses along the trajectories of neuro-muscular attention and coordination laid out by their earlier movements. Every movement made thus opens and forecloses possibilities for movement-making in the following moment. *I am the movement that is making me.*

So too, bodily movements do not occur in a vacuum. Every movement *is* always already a *relationship*, a *pattern* of both *sensing* and *responding* to something that appears.[5] If light bouncing off the vibrating tail feather of a bird interrupts the wiggling of my eye, I may turn my head to focus on the creature. Once I have made and practiced such patterns of visual, physical, and mental coordination, these patterns may occur spontaneously, priming my gaze as I walk through the woods or across a field. In this way, whatever movements a person makes as they think, feel, and act will fund a nexus of relationships where each relationship is marked by a capacity to perceive and respond to the other. Recent neuroscience research goes so far as to indicate that a heightened sensory awareness predicts a human's ability to empathize with others—that is, their ability to *move with*.[6] In all of these ways, the bodily movements a person makes become who that person is *in relation to* other persons, agents, and forces that comprise their worlds.

Moreover, the movements that make me and my relationships are not limited to bodily movements that "I" decide to make. The movements that make me are not just "mine." Whoever "I" am is constantly *being moved* from within and without, at various scales of shape and size, by forces that are organic and inorganic, social and personal. The movements that make me include the action of systems, organs, cells, and bacteria operating inside my skin, as well as the movements of people around me, and the cultural norms, social conventions, religious rituals, and elemental forces pressing upon me.[7]

The movements that make me extend, as well, to actions made by myriad organisms over billions of years of evolutionary experimentation. As evolutionary biologists note, the mere structure of a human arm, leg, head, or heart represents succinct sets of movement patterns that were made, remembered, and passed on by creatures who died long before humans existed.[8] Any human body is a unique individual that exists as a *potential to move* in specific, ever-evolving patterns of sensing and responding, with varying degrees of range and ability.

To go even further, the trajectories of sensation and response that bodily movements open are not confined to the individual or their immediate relationships. The movements we make create what appears to us *as the world*. Some neuroscientists are now arguing that the human brain evolved, not only to remember patterns of movements made, but to use those remembered movement experiences to establish a map of *what is*, that can predict the outcome of future movements.[9] In other words, the movements a bodily self makes form a repository of possible, imaginable experiences; they educate a person's sensory capacity for perceiving what they can then conceive as *real*. The implication here is that there can be no truth or world "out there" that exists independently of the bodily movements a person makes in perceiving it. Those movements may be as small as an eye scanning a page, or as large as an exuberant full body romp. Regardless, these movements *matter* to what people are able and willing to acknowledge as worthy of consideration, as having value.

Finally, although humans are never fully conscious of the movements making them, nor ever in complete control of their movements, they can and do learn to *participate with some degree of consciousness* in this generative, relational, open-ended rhythm of bodily becoming that they are. This participation is not (solely) a mind over body phenomenon. It occurs through the medium of sensory awareness. As noted above, human infants have a heightened capacity to *play* with movement patterns—to notice and recreate patterns of movement made by others.[10] Infants rely on this *kinetic creativity* in order to create life-enabling relationships with their caregivers. Yet, while other young primates mimic caregivers, they outgrow the desire to do so. Humans do not. Humans sustain a lifelong capacity to seek and find patterns of movement that will entice others to connect with them, and all for the pleasure of it.[11]

From the perspective of bodily becoming then, *to dance is to exercise a distinctively human kinetic creativity in cultivating specific patterns of sensory awareness that guide participation in the ongoing rhythms of bodily becoming.* Traditions and techniques of dance, in turn, represent collections of movement patterns that a group of people have discovered, codified, and passed on for their ability, not only to catalyze this kinetic creativity, but to help apply it in life-enabling ways within their personal, social, natural, and spiritual worlds.[12]

This phenomenological perspective on dance helps explain why many religious traditions, characterized by belief in an immanent, relational divine, also feature a significant commitment to dancing as a medium of religious experience and expression. A quick survey of three cases confirms that dancing serves as a primary means by which people in these traditions cultivate a *sensory awareness* of the whole world as dynamic, generative, and worthy of utmost care and respect—that is, as divine. So in these cases, dancing serves as the primary means by which people learn how to act accordingly.

Dance in/as Religion

The *Ju'hoansi* ("Real People"), or Bushmen of the Kalahari Desert, arguably the oldest extant culture on the planet, engage in an all-night healing dance as their primary ritual. According to elders, the action of dancing transforms the senses of those who dance, enabling them to perceive a luminous web of dynamic, ever-changing relationships that comprises all that is, including gods, humans, and the natural world.

In the dance, people of all ages and genders perform a step-shuffle pattern in a circle, most often around a fire, to the sound of clapping and singing. As they do, they stir and heat n|om—translated by anthropologists Bradford and Hillary Keeney, both recognized as healers by Bushmen elders, as "enhanced life force."[13] As n|om rises within the dancers, they enter what the Bushmen call "First Creation," an experience of reality in which every entity, in every shape and form, is *change and is changing* into every other. The dancers perceive all that is, including humans, animals, Earth, and gods, as sharing the same "nature," as constituting one another. Yet the elders are clear: when they dance, they do not enter a trance or an altered state of consciousness. They become "real people"—they become who they have the potential, as humans, to be.

Yet, as the elders explain, this shift in perception is not the point of the dance. The only seeing that matters is a seeing that acts—that heals. As the singing and dancing cause n|om to "boil," an experienced healer is able to see the places in participants where *change* is stuck and unable to move. This stuckness causes physical and mental illness, as well as conflicts with others. An experienced healer lays shaking hands on these persons, delivers "nails" of enhanced life force, and "pulls out" the sickness. A healer, in other words, releases the capacity of pain itself to change and become something else.

In this communal process of dancing, singing, and healing, the elders confirm that they not only *know* "God's Love," they know it *because* they are participating in it—doing its work. Here, the meanings of the words "God" and "Love" are not abstract; the meanings are given in the sensory experiences of moving and being moved by others that the dancing affords.[14]

Anthropologists who have studied the Bushmen report that the Bushmen frequently refuse to give straight answers to questions about why they dance. As elder =Oma Djo tells Richard Katz, "You just dance."[15] Some researchers have gone so far as to claim that the Bushmen have no religion, as they have no texts and their orally translated stories appear to make no sense.[16] However, the elders themselves—as translated by the Keeneys—claim that the action of dancing is itself its own end. Dancing creates its own reality. The movements of dance educate the senses of those involved to ever-evolving patterns of perception, meaning, and action concerning people's relationships to each other, the divine, and the natural world that the Bushmen have discovered, remembered, and passed down for centuries. These patterns have enabled their existence as a nomadic people in the African desert.[17] For the Bushmen, there is no other way to represent what the action of dancing enables a person to know other than the dancing itself: what dancing affords *is* a person's primary orientation within the world.

In a second example, an ocean away, Rosalie Daystar, of Pembina (Little Shell) and Chippewa ancestry, a pioneer in Native American modern dance, teaches that dancing is an Indigenous Way of Knowing—not because dancing involves learning a set of steps that has information symbolically coded within it, but because the act of dancing these steps provides a specific sensory education. As Daystar explains, Indigenous knowledge is not written down; it is spoken, sung, and danced. People rely on rhythmic bodily movements to awaken what they call "Intuition." To learn to dance then, is to learn *to give oneself* to the specific movements one is making and be transformed by them.

While widely diverse, Daystar explains that in Native contexts, the movements of the dancing are based on circular patterns that evoke the Medicine Wheel and trace the four directions. As one learns and repeats these patterns, the action of doing so shifts a person's lived experience of the world. Daystar describes this process in relation to a traditional women's dance, "The pulse of life vibrates through the body as we stand on Mother Earth, causing us to imitate its pulse: bouncing down and rebounding over and over again, without any thought of wanting to experience an end."[18] Dancing this dance, women learn to sense and respond to the sound of the drum as the pulse of life that connects human and Mother Earth. Over time, as people practice moving to—or being moved by—this pulse, what emerges in them is Intuition: a capacity to sense and respond to whatever is happening in ways that are guided by the sensory awareness of one's connection to the Earth. Actions that flow from this Intuition will promote the good life, *Mno-Bimadziwiin*, a life lived in "balance and harmony."[19] In sum, those who dance come to know *themselves* as able to move in synchrony with a divinity that is, as experienced through the dance, present.[20]

Referring specifically to the ceremonies of the Anishinaabe people in Michigan, dancer and dance anthropologist Gertrude Kurath confirms that the Anishinaabe dance events are "kinetic processes." The action of dancing and singing educates the senses of those who participate so that they are able to "create social and cosmic conditions conducive to sociality, healing, and well-being."[21] Kurath affirms that any conceptual distinctions between religious worldview and artistic action, or between idea and practice, are incapable of capturing what it is about the dancing that renders it religiously effective for the Anishinaabe: that is, the transformation of a human person's capacity for discernment.[22] For Kurath, as for Daystar, what this transformation yields is not a determined program of action, but an ability *to sense and respond* in the moment, in ways that respect and nurture the matrix of relationships that the community has come to value as life-giving. What changes is the world, and the dancer's orientation within it.

As the third of many possible examples, traditions of the African diaspora such as those found in the Caribbean and Central America also feature a regular, rhythmic dance practice at the heart of religion. Anthropologist Yvonne Daniel, in her book *Dancing Wisdom: Embodied Knowledge in Haitian Vodou, Cuban Yoruba, and Bahian Candomblé,* stresses time and again that traditions rooted in African religions privilege the action of learning and repeating rhythmic movement patterns—that is, dancing—as an effective conduit for the most precious and important values and knowledge.[23] Daniel, a dancer and scholar, describes the ceremonies of African-diaspora religions as "dance-dependent ritual structures with dance-initiating objectives." [24] Not only is a given ritual dependent upon the act of dancing for its efficacy, its goal is to initiate dancing as the medium in and through which participants come to know "divinities"— including animal and plant forces, ancestors, spiritual beings, and cosmic entities. As people dance, Daniel explains, new spheres of experience and meaning come into being. The dancing creates and nourishes relationships among participants and the sources of their well-being. By dancing, she explains, people imbibe "resilient patterns of human thoughts and time-tested human behaviours."[25]

In these three cases, *rhythmic bodily movement is the element of a ritual that renders the ritual effective in solidifying belief and guiding action.* A philosophy of bodily becoming explains why. The act of making and repeating specific patterns of bodily movement transforms people's sensory orientation according to the effort and attention that making those patterns demands. The movement patterns thereby open ranges of knowledge and possibilities of perception specific to a community and their social, cultural, and environmental contexts.

Specifically, in each of the cases above, this dance-enabled knowledge includes: 1) a *sensory awareness of one's own bodily self as moving and as being moved*; 2) a sensory awareness of *ever-changing relationships* as the constitutive "substance" of the universe and; 3) a sensory awareness of how one's own bodily movements in the dance, and beyond, *participate* in the ongoing life of these relationships. As a "sensory awareness," this knowledge can be indicated by words but not captured by or transmitted through words; it exists as an ability to mobilize one's kinetic creativity within a given context in line with what the community has deemed "good." While the specifics of the respective world-views differ, in each of these cases, one who dances is able to conceive of the world in all its material, spiritual breadth as a larger *dance* in which one's own bodily movement is an active, contributing part. One who dances is able to discern how to act in ways that honor that ongoing dance.

Divine Immanence and Planetary Thinking

Evoking dance as a critical category for examining cultures that embrace divine immanence sheds light on the role played by bodily movement in the gener-ation of those worldviews. Though sparsely mentioned in the other chapters of this volume, dancing plays a vital, constitutive role in a number of the tra-ditions described, including Native American, Hindu, Buddhist, African, Confucian, animist, Wiccan, and Christian contexts. Even in moments when practitioners may not appear to be dancing, thinking about their bodily actions as "dance" can illuminate how these movements—even when sitting still—serve to map a range of experience, knowledge, and action including, and especially concerning, a divine that presents in or as bodies and nature.

Many chapters featured in this book highlight vivid accounts of robust worldviews that may be described as examples of divine immanence. Words such as "cosmovision" (Grim); "cosmos-world" (Jefferson-Tatum); "anthropo-cosmic universe" (Tucker); "entanglement" (Keller), and "reciprocity" (Grim), to cite a few, point to worldviews that, despite striking differences, aspire to be inclusive, relational, and holistic. In these accounts, a human appears as one node in a dynamic, unbound web that may include living and nonliving entities, as well as natural, material, and spiritual phenomena. A human ap-pears as a creature within this web who can and should learn to respect all moments of it including, and especially, the natural world as (a manifestation of the) divine.

It is also evident in all of these examples that divine immanence is never merely an idea. In every case, material objects, both natural and human-made, come to life for the humans who worship them. Amulets, crosses, and pieces of wood; rocks, animals, and landscapes; icons, mandalas, and pots; and natural and animal phenomena appear to some humans as having their own *agency*. And in every case, the people who are able to sense and respond to material, natural objects as in some way divinely alive, engage in some kind of *bodily practice* that enables them to perceive and conceive these phenomena in this way. They caress relics, dress rocks, or meditate on their breathing; they practice aesthetic, somatic, and spiritual disciplines; they perform rituals.

As many of the authors acknowledge, these actions are not simply expressions of independently held beliefs about the objects. Rather, *these bodily movements matter to a person's experience of the object as having agency.* To one who does not, or cannot, or will not, make the movements of approaching, bending, holding, or touching, the rock, tree, or pot will not come alive. The movements do not guarantee that a seemingly material phenomenon will act; but without these movements, it will not. Through bodily movement, people cultivate a capacity to perceive and be moved by sources and forces beyond themselves.

Thinking about these ritual actions in terms of "dance" confirms that these bodily movements are neither accidental nor external to the beliefs in question: these bodily movements *are* the medium through which ideas of divine immanence become true for those who come to believe in them. *I am the movement that is making me.*

From the perspective of bodily becoming, the patterns of bodily movement enacted in a ritual or ceremony educate people's senses, training them to pay attention to a given relic or icon, tree, landscape, or image of God. The patterns of bodily movement guide people to perceive certain phenomena in greater detail, to notice particular qualities over others. Insofar as the experience of moving in relation to an object rewards people with sensations of pleasure, or connection, or belonging, they feel moved by that object. Their desire for this experience of being moved by that object may grow. As they repeat their movements with greater attention and focus, giving themselves to their movements, their capacity to be moved by it may increase. At that point, a propensity to sense and respond in similar ways to related natural and material phenomena as *divine* may emerge in a way that seems spontaneous.

Again, none of these developments are guaranteed; but the movements made create the opportunity for people to have sensory experiences of material and natural phenomena as animated—as moving them to think, feel, and act in life-enabling ways—and so support the belief that these phenomena are indeed vessels for, or manifestations of, a divine.

This observation carries implications for our planetary project. First, from the perspective of bodily becoming, the process of coming to *conceive* of a relational, dynamic worldview—and *believe* in it—is never just a process of learning to think about what is "out there," even if that "out there" includes "me." Rather, the process of coming to believe in divine immanence entails learning to move in ways that allow that reality to come alive for oneself—in one's own bodily self. It involves becoming someone for whom this perception is possible, and this idea makes sense.

Said another way, *insofar as any concept of divine immanence must include "me" as one part of the whole, that concept cannot be true for me unless I have a sensory awareness of my own bodily movements—however big or small or able they may be—as actively participating in this relational whole.*

From the perspective of bodily becoming, this sensory education is critical, not just because it enables the concept of divine immanence, but because it guides further action. As noted in relation to the three cases above, the dances that people pass on are those that guide them in aligning the actions of their kinetic creativity with what a given tradition knows as the "good life." Dancing trains people to pay attention—to create relationships—with sources that actually sustain life.

In this respect, *any idea of divine immanence will only prevail if it is rooted in practices that guide people to move in ways that heal, protect, and nourish what they perceive through those movements as healing, protecting, and nourishing them.*

Here then, the use of dance as a *critical* category, illuminates why the activists studied by Sarah Pike tell us to "do rituals to save the Earth." These rituals matter to our ability to sense and respond to the Earth in *mutually* life-enabling ways. Such sensory retraining—which is always shaped by history and tradition—is essential for changing our actions and not just our ideas.

Ongoing

What then are the practices and rituals that will tune our twenty-first century senses so that we learn to notice, desire, and follow through with Earth-friendly ways of being? What are the rhythmic bodily movements, in this time and place, which will allow ideas about the value of the natural world and our participation within it to become real for us?

The ubiquity of dance in the world's religious traditions is a reason to welcome these dances as theoretical and practical resources for shifting abusive patterns of human engagement with life on Earth. Humans cannot not dance. It might even be said that the Earth itself is a dance.[26] The question is not will we dance, but will we take responsibility for cultivating a sensory awareness that can guide us in dancing well.

Notes

1. Kimerer L. LaMothe, *Why We Dance: A Philosophy of Bodily Becoming* (New York: Columbia University Press, 2015); *Nietzsche's Dancers: Isadora Duncan, Martha Graham, and the Revaluation of Christian Values* (New York: Palgrave Macmillan, 2006); "Does Your God Dance? The Role of Rhythmic Bodily Movement in Friedrich Nietzsche's Revaluation of Values," in *Dance as Third Space: Interreligious,*

Intercultural, and Interdisciplinary Debates on Dance and Religion(s), ed. Heike Waltz (Gottingen, Germany: Vandenhoeck & Ruprecht, 2021), 81–94.

2. William McNeill, *Keeping Together in Time* (Cambridge: Harvard University Press, 1997).

3. Daniel Siegel, *The Developing Mind* (New York: Guilford Press, 2015), 13.

4. Sarah Hrdy, *Mothers and Others* (Cambridge: Harvard University Press, 2009), 23, 41, 49, and Alan Fogel, *The Psychophysiology of Self-Awareness: Rediscovering the Lost Art of Body Sense* (New York: W.W. Norton, 2009), 13.

5. Siegel, *The Developing Mind*. See also Maurice Merleau-Ponty, *Phenomenology of Perception*, trans. Colin Smith (London: Routledge, 2002), Part I, chapters 3–4.

6. Sandra Blakeslee and Matthew Blakeslee, *The Body Has a Mind of Its Own: How Body Maps in Your Brain Help You Do (Almost) Everything Better* (New York: Random House, 2007), 181.

7. While new materialists celebrate this distribution of agency across infinite material relations, they often neglect the agency of bodily movements that are specific to a human form. See my critiques, "Can They Dance? Towards a Philosophy of Bodily Becoming," *Journal of Dance and Somatic Practices*, Vol. 4 (1) (2012): 93–107, and "Becoming a Bodily Self," in *Religious Experience and New Materialism*, eds. Joerg Rieger and Ed Waggoner (New York: Palgrave Macmillan, 2016), 25–53.

8. Neil Shubin, *Your Inner Fish: A Journey into the 35 Million-Year History of the Human Body* (New York: Vintage Books, 2008), 149.

9. Rodolfo Llinas, *I of the Vortex: From Neurons to Self* (Cambridge: MIT Press, 2001), 5, and Sandra Blakeslee and Matthew Blakeslee, *The Body Has a Mind of its Own*, 170.

10. V.K. Ramachandran, *The Tell-Tale Brain: A Neuroscientist's Quest for What Makes Us Human* (New York: W.W. Norton, 2011), xv–xvi; chapter 4.

11. Chip Walter, *Thumbs, Toes, and Tears: And Other Traits that Make Us Human* (New York: Walker and Co., 2006), 32–5. Of course, because humans have this capacity to dance, we can create and become patterns of movement that lead us to believe we are minds living in bodies who cannot, will not, and do not dance.

12. In making this statement, I am not claiming that all dance techniques and traditions are inherently friendly to the Earth. There are, of course, techniques and traditions that require arduous denial of the senses and involve intense pain, as well as those that occur inside, apart from the natural world. This range supports the point: movement matters.

13. Bradford Keeney and Hillary Keeney, eds., *Way of the Bushmen as Told by the Elders: Spiritual Teachings and Practices of the Kalahari Ju/hoansi* (Rochester, VT: Inner Traditions, 2016).

14. Ibid. For an extended discussion of scholarship on Bushman dance and a contrasting approach, see Kimerer LaMothe, *A History of Theory and Method in the Study of Religion and Dance* (Leiden: Brill, 2018), chapters 11 and 12.

15. Richard Katz, Megan Biesele, and Verna St. Denis, *Healing Makes Our Hearts Happy: Spirituality and Cultural Transformations among the Kalahari Ju|'hoansi* (Rochester, VT: Inner Traditions, 1997), 134.

16. Guenther takes an innovative approach, claiming that the ambiguity of Bushmen stories is itself instructive. See Mathias Guenther, *Tricksters and Trancers: Bushman Religion and Society* (Bloomington: Indiana University Press, 1999).

17. For a thorough discussion of the sensory education the dancing provides, see LaMothe, *A History of Theory and Method*, chapter 12.

18. Rosalie Daystar, "Dancing the Four Directions," in *Dancing on Earth: Special Issue of the Journal of Dance, Movement, and Spiritualities*, ed. Kimerer L. LaMothe, Vol. 4 (2) (2017): 188.

19. Ibid.: 185.

20. Daystar gives an example of this process in an extended description of her own experience of choreographing "Dancing the Four Directions," a piece based on teachings she had learned from Edna Manitowabi, Anishinaabe elder and Professor of Indigenous Studies and Indigenous Performance Studies at the University of Trent. As she danced these teachings, they became *hers*.

21. Michael McNally, "Introduction," in Gertrude Kurath, *The Art of Tradition: Sacred Music, Dance, and Myth of Michigan's Anishinaabe, 1946–1955*, ed. Michael D. McNally (East Lansing: University of Michigan Press, 2009), xx.

22. Ibid., 46, 92, 138, 392. See also LaMothe, *A History of Theory and Method*, 60–63.

23. Yvonne Daniel, *Dancing Wisdom: Embodied Knowledge in Haitian Vodou, Cuban Yoruba, and Bahian Condomble* (Urbana: University of Illinois Press, 2005), 4.

24. Ibid., 4.

25. Ibid., 55.

26. Elisabet Sahtouris, *EarthDance: Living Systems in Evolution* (Bloomington, Ind.: iUniverse, 2000).

The Animist, Almost Feminist, Quite Nearly Pantheist Old Materialism of Giordano Bruno

Mary-Jane Rubenstein

Reattunement

As Joerg Rieger reminds us in his contribution to this volume, "Which Materialism, Whose Planetary Thinking?" a torrent of movements, philosophies, and consumer practices—many of them incompatible with one another—have either claimed or received the term "materialism." In the face of this semantic riot, the collaboration at hand has chosen to focus on the politically transformative potential of those manners of thinking and being that locate agency, creativity, and change within the animal-vegetal-mineral world, rather than in some spiritual sphere above, beyond, or even within it. Such *materialisms* affirm some of what has been gathered under that name while contesting others; in particular, they resist the impulse to reduce the world to information, subatomic particles, or the random firing of neurons. At their best, the vibrant materialisms that our loose collective finds most compelling do not merely collapse the tired old distinction between mind and body. Rather, they render it just that: tired and old, hardly even worth mentioning.

As this distinction and its scientifically reactionary reversals are all products of the intellectual tradition we incoherently call "Western," the clearest paths toward a truly generative materialism will likely run through traditions that have very little to do with mainline Greco-Roman-Abrahamism. In this spirit, some of the papers in this volume's first section attune us to counter-Christianities, while others turn us toward the variously agential materialities within Confucian, Daoist, Africana, Hindu, Jain, Buddhist, Native American, and other broadly "animist" Indigenous philosophies. Most of the contributions to this book's second section seek similarly to dismantle the tired old opposition by swerving around it, drawing our attention to contemporary theories of immanence that set themselves explicitly against the tired old distinction, whether they attribute it to Kant, Descartes, Platonized Christianity, or indeed, "the West."

As we pursue these non- or anti-"Western" vitalisms, however, I find myself wondering whether it might be instructive to revisit the source of the problem in order to recall its structure, contours, and conceptual cross-bracings. After all, the distinction between spirit and matter has traditionally mapped itself onto those old hierarchical privileges of light over darkness, male over female, mind over body, and of course God over creation. So, attending to its formulation means attending to the formulation of its racially, environmentally, sexually, and theologically toxic companions. To be sure, there is no singular source of these perennial privileges, yet they find paradigmatic articulation in the work of Aristotle. By returning to this articulation, and to its startling deconstruction in the work of one of Aristotle's heretical commentators at the dawn of the modern age, I am hoping to locate a structural incoherence at the heart of the "Western" tradition with respect to the status of matter. Ultimately, I am seeking unexpected, renewable, planetary resources that might recode matter's dark femininity as active, creative, transformative—perhaps even divine. This recoding brings us dangerously close to the perennial heresy known as "pantheism," but it could be that such heretical proximity might be a productive stance from which to mobilize the material against the disembodied, the planetary against the global.

Matter in Waiting: Aristotle

Aristotle's *Physics* encodes matter as passive and inert, lying in wait for another principle to discipline, order, and shape it. It is matter's "own nature," Aristotle explains, "to desire and yearn for [form]."[1] Form, he continues, provides unity, order, and animacy, transforming matter's empty potential into actuality.[2] This means that matter itself has no qualities apart from "privation."[3] And lest we think the persistent gendering of these terms is merely implicit or accidental, *Physics* clearly states that "what desires the form is matter, as the female desires the male, and the ugly the beautiful."[4]

At first glance, Aristotle seems to give us the stark opposition of matter and form that our contemporary immanentisms so fastidiously contest. Yet even in the work of this paradigmatic dualist, the dualism is not so simple; rather, Aristotle insists that form is "not independent of matter."[5] Indeed, as one commentator glosses a famous example, "all natural forms are like something which is 'snub,' where something is snub only if it is concavity-realized-in-a-nose."[6]

Unlike the Platonic Forms, which dwell in a realm prior to and independent of the material sphere, Aristotelian form is totally bound up with matter; in fact, matter allows form to come into being in the first place. Matter, in Aristotle's words, is the "ultimate substratum"—that which precedes, underlies, and follows each evanescent configuration of matter-and-form.[7] So when a tree becomes logs or mulch, the forms change dramatically, but the material of the wood persists. And when a log becomes fire, the "proximate material" of the wood disappears, but *matter itself* persists as fire, smoke, and ash.

In sum, matter is the condition of possibility of all substance. It is that ineffable primordium of which nothing can be properly predicated because it enables predication itself. Those who have ears to hear might pick up traces in this primordium of some feminized, apophatic divinity. Indeed, the medieval philosopher David of Dinant (1160–1217) took Aristotle's "ultimate substratum" to mean that the divine intelligence was identical to primal matter—or in a Christian register, to the "deep" or *tehom* of Genesis (1:2).[8] Such divine materiality led him to proclaim the equivalence of creator and creation. If, as David reasoned, "the matter of the world is God himself, and the form that comes to animate matter is nothing other than God making himself sensible," then "the world is therefore God himself."[9] For this crime, which *The Catholic Encyclopedia* continues to brand "the most thoroughgoing pantheism,"[10] David's books were burned, his followers executed, and his ideas given a particularly uncharitable treatment by Albertus Magnus and his star pupil, Thomas Aquinas.[11]

David's theo-materialist interpretation of Aristotle did not, therefore, become the received reading of Aristotle. This is not to say that David was the only person to identify the divine intelligence with prime matter; to the contrary, a similar position has been ascribed to the Islamic philosopher Ibn Rushd (Averroes), who asserted the eternity of matter against the doctrine of *creatio ex nihilo*. It has also been ascribed to the Jewish Neoplatonist Ibn Gabirol (Avicebron), whom early modern Christians often mistook for a Muslim Aristotelian.[12] The extent to which these philosophers actually divinized matter is a question of ongoing debate,[13] but they certainly held it in higher esteem than Aristotle himself. Despite matter's interiority and anteriority to form, and despite its resistance to all conceptualization, Aristotle hardly divinizes it. Rather, he ascribes divinity to a (sometimes singular, sometimes plural) "Prime Mover" positioned sufficiently beyond the fixed stars to give them a cosmogonic first push.[14] This Prime Mover is pure actuality, which is to say, form uncontaminated by matter. Matter, in the meantime, continues throughout the authorship to embody pure passivity, privation, and longing. In relation to form, it is unquestionably the inferior term—the ugly, womanly, shapeless gunk that needs something manly to bring it to order and life. And yet, it nevertheless precedes and conditions the possibility of form.

In sum, matter occupies a position of raging ambivalence in the work of Aristotle and his interlocutors—an ambivalence that rarely finds elaboration in contemporary critiques of Western dualisms, but which would certainly assist their cause. With this aim in mind, I would like to turn to the brilliant exploitation and deconstruction of this Aristotelian ambivalence in the work of the under-attended Renaissance philosopher Giordano Bruno, a Dominican friar-turned-pantheist whom the Inquisition executed at the dawn of the seventeenth century.

Matter Reimagined: Giordano Bruno

At the beginning of his dialogue in *Cause, Principle, and Unity* (1584–85), Bruno's mouthpiece, Teofilo (one of several including Filotelo, Discono, and Gervasio), denounces those Aristotelian pantheists who teach that "matter alone is the substance of things, and that it is also the divine nature, as the Arab named Avicebron has said."[15] Although Teofilo will eventually assert this very position, he uses this racialized aside about Ibn Gabirol to placate his European-Christian audience and throw them off Bruno's track. Teofilo even goes so far as to insist that true philosophy must make absolute distinctions between form and matter, "active potency" and "passive potency," and "the power to make" and "the power to be made" (55). In short, he says, true philosophy must distinguish creator from creation.

As the reader will no doubt notice—perhaps with a bit of consternation, considering the iconoclasm I have been promising from this text—these are traditional, Aristotelian categories, mapped in raced and gendered opposition, under the distinction between God and the world. The likeliest explanation for Teofilo's beginning with the very dualisms he will go on to abolish, is that he is meeting his readers at their own level. The universities of the late sixteenth century were filled with neo-Thomist Christians (scholars who Bruno had ridiculed in an earlier, more audacious dialogue as "Peripatetics who get angry and heated for Aristotle"[16]), and Teofilo is staking his eventual implosion of these terms on an analogical premise that his interlocutors will find unshakeable: *form* is different from *matter*, as (light-masculine) *activity* is different from (dark-feminine) *passivity*, and as *maker* is from *made*.

Even though these categories traditionally line up under the headings of "God" and "creation," respectively, Teofilo makes it clear from the beginning that his investigation into the "cause and principle" of the universe will have nothing to do with God. He is only a natural philosopher, he explains, and as such he is dealing with only natural causes (34). Bracketing the question of the *first* principle and cause, he will only "look into the principle and cause insofar as . . . either it is nature itself or it shines in the elements and the bosom of nature" (36). In other words, he will examine the source of all things in as far as it either *is* or *animates* the material world itself. A source-of-all-things, he strategically assures us, that certainly is not God.

In line with Greco-Christian tradition, Teofilo begins his cosmogony with the "form" of the universe, which he calls "the world soul:" "a vital, vegetative, and sensitive principle in all things which live, vegetate, and feel" (6). The chief faculty of the world soul is what Bruno calls "the universal intellect," which he designates as "the universal physical efficient cause" (37/39). The world soul is therefore the *principle* of the universe, meaning it precedes, contains, and fills everything that exists; whereas the intellect is the universal *cause*, meaning it brings everything into being.[17] Matter, by contrast, is the stuff *on* which the world soul works. Matter "has no natural form by itself but may take on all forms through the operation of the active agent which is the principle of nature," the world soul (56). But precisely because matter is, in this sense, the "receptacle of forms" (61),[18] matter is indispensable to the emergence of anything that is. After all, Discono asks, "how can the world soul . . . act as shaper, without the substratum of dimensions or quantities, which is matter?" (55). Since form cannot exist independently of matter, it must be internal to it, "forming [matter] from inside like a seed or root shooting forth and unfolding the trunk" (38).

At this point in the dialogue, we have come as far as Aristotle will go, with matter figured as the surprisingly formidable "universal substratum"—the stuff that remains even as accidental forms arise in it and fall away. As we have seen, Aristotle nevertheless persists in denigrating matter as sheer passivity, as possessing neither powers nor qualities, and as "yearning" for a masculinized form to come and make it into something. The contradiction is enough to prompt the character of Discono to cry, "Why do you claim, O prince of the Peripatetics, that matter is nothing, from the fact of its having no act, rather than saying that it is all, from the fact that it possesses all acts?" (82). Why does Aristotle fail to adhere to his own insight? If form does not exist without matter, but is rather preceded, followed, and even generated by it, then one presumably ought to say that matter is not empty of all qualities but rather full of them, containing *in potentia* all the forms it actualizes over time. This, says Filoteo, is what David of Dinant knew (7), and what Averroes almost knew, ". . . he would have understood still more," he laments, "had he not been so devoted to his idol, Aristotle" (80): *matter does not lack form* and so cannot desire it. Rather, matter gives rise to form, and as such can only be said to be "deprived of forms" in the same way that "a pregnant woman lacks the offspring which she produces and expels forth from herself" (81).

According to all of the characters that the dialogue presents as respectable interlocutors, the reason that so few people have reached the insight that matter contains and gives rise to all things, is that Aristotelians hate women. As we have seen, the *Physics* explicitly aligns matter with femininity—an unshaped ugliness lying in wait for masculine form to bring it to order and beauty. Bruno gives this position a comical, exuberant spokesman through the dopey character Poliinnio, "one of those stern censors of philosophers . . . reputed to be a follower of Socratic love, an eternal enemy of the female sex" (29).

The fourth dialogue opens on Poliinnio alone, who in the absence of his quicker-witted colleagues is free to deliver his thoughts on the manifold ills of matter in an uninterrupted, verbose, and increasingly ridiculous rant. "And the womb never says 'enough,'" Poliinnio begins, likening the operations of matter to the hysterical longings of a sex-crazed woman (70). Matter, according to Poliinnio, displays "the insatiable craving of an impassioned female" (10) inasmuch as "she" is "never sated with receiving forms" (70). For this reason, he explains, matter is:

> . . . called by the prince of the Peripatetics . . . *chaos*, or *hyle*, or *sylva* [abundant material], or . . . cause of sin . . . disposed to evil . . . not existing in itself . . . a blank tablet . . . unmarked . . . litter . . . field . . . or *prope nihil* (almost nothing) . . . finally, after having taken aim with several comparisons between various disparate terms . . . *it is called "woman"* (70).

Citing Helen of Troy, Delilah, and Eve, Poliinnio goes on to charge women with having caused the downfall of all great men and nations. Similarly, he reasons, matter is the ruin of all form, which on its own "does not sin, and no form is the source of error unless it is joined to matter" (71). It is, therefore, no accident, Poliinnio concludes, that the *Physics* compares matter to femininity. For it cannot be denied that matter shares all the qualities of:

> . . . the female sex—that sex, I mean, which is intractable, frail, capricious, cowardly, feeble, vile, ignoble, base, despicable, slovenly, unworthy, deceitful, harmful, abusive, cold, misshapen, barren, vain, confused, senseless, treacherous, lazy, fetid, foul, ungrateful, truncated, mutilated, imperfect, unfinished, deficient, insolent, amputated, diminished, stale, vermin, tares, plague, sickness, death (72).

This lengthy and progressively absurd monologue ends up serving three purposes in the text at hand. First, it exposes the traditional philosophical denigration of matter as a product of untrammeled sexism. Second, it exposes such sexism as baseless and anti-intellectual, coming as it does from the mouth of a character who Gervasio calls "the biggest, most bumbling beast that exists in human form" (34). And third, it provides Teofilo with the metaphorical basis of a radical transvaluation of matter itself. Turning the Peripatetics' own associations against them, Teofilo provokes them, unwittingly, to demonstrate the preeminence of *hyle*, which "sends all forms forth from its womb" and as such, is the origin of all that is (82). In effect, Teofilo's strategy is to retain the traditional gendering of matter while shifting our focus from the heteronormative sex act to the act of giving birth. From this vantage point, he is able to assert that far from lacking, desiring, or indeed receiving anything, matter already "possesses" within itself everything it eventually brings forth (82). As his dialogic twin Filoteo suggests in his framing summary, matter is "not a *prope nihil*, an almost nothing, a pure and naked potency, since all forms are contained in it, produced by it, and brought forth by virtue of the efficient cause (which . . . can even be indistinguishable from matter)" (9).

At this point, Teofilo is finally able to unify all the distinctions he has taken such pains to separate. While it is the condition of possibility of all things, this material cause is the principle of creation, which is to say the world soul itself. And in as much as matter brings all things forth, it is also the efficient cause of creation, which is to say the universal intellect. Hence the coincidence of corporeality and intellect, body and soul, principle and cause, activity and passivity, and—most centrally for our purposes—matter and form (8, 66). Crucially for Bruno, however, this coincidental cascade does not erase the distinctions it holds together. Rather, as Filoteo explains, the assertion that "all is one" means "there is unity in the multiplicity and multiplicity in the unity . . . being is multi-modal and multi-unitary" and, as Teofilo puts it, "multiform and multifigured" (10, 90).[19]

Especially for those of us with an eye on pantheism, the question becomes whether this many oneness of form and matter, act and potency, and intellect and material also amounts to a differential coincidence of *God* and *world*. Filoteo tempts us with this possibility when he suggests that "what is supreme and divine is all that it can be" and that likewise, "the universe is all it can be" (8). Perhaps this means (by virtue of the transitive principle) that the universe itself is "supreme and divine?" Teofilo is understandably reluctant to assert this particular identity, and so he qualifies it with a move he learned from Nicholas of Cusa: the distinction between "contracted" and "uncontracted" infinities.[20] "The universe is all that it can be, in an unfolded, dispersed, and distinct manner," he explains, "while its first principle is all it can be in a unified and undifferentiated way" (66). Therefore, he implies, the two do not coincide. If, however, the divine first principle relies upon the universe that incarnates it as form relies upon matter, then creator and creation would coincide after all. Discono tries numerous times to get Teofilo to extend his dialectics in this manner, but Teofilo keeps reminding him that their conversation has deliberately excluded "the supreme and most excellent principle" (81), restricting itself to physical causation. None of this, he repeatedly insists, has anything to do with God.

Rhetorically and strategically speaking, Teofilo's restraint here is perhaps well-advised. Logically speaking, however, there is no reason to limit the coincidence of opposites to physical causes—especially in as much as the physical and the metaphysical presumably coincide in the unity of sensible and intelligible matter. By leading us to this possibility without quite entertaining it, Bruno allows his reader to contemplate the notion of God's identity within the universe, should she be so inclined—while himself stopping a hairsbreadth short of heresy. Even so, Bruno does allow Teofilo to conclude that if matter indeed contains all forms, then it "must, therefore, be called a divine and excellent parent, generator, and mother of natural things—indeed nature entire in substance" (83–84). At this point, one starts to wonder just what use the perennially bracketed "supreme first principle" might ultimately be. If nature is itself divine, if it generates all sensible and intelligible things from itself and is, as such, an omni-gendered parent (both "generator" and "mother"), then what on Earth would we need from a God above, beyond, or before this spiritual-material divinity? One might suggest, perhaps, that such a God is required to give the universe a first push at the beginning of time, but this would limit, rather severely, the function and continuing relevance of God. Besides that, there *is* no "beginning of time" for Bruno; the universe is eternal and so needs no first push.

Theologically speaking, then, what this "strictly physical" dialogue has done is to call each of the divine faculties down into nature itself—all the while pretending not to speak of God. It is precisely by bracketing the "supreme first principle" that Bruno goes on to render such a principle irrelevant, leaving us with an omni-formed, ensouled matter as the creator and end of all things. Insofar as this created creator is both intellectual and extended, Bruno's "cause and principle" of the universe looks remarkably like the single Spinozan "substance" it goes on to influence.[21] For Bruno as for Spinoza, the manifold animals, vegetables, and minerals around us are all physical and ideational expressions of the same substance, which is composed substantially of its expressions and is for that reason many in its oneness—or in Filoteo's words, "multi-modal," "multi-unitary," "multiform," and "multifigured."

This multi-unitary substance leads Bruno to proclaim the vitality of all things. Because everything in existence is an expression of the world soul, Bruno reasons, *everything has a soul*—and everything is therefore animated. The logic seems to him so sound that he asserts, in a prefatory summary of the dialogue, "It is . . . unworthy of a rational subject to believe that the universe and its principal bodies are inanimate" (6). Although this statement seems to limit the scope of animacy to the world as a whole (and the Sun, Moon, and stars), Teofilo proceeds, over the course of his instruction, to extend animacy to all inner-worldly beings. The teaching, his interlocutors object, is a strange one, "Common sense tells us that not everything is alive," cautions Discono, only to be immediately countered by Teofilo's reply, "The most common sense is not the truest sense" (42).

This exchange stirs the ire of the embattled Poliinnio who attempts to force the argument into absurdity. "So my clogs," Poliinnio asks, "my slippers, my boots, my spurs, as well as my ring and my gauntlets are supposedly animated? My robe and my palladium are animated?" (43). Teofilo's remarkably measured answer is that "the table is not animated as table, nor are the clothes as clothes, nor is leather as leather . . . but . . . they have within them matter and form. All things, no matter how small and miniscule, have in them part of that spiritual substance which, if it finds a suitable object, disposes itself to be plant, or to be animal" (44). The omni-creativity of this multi-unitary spiritual substance—which, we will recall, is also a material substance—means that all things, "even if they are not living creatures, are animate" (44). Nothing is inert, dead, mere, or for that matter, exploitable matter.

For Teofilo, this universal animacy means that the pre-Socratic philosopher, Anaxagoras, was right when he said that "all things are in all things."[22] After all, the same spiritual-material world soul that animates the cactus also animates the polar bear, so the whole universe appears in contracted form in each of them. As Teofilo puts it, "each thing in the universe possesses all being" (89). Far more recently than Anaxagoras, Nicholas of Cusa had taught this same precept as a theological principle: God is present everywhere throughout the boundless universe, he argued, and as such God is as fully present in a mustard seed as in a man.[23] This radical indwelling was, in fact, what it meant for Cusa to call God "creator" in the first place. "Creating," he ventured, "seems to be not other than God's being all things."[24] And, to the extent that God is the being of all things, and all things dwell reciprocally in God, it can in fact be said that "all are in all and each are in each."[25]

Although the logic is nearly indistinguishable from Cusa's, Teofilo does conceive two major, but subtle, departures from his more orthodox predecessor. First, he effectively eliminates the Cusan difference between God and the universe, entreating us, by virtue of this entangled animacy, not to "look for the divinity outside of the infinite world and the infinity of things, but inside that world and those things" (82). Second, he qualifies the Cusan-Anaxagoran proclamation of "all things in all things" with a pre-Spinozan principle of particularity. "Everything is in everything," Teofilo affirms, "but not totally or under all modes in each thing" (90). So, this piece of toast has carbon, wheat, yeast, salt, fire, human labor, mechanical production, time, space, and, most likely, traces of polycarbonate or polyvinyl chloride in it—indeed, it has the *substance* of the whole universe within it—but it does not, for all that, contain a teabag. To be sure, the toast contains and reflects the same "being" (and earth, vegetality, water, air, and probably trace plasticity) that also finds itself expressed as a teabag, but the teabag as teabag is not in the toast as toast. Hence, the universal interrelation *and* the irreducible particularity of all things—a differential holography enacted through the divine generativity of matter itself. In his bold sort of qualified stutter, Teofilo is therefore led once again to conclude that matter is what we have meant by the origin, end, and life of all things: matter, he suggests, is indeed "so perfect that, if well pondered, [it] is understood to be a divine being in things, as perhaps David of Dinant meant, who was so poorly understood by those who reported his opinion" (86). This, at least, is the position of Teofilo, proponent of "what the Nolan holds," reaching backward and forward across the centuries from one heretic to another to proclaim a transvalued, still feminine *matter*: an omnicreative, omnitransformative, multimodal coinherence of intellect and extension that is not not divine.

Notes

1. Aristotle, "Physics," in *The Complete Works of Aristotle*, trans. R. P. Hardie and R. K. Gaye, ed. Jonathan Barnes (Princeton: Princeton University Press, 1984), 1.9.192a, 16–17.

2. Ibid., 2.1.193b, 7–8. The essence or "nature" of any particular thing, he insists, is "the form . . . rather than the matter, for a thing is more properly said to be what it is when it exists in actuality than when it exists potentially."

3. Ibid., 1.9.192a, 26; Aristotle, "Metaphysics," in *The Complete Works of Aristotle: The Revised Oxford Translation*, trans. W. D. Ross, ed. Jonathan Barnes (Princeton: Princeton University Press, 1971), 7.3.1029a, 24–6.

4. Aristotle, "Physics," 1.9.192a, 23–4.

5. Aristotle, "Metaphysics," 6.1.1026a, 6.

6. "Form Vs. Matter," in *Stanford Encyclopedia of Philosophy* (https://plato.stanford.edu/entries/form-matter/: February 8, 2016); quotation marks added. See Aristotle, "Metaphysics," 6.1.1025a, 30–32.

7. Aristotle, "Metaphysics," 7.3.1029a, 24.

8. Enzo Maccagnolo, "David of Dinant and the Beginnings of Aristotelianism in Paris," in A History of Twelfth-Century Western Philosophy, ed. Peter Dronke (New York: Cambridge University Press, 1992), 435.

9. David's fragment is published in Tristan Dagron, "David of Dinant–Sur Le Fragment [Hyle, Mens, Deus] Des Quaternuli," Revue de Métaphysique et de Morale, Vol. 40 (2003): 424–5; translation mine.

10. William Turner, "David of Dinant," in The Catholic Encyclopedia (New York: Robert Appleton Company, 2017).

11. On David's alleged conflation of "active" divinity with "passive" matter, see Albertus Magnus, "Summa Theologiae Sive Scientia De Mirabili Scientia Dei," ed. E. Borgnet (Paris: Vives, 1894), 2.12.72.1, 2.12.72.4.2, and Thomas Aquinas, Summa Theologiae, 5 Volumes, trans. Fathers of the English Dominican Province (Allen: Christian Classics, 1981), 1, 3, 8.

12. Giordano Bruno, "Cause, Principle and Unity," in Cause, Principle and Unity and Essays on Magic, trans. Richard Blackwell, eds. Richard J. Blackwell and Robert de Lucca, Cambridge Texts in the History of Philosophy (Cambridge: Cambridge University Press, 1998), 80, 55, 61. **Subsequent references to this text will be cited internally.**

13. The controversy with respect to Ibn Rushd poses the question of whether calling matter eternal renders it equal to God. Believing that it did, Bishop Etienne Tempier condemned two hundred and nineteen "Averroist" theses in 1277. Ibn Gabirol goes further than Ibn Rushd, not only identifying matter with form, but occasionally privileging the former over the latter. See Ibn Gabirol, The Font of Life (Fons Vitae), Mediaeval Philosophical Texts in Translation (Milwaukee: Marquette University Press, 2014), 5.42.334, 3.9.99.

14. On the numerically shifting movers, see Aristotle, "Metaphysics," 7.8.1073a-1074b and Mary-Jane Rubenstein, Worlds without End: The Many Lives of the Multiverse (New York: Columbia University Press, 2014), 35.

15. Bruno, "Cause, Principle and Unity," 55. Cf. p. 64: "The Epicureans have said some good things, although they have not risen beyond the material quality."

16. Giordano Bruno, The Ash Wednesday Supper, trans. Edward A. Gosselin and Lawrence S. Lerner, Renaissance Society of America Reprint Texts (Toronto: University of Toronto Press, 1995), 95.

17. On the distinction between principle and cause, see Bruno, "Cause, Principle and Unity," 36–7.

18. Here, Bruno is equating matter with Plato's khôra, which is propertyless "space" in Plato's Timaeus.

19. On the manyness of Bruno's "one," see Antonio Calcagno, Giordano Bruno and the Logic of Coincidence: Unity and Multiplicity in the Philosophical Thought of Giordano Bruno, ed. Eckhard Bernstein, Renaissance and Baroque Studies and Texts (New York: Peter Lang, 1998).

20. Nicholas of Cusa, On Learned Ignorance, trans. H. Lawrence Bond (New York: Paulist Press, 1997), 2.4.113.

21. There is no incontrovertible evidence that Spinoza read Bruno, but commentators often deduce that he either may or must have done so. For these positions, respectively, see Steven Nadler, *Spinoza: A Life* (Cambridge: Cambridge University Press, 2001), 111, and J. Lewis McIntyre, *Giordano Bruno* (London: Macmillan, 1903), 337–43. For an analysis of the dialectical similarities between the two thinkers, see Arthur O. Lovejoy, "The Dialectic of Bruno and Spinoza," in *The Summum Bonum*, ed. Evander Bradley McGilvary (Berkeley: The University Press, 1904).

22. Anaxagoras taught that the milk a child ingests, for example, can only become bone and blood, "if there is already bone and blood in the milk." Patricia Curd, "Anaxagoras," *Stanford Encyclopedia of Philosophy* https://plato.stanford.edu/entries/anaxagoras/(October 1, 2015).

23. Mary-Jane Rubenstein, "End without End: Cosmology and Infinity in Nicholas of Cusa," in *Desire, Faith, and the Darkness of God: Essays in Honor of Denys Turner*, ed. Eric Bugyis and David Newheiser (Notre Dame: University of Notre Dame Press, 2016), 21–2, 27–8.

24. Nicholas of Cusa, *On Learned Ignorance*, 2.2.101.

25. Ibid., 2.5.117. For a contemporary cosmo-theological exposition of this principle, see Catherine Keller, *Cloud of the Impossible: Negative Theology and Planetary Entanglement* (New York: Columbia University Press, 2015), 114–5.

Emergence Theory and the New Materialisms

Kevin Schilbrack

The Idea of Emergence

Emergence theory is a recently revived attempt by philosophers to provide an ontological account of what it means to say that a whole is greater than the sum of its parts. Entities with their own distinctive properties often come to be connected to each other in particular ways—atoms become molecules, athletes become teams—and "emergence" names the phenomenon when the composite whole has properties that cannot be found in any of its constituent parts, except when they are connected in that particular structure. In this chapter, I argue that one can use the concept of emergence to give a non-reductive account of how material entities combine to make new composites, including how people combine to make social groups so that "a social group" can refer, not merely to a heuristic device, but also to an entity operating in the world. In short, this paper recommends emergence as a conceptual tool that helps us see how the entities in the world are shaped by real—though sometimes invisible—material structures.

Emergence is a valuable conceptual tool because it avoids the dualisms that hinder our ability to explain behavior naturalistically, and this is the case both when we study behavior in the natural world and when we narrow our focus to the human part of it. Emergence offers an account of living entities without introducing any mysterious entelechies or *élan vital*, an account of human entities without speaking of immaterial souls or minds; and an account of cultural entities that does not invoke some Hegelian Geist. Instead, this tool treats living organisms, human minds, and social groups as material realities that are nonetheless more than all the simpler entities that make them up. Given this approach, therefore, social groups would not float above material forces but would instead be seen as grounded in the character of the group's members and the relations between them, which are grounded in human biology, grounded in chemistry, grounded in physics.

This chapter has three parts. The first offers an overview of the idea of emergence. The second applies it to the case of social groups, along with a sketch of what I consider the most significant challenges to an emergence theory and a proposal about the best way for emergentists to navigate them. The third identifies where this way of thinking supports and dissents from other forms of planetary thinking discussed in this volume.

The central idea of emergence is that a whole is greater than the sum of its parts and, for this reason, an account of the whole only in terms of its parts will be incomplete. A classic example of emergence, cited repeatedly since it was proposed by John Stuart Mill (1988 [1872]), is the fact that hydrogen and oxygen atoms can combine with each other to form a molecule, H_2O, which has properties that the component hydrogen and oxygen atoms did not have. The water molecule as a composite whole is not composed of anything other than its constituent gas atoms, but when enough of those parts connect to each other in a particular structure, properties not possessed by the parts—like H_2O's drinkability, solvent abilities, and surface tension—emerge. These novel properties mean that water can have effects on the world that hydrogen and oxygen cannot. Mill and other British emergentists, argued that the existence of the novel properties found in molecules justified the distinction of chemistry from physics, so that even though there is only one kind of matter—physical matter—the increasing complexity of structured composites meant that different sciences were needed to respond to different kinds of properties operating in the world.

One can repeat this emergence theory for living organisms. That is, water molecules combined with other kinds of molecules build higher-level entities, living cells, which have new properties that their constituent molecules lack. The molecule that was treated as a whole in chemistry is now treated as a part in biology, and the emergent properties of these organic entities then justify biology as a discipline distinct from chemistry and physics. Continuing this logic "upward," so to speak, cells can combine into organs, with their own emergent properties, and organs into bodies, again with new properties. Once we get to this domain of living entities, one discovers that there are emergent properties that abet the life of the organism of which they are a part: a heart, for example, has the property of circulating blood through the organism. Properties that are functional like this do not exist in physics, and so new kinds of explanation for the entity's behavior become available.

For many philosophers, the Holy Grail that emergence theory might be able to reach is an account of the mind that is both materialist (with no appeals to non-physical realities) and yet non-reductive (so that mental properties such as beliefs and desires have effects that individual neurons do not). The emergentist argument is that, just as molecules are composed of atoms in a certain structured relationship, but have chemical properties not reducible to their atomic parts; just as cells are composed of molecules in a certain structured relationship, but have biological properties not reducible to their molecular parts; animals are composed of their physico-chemico-biological parts in a certain structured relationship, but they have psychological properties not reducible to those parts. An emergence theory of mind seeks to account for mental properties as emergent from the structured components of the material bodies of the organisms made up of them, especially the nervous systems, without proposing the arrival of a second, non-material substance of some kind. As a non-reductive materialist philosophy of mind, emergence theory treats the mind as the property of a material entity, but it also treats that material entity as an astoundingly complex, composite whole that possesses novel properties not possessed by its constituent parts. If an emergence theory is successful, the result is a philosophy of mind that includes consciousness, intentions, beliefs, and desires as psychological properties, as effects of the organization of the animal's physical, chemical, and biological components, but not found in them; save for when they are arranged in that particular way. The hope here is that emergence can provide a non-Cartesian theory of minds: a materialist account of mind that would not contradict the goals of immanent religiosities.

Emergence theories are promising in all of the above disciplines in that they seem to offer a "third way" between two problematic positions. On the one hand, there are eliminative reductionist views according to which understanding an entity means breaking it down and explaining its behavior in terms of the parts that make it up. Although emergence agrees that entities in the world are constituted by nothing other than their material components, it also insists that those entities only possess the properties they do when those components are properly organized. As a consequence, the emergent, higher-level entity cannot be eliminated from the explanation of its properties. Form matters. In this way, emergence theory seeks to avoid eliminative reductionism.

On the other hand, there are dualist views according to which the behavior of parts and that of the whole is so different, that explaining the difference requires, not only material entities in increasingly complex relations, but also another, very different type of reality. By contrast, emergence theories attempt to attribute the causal properties of higher-level realities to the structured relations of the entities that compose the whole without ever "jumping" to something non-material. Atoms with their properties are material realities; molecules have properties not possessed by the atoms when those atoms are not part of the molecular structures; but the molecules are composed of nothing other than those material realities in a certain structured relation, and so molecules are also material realities. The same is true of cells, and then of multicelled animals, and then of animals who are persons. In this way, emergence theory seeks to avoid dualism.

If emergence theory is able to successfully avoid these two pitfalls, the consequence would be a single-substance picture of the world that does not treat life or mind as eliminable from a complete account of things. In this view, each entity that exists is "laminated" in that it is composed of material parts that are composed of material parts, and the question of why an entity behaves as it does would be amenable to answers at multiple levels (Elder-Vass, 49–53). The promise of emergence theory, then, is a coherent version of non-reductive materialism.

Social Emergence, Strong and Weak

Even though John Stuart Mill championed the idea of emergence, he hesitated to apply it to social groups. Nevertheless, those interested in social realities can go one level "above" an emergent theory of individual persons to also think of groups as emergent entities. From the emergentist perspective, even when a social group is composed of nothing but individual people, the enabling and constraining structure of the group endows some of them with new powers, while inhibiting other powers, so that group members possess properties that they do not and could not have outside social relations of that type. These emergent properties are distinctively social in that they depend on the particular relationships between those individuals. Examples of such properties include those of a landlord, a lead singer, a girlfriend, a senator, a goalkeeper, or an assistant professor. In short, emergence theory treats groups as structured composite entities in the world, and the existence of these groups then justifies explanations that employ properties that depend on the existence of those human interactions.

The crucial step for an emergentist account is the distinction between resultant and emergent properties.[1] A social group's resultant properties are those that the set of people would have, regardless of the group's structure. For example, the weight of the group is the same as the sum of the weight of the individuals in the group, so the group's weight is a resultant property. The group's ability to walk and its average IQ are also resultant. Resultant properties are simply aggregative, in that the whole would have them however their parts were organized, and each member would have their share of these properties, whether or not they were part of this particular whole, even if the set of people had no structure. By contrast, the emergent properties of the group are those that members possess by virtue of their relations with each other. The group members would not possess the properties of *being a jury* or *being an orchestra*, even aggregately, unless they participated in the structured relations of the group.

Emergence theory is not a settled position, but one still entangled in thorny debate. The central challenge is that because emergence is supposed to offer a third way between a reductive eliminativism and a non-reductive dualism, its proponents have to perform a balancing act to clarify how an entity can be simultaneously dependent and independent of its parts. The entity has to be dependent in that it consists of those parts and nothing else, but also independent in that it has novel properties they lack. If one's account of emergence leans too far in the first direction, it slips back into an eliminationist view according to which chemicals, living things, consciousness, and social groups make no difference in the final analysis in how we explain the world. If one's account leans too far in the other direction, and one argues that the properties of wholes cannot be derived from their parts, then the whole and the parts seem to occupy two distinct realms. This answer suggests that emergent properties are simply brute, inexplicable facts, requiring their own explanatory laws, and the theory slips back into dualism. In this section, I want to sketch how an emergence theory might navigate a path through this two-sided challenge.

The central topic on which emergence theory has been challenged is its account of causality. Emergence is the view that from the assembly of parts in a given structure emerge systemic properties that have new effects in the world—but what is it that *wields* these effects? It seems natural to point to "the whole" as the agent: what emergence refers to, on this view, is a new, higher-level composite entity that possesses causal powers. A reductive physicalist critique of this holism is that since the whole is composed of parts, anything the whole does could be explained by the parts, and the whole is explanatorily superfluous.

A closely related topic on which emergence theory has been challenged is that of explanatory reduction. Emergence theory holds that emergent properties, by definition, cannot be found in their constituent parts—but how unpredicted is the appearance of these properties? Can the emergent properties of the whole be explained by the properties of, and the relations between, the constituent parts? For example, if one studied hydrogen and oxygen atoms, could one predict that, when combined in a certain structure, the composite whole would be a liquid with solvent properties? This question about explanation presents a dilemma that threatens the coherence of the very idea of emergence. If the emergentist answers no, that the emergent properties of the whole *cannot* be reductively explained by its constituent parts, then a whole and its parts are two completely different kinds of realities and, despite claims to the contrary, emergence seems to reintroduce a dualistic ontology. However, if the emergentist answers yes, that the emergent properties of the whole *can* be explained by its constituent parts, then natural and social scientists can continue to explain the world using reductive methods that break composites down to the elements that make them up, and to say that the whole "emerges" does not add anything.

The answer that emergentists should give to the question about explanatory reduction is yes: one *can*, in principle, explain the existence of emergent properties from their parts, their properties, and their relations. For instance, given the charges of the ions in hydrogen and oxygen atoms, one could predict that an H_2O molecule would have a bent shape and that a mass of them would be slippery, they could dissolve salt cubes, and so on. To hold that some property is emergent does not rule out a scientific explanation. However, even though higher-level properties can be explained by their parts, and they are reducible in that sense, it does not follow that the whole can be eliminated from one's explanations. Unlike the effects of the resultant properties of the whole, the effects of emergent properties are not simply summations of the effects of lower-level mechanisms, because their existence depends upon the particular relations between the parts. In principle, emergent properties can be explained, but their explanation depends upon the structured relations between the parts when they constitute the whole. The answer to the question about reduction, then, is that emergence leads to a distinction between "explanatory reduction" but not "eliminative reduction" (Elder-Vass, chapter 3). On this account, emergent properties, in principle, are derivable, predictable, and explicable in terms of their parts, but this is a nonreductive theory in the sense that the existence of higher levels continues to be salient for, and cannot be eliminated from, how one explains the behavior of wholes.

This distinction can also help emergentists answer the challenge regarding causality. The claim that has been criticized is that emergent wholes are new causes in the world, in addition to their organized parts. That view is sometimes called "strong emergence." An alternative answer, "weak emergence," holds that by joining in a whole, *the parts* gain new powers and constraints by their relationships with the other parts. To say that "the whole did it" is not to introduce an additional cause but rather a shorthand way of saying that the parts in their organization did it. Although the weak emergence answer does not introduce the whole as an additional cause in the world, it is still an emergentist view, since the configuration of the parts is essential, and the parts have effects they could not, or would not, have if they were not in that form.

To be explicit, this essay recommends a weak emergence theory of society. This lets us recognize that participating in a group has *effects* on the individuals in the group without reifying the group or treating it as a *cause* in addition to its members. Weak emergence permits one to speak of a group of people having an intention, believing something, or doing an action, because a plural agent is ontologically different from a mere aggregate of independent people. Nevertheless, the causal powers of a whole are both dependent on its parts and instantiated in them. The slipperiness of water will be tied to the properties of the elements and how they are arranged; the success of the championship team will be tied to the talents of the players and how they are organized. This means that the resultant properties and the emergent properties of a group are similar in that one can explain both by pointing to the members of the group, but they differ in that the emergent properties would not exist unless those members were properly structured. This is what it means to say that a whole can be reductively explained, but not reductively eliminated.

Emergence and the New Materialisms

Emergence theory is an ontological proposal and its ethical and political implications have not yet been spelled out.[2] As a proposal that offers to root human subjectivity and agency in people's bodies, their social relations, and their material conditions, however, emergence may provide a tool useful for multiple forms of new materialism. Emergence theory can be useful for materialist thinking, in the first place, because it is a form of realism: the emergent properties of atoms, molecules, cells, animals, groups, and perhaps even God, exist and operate whether or not human beings are aware of them. Secondly, it can be useful because it avoids anthropocentricism: for emergentists, human animals are wholes, and they are astoundingly complex wholes, but they are not the telos of the world and they are, in turn, parts of larger social, ecological, and perhaps divine wholes. Moreover, as an activist or powers-based ontology, emergence theory recognizes agency throughout the nonhuman universe, and this willingness to talk about causal mechanisms whose effects are independent of human knowledge, means that emergentists reject both empiricist skepticism about causal powers and the correlationism that attributes those powers to the work of the cognizing subject.

Given this non-anthropocentric realism, emergence theory could be allied to—or perhaps is simply an example of—some of the positions proposed in this book. For instance, emergence theory is close to, if not exemplary of, what

Kimerer LaMothe calls "relational immanence," that is, the idea that the relations between entities give rise to genuinely novel properties—including energy, life, and consciousness—without ever "jumping" to a non-immanent source (see, "Dancing Immanence: A Philosophy of Bodily Becoming"). For the same reason, a composite emergent entity is close to, if not exemplary of, what Catherine Keller calls "intercarnation" rather than incarnation since the emergent properties of human and other agents arise out of the organized contributions of their parts and do not enter "in" or "down" from some other realm (see, "Amorous Entanglements: The Matter of Christian Panentheism"). Despite how "magical" the powers of a chromosome or the music of an orchestra may seem, emergence theory understands their agency in terms of their interacting parts and how they are arranged. Emergence theory is also close to, if not exemplary of, what Mary-Jane Rubenstein calls "differential holography" (see, "The Animist, Almost Feminist, Quite Nearly Pantheist Old Materialism of Giordano Bruno"). This label refers to a pluralistic ontology that combines "universal interrelation," the notion that no entity exists outside the web or tapestry of relations it has to others; and "irreducible particularity," the notion that the specific elements constituting any entity and the entity's location in the web or tapestry make that entity ontologically unique. Emergence theory and differential holography both eschew natural/supernatural and immanent/transcendent dualisms, but nevertheless make distinctions within the one natural and immanent plane.

The feature of emergence theory that is most promising for new materialists is that it offers a way to speak of composite wholes, even when invisible, not simply as abstractions or heuristic devices, but rather as real entities in the world. This feature could provide conceptual support to those working on: Latourian networks; Deleuzian assemblages; Bourdieusian fields; Foucauldian dispositifs; Wittgensteinian forms of life; or Bertalanffyian systems—not to mention environmental niches, ecologies, and nature as a whole—and it could have tremendous value for the analysis of plural subjects. Emergence theory seeks to explain the nature, not only of physically contiguous composite entities (like a neuron or a fish), but also that of collectives. It follows that emergentist vocabulary can help us speak, as Joerg Rieger hopes, of "the Earth" or "the environment" as a real whole, with top-down effects for everything within it (see, "Which Materialism, Whose Planetary Thinking?"). Human social groups are also composite wholes, stitched together by intentional relations, and emergence theory, therefore, can help us speak, as Rieger also hopes, of a political movement or a religion as a real thing in the world. In this spirit, emergence could also contribute to a realist account of Timothy Morton's "hyperobjects," such as the climate or the economy, or even to a realist account of pantheism or panentheism. Because emergence theory recognizes that wholes have effects on their parts, it rejects the methodological individualist claim that the behavior of social groups can be fully explained by attention to the individuals within them.

Despite these potential alliances, there remains a significant tension between emergence theory and at least some of the new materialist positions in this volume. The tension arises from the fact that new materialism is inherently committed to undoing harmful hierarchies—and emergence theory is inherently hierarchical. New materialism critiques, in particular, the hierarchy of spirit and matter, an ontological distinction that undergirds not only the religious hierarchy of God over God's creation, but also the hierarchies of human beings over the natural world, and of those human beings said to be more spiritual over those said to be less so. On the other hand, emergence holds that any entity is composed of parts that are themselves composed of parts: a herd is composed of animals; each of which is composed of cells; composed of molecules; composed of atoms; composed of particles. For emergence theory, each composite entity is stratified into higher and lower levels, and nature itself is plausibly interpreted as hierarchical. It may therefore seem that emergence could provide ontological support for reactionary politics.

Above all, one sees this tension between emergence in the way that some new materialists seek to undo the dualism of subjects and objects. Emergence theory holds that entities have properties not possessed by their parts, and living entities have organic properties not possessed by their inorganic parts. In this way, emergence theory distinguishes between those entities that are enabled by their parts to have subjective experiences and their non-living parts that lack those properties.[3] Emergence theory does recognize that all material things have some form of agency. However, not all agency is of the same type. Emergence theory treats the properties of living things as emergent, and it does not support the claim that life or mind permeates all things. Emergentists would therefore not agree that "everything is alive" (Rubenstein) or that "ideas and matter [are] on a single plane" (Whitney Bauman, "Developing a Critical Romantic Religiosity for a Planetary Community").

One also sees this tension between emergence theory and the idea, endorsed by some new materialists, of a flat ontology. A flat ontology and opposition to the emergentist claim that a whole is greater than the sum of its parts pervade Sam Mickey's chapter on object-oriented ontology (see, "Solidarity with Nonhumans: Being Ecological with Object-Oriented Ontology"), but they can also help us spell out how the two approaches might be reconciled. Mickey quotes Timothy Morton's statement that a whole is actually dependent on, more fragile than, and ontologically "smaller" than the sum of its parts. Although Morton contradicts the emergentist claim verbally, the observations that a composite whole depends on and is more fragile than its parts are central to emergence theory. They do not contradict the key idea that a whole is greater than its parts in the sense that it possesses novel properties that they lack. We see opposition again in Graham Harman and Morton's statement that a whole always "withdraws" into its parts. As Mickey explains, "When you look for any whole object, all you find is its enmeshed parts, which interrelate to make up the whole; the whole remains withdrawn." This explanation flirts with the category mistake made by someone who says that they have toured every part of the campus but have never seen the university. It also resembles the eliminationist assertion that only parts exist, not wholes, and then each part also "withdraws" into its parts, and at every level the putative whole can be eliminated; this is the argument given by reductive physicalists. I take it, however, that this critique of wholes reflects Harman and Morton's anti-reductionist insistence that no account ever captures an object in its entirety. But the claim that a whole is greater than the sum of its parts does not imply that any particular object is fully known. On the contrary, a stratified ontology that goes beyond Humean empiricism insists that what is real is more than what anyone sees.

The most sustained opposition to the emergentist claim in Mickey's essay comes from Terrence Deacon's slogan that a whole is *less* than the sum of its parts. Mickey explains Deacon's idea that, "a living organism does not add *something more* to *nothing but* matter" but rather takes something away, and Mickey illustrates this by comparing a water molecule embedded in a chimpanzee's skin cell to the animal as whole. However, we should keep in mind that Deacon is not a critic of emergence theory, but one of its foremost contemporary proponents. His version of emergence is noteworthy as he stresses that an emergent whole is based on "absences" or "holes" and, as Mickey notes, for Deacon, parts are limited by their participation in a whole in a way that they would not be if they were outside it. Deacon's point is that energy, forced by an emergent structure into a smaller range of possible paths, is more potent than energy that diffuses in any direction, and that an emergent whole's powers are increased as the organization of the whole works to channel energy.[4] But recognizing the role of constraints in building a composite whole undermines neither the idea that emergent wholes gain powers not present before, nor the distinction between living wholes and their nonliving parts. As Deacon says of inanimate bits of matter that were the product of stars exploding and now temporarily constitute one's body: "Together they are alive; apart they are not" (Deacon, 113).[5] It is true that a water molecule embedded in a chimpanzee's skin cell cannot flow down a river or fall as rain. But being located in the chimp removes none of the properties of H_2O. On the contrary, that location enables the molecule to help make possible the chimp's grooming, eating, tool-use, and so on—a range of higher-level activities unavailable in the inanimate world. In short, Deacon's rhetorical focus on absences does not undermine emergence theory's stratified view of the world.

Politically speaking, some may worry that emergence would bring a form of what Mickey calls "explosive holism." This is the danger when a philosophy teaches that the value of the whole outweighs the value of its individual parts. Mickey illustrates this danger in Mill's utilitarianism and Kant's transcendental ego, and it is a legitimate problem that one also sees in Hobbes's leviathan, Marx's communism, Leopold's land ethic, or any other political ethic that refers to a political whole. To avoid this danger, perhaps emergence theory and a flat ontology can be reconciled in the following way.

An ontology is "flat" when it holds that no entity that exists is more real than any other, but a flat ontology need not (and should not) deny that these equally real entities build structures of unequal dependencies and enablements. One can therefore combine a flat ontology with an emergence-inspired stratified ontology. Consider the following. A flat ontology rejects the idea that the parts of which all things are made are more real than the composite wholes they make (as claimed by reductive physicalists). It also rejects the idea that sensory appearances are more real than the mechanisms in the world that allegedly cause them (as claimed by Humean empiricists). Rejecting these two ideas leads to the view that one's ontology includes both parts and their wholes, both appearances and their causes. If a flat ontology can permit the existence of these levels of being, then it can, like emergence theory, be a version of LaMothe's "relational immanence" or Rubenstein's "differential holography." A flat ontology can hold, with emergence theory, that all things exist on the one "flat" plane of immanence, but that they are nevertheless differentiated by their relations. Within the one immanent plane, being would be stratified, not flat.[6]

For the past century, philosophers have taken a linguistic turn to focus on the ways that language mediates what one can know, a focus that encouraged skepticism that one could speak of real causal powers "out there" in the world. By contrast, emergence theory contributes to the now-growing retrieval of realism and the recognition that the world includes mechanisms whose effects do not depend on being conceptualized by human beings. The ethical and political dimensions of emergence are still relatively undeveloped, but emergence is at least allied to the ethically and politically transformative potential of ways of thinking and being that locate agency throughout the natural world, "rather than in some spiritual sphere above, beyond, or even within it" (Rubenstein). The most important contribution of emergence theory to this project is its argument that it is precisely the cooperation between the interrelated parts of an emergent whole that explains why agency, creativity, and change are real effects in the world.

Notes

1. All the good ideas in this paragraph come from Dave Elder-Vass, *The Causal Power of Social Structures: Emergence, Structure and Agency* (Cambridge: Cambridge University Press, 2010). Further references to this text can be found within the body of this essay.

2. The theological implications of emergence are also relatively undeveloped, although the British emergentist, Samuel Alexander, gave the 1916–18 Gifford lectures arguing for an emergent God. See Samuel Alexander, *Space, Time, and Deity*, Vol. 1–2 (London: Macmillan, 1920), and see Philip Clayton and Paul Davies, eds., *The Re-Emergence of Emergence: The Emergentist Hypothesis from Science to Religion* (Oxford: Oxford University Press, 2006), chapter 1, 12–14.

3. For a pointed debate on this subject between panpsychists and emergentists, see Demian Wheeler and David E. Connor, eds., *Conceiving an Alternative: Philosophical Resources for an Ecological Civilization* (Anoka: Process Century Press, 2019), especially chapters 10 and 12; and John Heil, "Emergence and Panpsychism," in *The Routledge Handbook of Emergence*, eds. Sophie Gibb, Robin Findlay Hendry, and Tom Lancaster (London: Routledge, 2019), 225–34.

4. Terrence Deacon, *Incomplete Nature: How Mind Emerged from Matter* (New York: W.W. Norton, 2012).

5. Terrance Deacon, "Emergence: The Hole at the Wheel's Hub," in Philip Clayton and Paul Davies, eds., *The Re-Emergence of Emergence: The Emergentist Hypothesis from Science to Religion* (Oxford: Oxford University Press, 2006), 11–150, quote on 113.

6. Ray Brassier, "Deleveling: Against 'Flat Ontologies'" (https://uberty.org/wp-content/uploads/2015/05/RayBrassierDevelevingAgainstFlatOntologies.pdf); accessed July 3, 2023.

New Materialisms and Planetary Persistence, Purpose, and Politics

Heather Eaton

Introduction

The appeal, power, and presence of new materialisms have become irresistible, intellectual tidal swells of novel insights, disruptions, and creativity. Past, present, and post-European intellectual currents are involved in these innovative cultural theories that have broad implications. New Materialism wants—and claims—to be initiating a revolution of thought.[1] It is an interesting phenomenon to witness the new materialisms seeping, or crashing, into established forms of knowing, thinking, and imagining. Countless disciplines are (re)formulating themselves in light of the flood of critiques and perceptions. The term now refers to multiple streams and rivers of thought, at times moving in different directions. There are distinct emphases and interests. Ideas from new materialisms are proliferating, expanding their influence, and incorporation, into an increasing plethora of disciplines and perspectives.[2] In turn, these efforts are establishing other streams of ideas and ingenuity. There are numerous cartographies and genealogies of the new materialisms.[3]

The waters between religion and new materialisms are also flowing. Publications such as Clayton Crockett and Jeffrey Robbins', *Religion, Politics, and the Earth: The New Materialism*; Whitney Bauman's *Religion and Ecology: Developing a Planetary Ethic*; *Religious Experience and New Materialism: Movement Matters*, edited by Joerg Rieger and Edward Waggoner, and countless articles and chapters attest to new sources for the ecological efforts in religious studies and theology.[4]

The focus here is on materialist turns in religion and ecology. This is creating spaces for new perceptions and sensitivities among planetarity, religion and ecology, and new materialisms. My primary concerns are with the biosphere, and where religion, planetary thinking, and new materialisms converge so that there is a future for lifeways of the biosphere. I triage the vast and rich expanse of new materialist discourses with respect to these concerns. My contribution has three parts. The first section names a few new materialist characteristics or themes that are relevant to this chapter. These include some issues within academia and disciplinary intellectual training that can inhibit plunging into the confluence of disciplines involved in new materialisms and planetary thinking. The second focuses on symbolic consciousness, and how these processes reveal infinite entanglements. The third considers evolution and cosmology and the role they could play in planetary thinking, infused with insights from new materialisms. The suggestion is that there are lacunae within new materialisms and planetary thinking, and understanding symbolic consciousness, as well as incorporating the significance of evolutionary processes, are proposed as correctives and expansions to both.

Themes

The three themes mentioned below represent a few aspects of new materialisms that intersect with projects geared towards planetary thinking and religions.

1) *New Materialisms: new?*: At times, there is a tendency within the new materialisms towards an intoxication of a self-endorsed exceptionality. The certitudes, special lexicon (agential realism, transversality, performativity, quantum affectivity, algorithmic situations, affirmative and creative entanglements), and bold claims about putting in motion revolutions of thought, can be exasperating. One can get lost in the waterways of new materialisms, while the planetary possibilities rapidly diminish. Still, new materialisms are a gush of fresh perspectives, innovative options, and vast collaborations among disciplines; defying the categories of inter, multi and transdisciplinary. The insights can indeed be intoxicating.

It is important to note that new materialisms—with their anti-Cartesian and Newtonian stances, and emphases on non-dualistic immanence, dynamic connectivity, and differentiated agencies *entangled all the way up and down*— are indebted to certain (perhaps minor) continental philosophies, now revised and refurbished.[5] Furthermore, these emphases are not *new* to some of the world's religious and cultural traditions. Themes within Confucianism, Daoism, Hinduism, Buddhism, Jainism, and Indigenous and animist traditions, have discerned, emphasized, and developed these insights, in distinct modes, for centuries or longer. While new materialisms' perceptions are fresh and powerful currents of thought and sensitivities in what are, loosely, Euro-Western analytic frameworks and epistemologies, such frameworks and epistemologies are not entirely new to the world.

For example, Pierre Teilhard de Chardin, Alfred North Whitehead, Thomas Berry, and many others, with their own linguistic innovations and blends of religion and science, also perceived the deep interconnectivity, relationality, and vital flow from subatomic physics to complex life; to dynamics of consciousness; to planetary and cosmic processes. These connections informed their teleologies. However, it should be noted that teleology is rarely a pertinent theme across the new materialisms' spectrum, which is usually focused on processes, developments, emergence, and undetermined becomings. Although a discussion of the pros, cons, relevance, or beliefs in teleology is germane to some planetary thinking projects, it is not a central theme here.

2) *Academic Formations:* The fluid intermingling among sciences, humanities, social sciences, and environmental humanities is crucial to maintain. The need to move out of academic silos, to lower barriers across specializations, to expand methodologies, and to reconceive education outside of disciplinary borders is decisive for planetary thinking. However, such blends are difficult to manage intellectually with precision and carefulness, for a few reasons. First, new materialisms are vast, and planetary thinking is strenuous. Both represent confluences that are challenging to articulate clearly and accurately. Second, disciplinary frameworks can be ideologically or methodologically inflexible. Third, intellectual habits and conceptual pathways may become intractable. Academic studies form minds to appreciate order and strictness of thought, yet often with a simple determinism and linear causality.

Most intellectual work requires some reductionism to communicate specific topics. What new materialisms and planetary thinking are attempting is to open academic flood gates, and enlarge frameworks, methods, and modes of analyses with the fluid intermingling of knowledge and insights. To represent the interplays of the material and the discursive are not readily conveyed in customary discourses. In addition, academics are routinely trained to deconstruct, expose fault lines, take a stance against something/one else, and remain in the realm of critique. To construct and/or propose viable alternatives; or invent, imagine, or create fresh and vibrant options, are far more difficult.

The multidisciplinary appeals allow for innovative modes of inquiries and contributions from distinct vantage points which, at times, include political and ethical purposes. However, the norm in academia is to try to evade overt political stances, ethical certainties, and ideological commitments. Planetary thinking is, overall, about scientific, ethical, and political projects. The desired ethos is to propose novel planetary thinking and visions, without losing perspicacity. Planetary thinking includes social, political, and ecological transformations.

Planetary thinking is most potent, and most relevant, when infused with political relevance. One goal of planetary thinking is to develop a robust sense of common ground (planetary, conceptual, ethical) which strengthens a sense of contributing to communal, planetary projects. The political significance between materialisms and social movements is considered in Rieger and Waggoner's *Religious Experience and New Materialism: Movement Matters*. These combinations are germane to the key interests of this chapter.

3) *Composites and Relationality*: The emphasis on relational immanence, reciprocity, mutuality, embodiment, and embodied subjectivities instructs approaches to planetary thinking. Intellectual work in planetary thinking is subject to, and entangled with, multiple dynamics of consciousness, cognition, affectivity, and somatic subtleties that are in flux. This theme flows into that of aesthetics. The arts, media, literary studies, and myriad approaches to aesthetics are intersecting within new materialisms. For example, James Miller reflects on how aesthetics with bodily and ecosystem permeability can produce new ethical and political paradigms in *China's Green Religion: Daoism and the Quest for a Sustainable Future*.

From another angle, Joerg Rieger writes that matters of culture, nature, and religion need to be integrated with matters of politics and economics, as these, together, are fashioning whatever is considered to be religious experiences and cultural reasoning. Usually, power and politics hold sway over the religions and religious experiences of the day. Nonetheless, human concerns shape religious experiences. He adds that the influence of the material production of work, the social structure of labor, and the international neoliberal machinery remain notably absent, and should be included in the discussions about all facets of religion.

Thus, to ponder new materialism, planetary thinking, and religions together represents multiple composites of interrelated topics. Relationality is a consistent theme. The metaphor of water is useful, noting the ways water encompasses, flows through and around matter, and is molecularly flexible to bond with most elements and molecules—making it the universal solvent. The composites within new materialisms and planetary thinking are important to keep in mind. The themes mentioned are also active in the background, as the topic shifts to symbolic consciousness.

Symbolic Consciousness and Planetary Thinking

While new materialisms offer refreshing waters in which to cleanse our minds and rejuvenate our spirits, there are areas that are yet to be developed and connected to larger waterways. One such area is that of symbolic consciousness.

The dynamics of symbolic consciousness are, in my view, woefully understud-
ied and overlooked, and yet are utterly related to new materialist assertions and
planetary thinking.

I am grateful for the work of biological anthropologist Terrance Deacon.
His work is unique and profoundly important in illustrating the precepts of
relationality, interconnections, immanence, and emergence. My comments
are intended only to indicate the importance of symbolic consciousness and
to mention some implications.

Homo sapiens and other hominins navigate existence symbolically, having
evolved with the capacities to live by means of a symbolic consciousness.[6] It is
this mode of consciousness that allows for representations of the world to form,
eventually as worldviews or social imaginaries.[7] Within the evolution and development of hominin species emerged the capacity to navigate the world
symbolically and then to live via a symbolic consciousness. Symbolic psychic
structures are a complex weave among active imaging, engagement with life's
exigencies, and symbolic rendering of experiences.

Symbolic consciousness, artistic representation, and tool making are older,
more complex, and involve more species than previously assumed.[8] Of language
alone, Deacon suggests that "the thousands of symbolic units comprising the
lexicon of a language (e.g., words and morphemes) effectively 'point' to one
another as though comprising a complex interconnected network."

The formation of a self-reflexive and supple consciousness that could function
symbolically and sustain the capacity to coordinate images, thoughts, emotions,
intuitions, insights, retention, and somatic memory was acquired over millennia.
In addition, the evolutionary, emergent, and interactive processes that led to
signs, representations, and imagery remain opaque. Nonetheless, over millennia
there emerged systematic renderings of experiences, which became codified.
They could be shared, taught, ritualized, and embedded in individual and
collective identity formations. Contemporary studies of social imaginaries,
worldviews, reality maps, and other terms indicate these aspects of symbolic
representations of the world.

This symbolic mode of being is the *modus operandi* of humans. A symbolic
consciousness is the way humans process and navigate the world. It is not
through or with symbols or images that humans navigate and represent the
material and ineffable realms. It is *within* symbols. John Dixon asserts that
remnants of ancient symbols and artifacts indicate that experiences were trans-
muted into systems of images to cope with, and delineate, the interior and
exterior exigencies of life.[9]

Although the dynamics of symbolic functioning can be dissected into aspects
involving external realities of culture, context, representations, and internal

realities of emotions, cognition, and ideation—all of which are embedded in identity formation, a sense of self, social imaginaries, and bonding patterns—this renders a superficial, even false, understanding. The division of exterior and interior is false, as are the differentiations. These composites are interrelated in ever-moving exchanges. Material inter- and intra-actions, contexts, events, and symbolic processes are inseparable, interwoven, and enmeshed within the very structures of human consciousness and behaviors. They operate within indivisible personal and social fabrics. They are shaped by, interact with, and impact material realms, revealing another form of reciprocity, subjectivities, and relational immanence.

Current work in mind-brain associations, imagination and cognition interactions, somatic studies, language acquisition, and biosemiotics are addressing these aspects of humans as a symbolic species. The interactive fluidity and mutability of symbolic consciousness seems to me to be an essential adhesive to new materialist insights.

For example, some suggest that the rapport between Earth dynamics and the development of symbolic self-consciousness is an evolutionary impetus to enhance symbolic consciousness. The mutual incursion between humans and the natural world would have been formidable, evoking by necessity, representation. Caves, vistas, storms, seasons, other animals, and the elements of air, water, fire, and earth formed human sensibilities, consciousness, and self-consciousness. Humans had no *techne* to control, and minimal ability to avoid, such powers and immensities.

Poignant experiences of the natural world evoke a blend of material, mythic, emotional, and psychic interfaces. These, then, require mediation, representation, and expression. For example, experiences of caves are often described in terms of intimacy, intensity, envelopment, or interiority. To experience the immensity of the forest is not unusual. Yet it is a multilayered and perplexing interior involvement. Gaston Bachelard devoted much of his life to analyzing such occurrences.[10] He perceived that it is an immensity felt while in the forest, described as "of the forest," yet experienced within our self-consciousness. We *feel* this immensity within ourselves, interpreting the forest as emanating more than the material. The imagination and associated symbolic expressions are able to amplify, indefinitely, the images and sensations of immensity, increasing their interior presence and power. Subsequent meanings are entangled with emotional, imaginative, symbolic, and cognitive apparatus and responses. Bachelard used the term the *material imagination* to show the dialectic between the material world and the limitless capacity of the human imagination and intensities of experiences. These elements seem to be missing from new materialisms and planetary thinking discourses.

The immeasurable interconnectedness affirmed in new materialisms have not developed these correlations. Jane Bennett's terms of *vibrant matter* and *vital materialism* point in this direction. However, appreciating *the force of things*, and *differentiated agencies of materiality* are not enough to grasp the composites of vitality between materiality and human engagement, and the symbolic assemblages involved. As human life is continuously interacting with countless composites, in immanent systems of images; then the infinite interactions among symbolic consciousness and material realities manifests connectivity as much as quantum dynamics does.

To add the realm of symbolic consciousness to the exchanges between planetary thinking, new materialism, and religions reveals a vast array of connectivity and porous boundaries. Symbolic consciousness offers insights into a depth of relatedness, a material immanence and intimate immensity, and both function as antidotes to binaries, dualisms, and abstract dissections. The continuity between symbolic consciousness and this materialist turn is of interest, most importantly the fluidity, impermanence, and symbolic nature of thought. The metaphor of water in this chapter helps portray that, whatever planetary thinking and new materialism involve, it is flowing, powerful, refreshing, and never static.

Nature, Evolution, and Cosmology

Studies of religion, ecology, and nature have, for decades, established and developed perspectives on multiple rapports between the natural world, an Earth community, and human evolution and endeavours. These fields have creatively and indirectly dabbled in new materialisms with the persistent reevaluation of partitions between religion/science, human/nature/culture, humans/other animals, transcendence/immanence, spirit/matter, and more. Planetary thinking has been developing through many initiatives, such as Gaia hypothesis research, *The Journey of the Universe* projects, *Big History, Epic of Evolution*, and other forms of planetary thinking. Some are allergic to religion while others see links between religious worldviews and planetary thinking.[11] Engaging explicitly with new materialisms, and seeing the repercussions to studies of religions, ecology, and nature is relatively new. There are alliances to be developed that can also incorporate awareness of symbolic consciousness, as well as evolutionary sciences. A brief look in two directions will reveal such alliances.

The first is the micro evolutionary realm of Deacon's book *Incomplete Nature: How Mind Emerged from Matter*. With deft detail, using theories of emergence and complexities, Deacon considers thermodynamics, morphodynamics, and teleodynamics in his quest to understand the rise of life and mind. He probes multiple connections between matter, life, mind, and energy, asking how greater complexity emerges from lesser. He dwells on the incompleteness of processes, on absences and constraints, and what he refers to as the *ententional*. He follows lines of evolutionary thought that include autopoiesis, auto-catalysis, self-organization, self-assemblages, and the omnipresent, vital interdependence of mind and matter.

In some respects, *Incomplete Nature* is a worthy partner to Karen Barad's inspiring book, *Meeting the Universe Halfway*. Deacon's focus is more on mind from matter rather than on quantum physics, although there is overlap. He studies absences while Barad writes of exclusions and possibilities. Both note that nothing is ever still: the flow of reconfiguring is constant, although there are both probabilities, as well as constraints, to possibilities. Suffice it to say that Deacon is demonstrating the tenets of new materialisms by pondering the emergence of more from seemingly less, within an utterly intra-connected/interconnected/entangled, immanently dynamic, vibrant, and vital matter.

If everything is *entangled all the way up and down*, then there is a flow, of sorts, between inorganic and organic matter. To understand this, even minimally, it is necessary to incorporate evolution and planetary processes, as there is continuity, integrity, and interaction among these. Given this, it seems odd that little attention is paid to evolution. Evolutionary biologists bring material substance and strength to the persistent, yet often abstract, claims of processes and dynamics of some new materialist disciples. Planetary thinking should also attend to evolution. Ernst Mayr eloquently stated, "What is Evolution? Evolution is not merely an idea, a theory, or a concept, but is the name of a process of nature."

Adding *the Earth as subject* to new materialist discourses is woefully incomplete, and inaccurate. Although some are considering *the vital and material contributions of the Earth,* I find these perspectives limited. Neither phrase reveals the limitless interconnections within evolutionary processes, emergent complexities, the vibrant matter of animal life and communities, or the countless entangled and vital systems from molecules, to matter, to mind. The point here is to indicate that new materialisms and planetary thinking must go farther into the connections they claim: to embed evolution, human emergence, and symbolic consciousness into the mix. This is not to promote or support anthropocentrism, but rather to add more dimensions into new materialisms and planetary thinking realms. Here is where the fields of religion, ecology and nature, and the project of planetary thinking have additions, even correctives, to new materialisms.

The second direction is from palaeontologist and priest, Pierre Teilhard de Chardin. His approach illustrates the inextricable and embedded entanglements that some new materialists are explaining. For example, Teilhard pondered the depth of affinity between what he called the *without from the within* of things. The without is the observable, as seen with the forms, transformations, and bonding arrangements from atomic structures to the formation of molecules and mega molecules, out of which arose and evolved all matter and life. To appreciate the transformations of atoms to molecules, there must be a commensurate theory of energy. Furthermore, and similar to Deacon, there needs to be a theory that connects the structures and activities with the processes and purposes of these transformations. Teilhard pondered these as a whole, meaning he would not separate anything from its structures, activities, developments, and directionality.

Teilhard studied bacteria cultures in the same manner, and then plants. The without and observable cannot explain the life dynamics of plants. With insects it is more difficult, with vertebrates futile, and breaks down completely with humans. As life evolves, the without of things—the observable—becomes increasingly incapable of explaining behavior, developments, intensifying complexities, and evolutionary directionalities. Of course, there is no ontological divide between the within and without of things. This is not a binary, but a manner of addressing intimate connectivity and performativity. Karen Barad used the phrase *things in phenemona*. Furthermore, if the interconnections are taken seriously *all the way up and down*, then all reality represents forms of psychic, spiritual, and material phenomena in intricate patterns of agencies, unities, and differentiations.[12]

Evolutionary studies support the notions of subtle, nuanced, and active unions of matter and energy which unite the interior dynamics with the transformations to increasing levels of complexity, as seen in planetary processes. Evolution is the primary context, process, and dynamic out of which everything planetary emerges and to which it needs to refer. Evolution is a necessary dimension of planetary thinking. As such, symbolic consciousness would be appreciated as a process of evolution, and a vibrant element of a living Earth. For some, evolution and cosmology must be understood together.

I end this chapter with a final comment about the infrequent and cursory mention of cosmology in the tomes of new materialisms. This is odd, given that virtually everything that is now understood, or perceived, about the dynamics of the cosmos is both resonant and consonant with new materialist assertions. It is beginning to dawn on, or emerge within, human consciousness that the expansiveness and essence of the world—cosmos, Earth, time, space, and processes—are central to knowing anything meaningful about planetary vitality, as well as being human.

The universe is being perceived with increasing clarity, detecting the dynamics, processes, developmental sequences, transformations, interconnections, and expansion of the universe. What is astonishing is that everything is so much more than was ever assumed or imagined. Furthermore, it is apparent that in spite of a capacity to parse knowledge into different physical processes—subatomic physics, astrophysics, nucleosynthesis, structures and formation of dark matter and dark energy, planetary formations, energy, relativity, etc.—there is coherence. The universe is integral: unity without uniformity. There is a cohesiveness within the great diversity found in how the universe functions, including in the birth and death of stars, and galaxy and planetary formations, including planet Earth. What is planetary thinking without integrating these aspects?

Coherence and *integrality* are also seen in the evolution and functioning of the biosphere. What is currently being learned about the biosphere is correspondence, mutual influence, and communication from the molecular and cellular, to the planetary processes, to complex biospheres (re)emerging.[13] Sciences use the terms of emergent complexity, entanglement, coherence, correspondence, congruence, or intelligibility to describe the overall orientation of the universe, and planetary activities. The language mirrors that of new materialisms and should also inform any planetary thinking.

Conclusion

This chapter has explored some characteristics of new materialisms, and how the multiple composites expand understandings and revitalize awareness of the breadth and depth of the realities in which we are immersed. New materialisms and planetary thinking challenge customary academic disciplines and the habits of Euro-Western thought. The insights of both are also seen as not necessarily "new" to some religious worldviews. The discussion then focused on correcting two lacunae by proposing increased knowledge of the dynamics of symbolic consciousness as well as integrating evolutionary frameworks. The hope is that these will strengthen and add potency to the transformative capacities of new materialisms and renewed planetary thinking for a planet in peril.

Planetary thinking, new materialisms, and religions are active, flowing waterways, filtering through our symbolic consciousness. If any planetary thinking can be a crucible for experiences of immanent and intimate immensities—from the cosmic to the atomic—then we have an "infinite plentitude of openness."

Notes

1. Rick Dolphijn and Iris van der Tuin, *New Materialism: Interviews & Cartographies* (Ann Arbor: Open Humanities Press, 2012), 85.

2. Information on previous and current training schools, publications, conferences, working groups, and more can be found at http://newmaterialism.eu. Accessed April 15, 2021.

3. Vera Bühlmann, Felicity Colman, and Iris van der Tuin, "Introduction to New Materialist Genealogies: New Materialisms, Novel Mentalities, Quantum Literacy," *Minnesota Review*, Vol. 88 (2017): 47–58; Monika Rogowska-Stangret, "Corpor(e)al Cartographies of New Materialism: Meeting the Elsewhere Halfway," *Minnesota Review*, Vol. 88 (2017): 59–68; Dolphijn and Tuin, *New Materialism: Interviews & Cartographies.*

4. Clayton Crockett and Jeffrey W. Robbins, *Religion, Politics, and the Earth: The New Materialism* (New York: Palgrave Macmillan, 2012); Whitney A. Bauman, *Religion and Ecology: Developing a Planetary Ethic* (New York: Columbia University Press, 2014); and Jorge Rieger and Edward Waggoner, eds., *Religious Experience and New Materialism: Movement Matters* (New York: Palgrave Macmillan, 2015).

5. Many explore influences and insights from a plethora of Continental philosophers and themes. The genealogies and cartographies present these in depth, as do Karen Barad, Jane Bennett, Bruno Latour, and others. See Karen Barad, *Meeting the Universe Halfway: Quantum Physics and the Entanglement of Matter and Meaning* (Durham: Duke University Press, 2006) and Jane Bennett, *Vibrant Matter: A Political Ecology of Things* (Durham: Duke University Press, 2010).

6. See Jonathan Gottschall, *The Storytelling Animal: How Stories Make Us Human* (Boston: Mariner Books, 2013); Terrance W, Deacon, *Incomplete Nature* (New York, W.W. Norton, 2012); Terrance W. Deacon, *The Symbolic Species: The Co-evolution of Language and the Brain* (New York: W.W. Norton, 1998); Richard D. Klein and Blake Edgar, *The Dawn of Human Culture: A Bold New Theory of What Sparked the "Big Bang" of Human Consciousness* (Hoboken: John Wiley & Sons, 2002); John W. Dixon, *Images of Truth: Religion and the Art of Seeing* (Atlanta: Scholars, 1996); Wentzel van Huyssteen, *Alone in the World?: Science and Theology on Human Weakness* (Grand Rapids: Eerdmans Publishing, 2004); David Lewis-Williams, *The Mind in the Cave: Consciousness and The Origins of Art* (London: Thames & Hudson, 2002); and David Lewis-Williams, *A Cosmos in a Stone: Interpreting Religion and Society through Rock Art* (Walnut Creek: Altamira Press, 2002).

7. I have written frequently on this topic: See Heather Eaton, "The Human Quest to Live in a Cosmos," in *Encountering Earth: Thinking Theologically with a More-Than-Human World*, eds. Trevor Bechtel, Matthew Eaton, and Timothy Harvie (Eugene: Wipf and Stock Publishers, 2018), 227-247; "Global Visions and Common Ground: Biodemocracy, Postmodern Pressures, and the Earth Charter," *Zygon: Journal of Religion and Science*, Vol. 49 (4) (2014): 917–937; "The Challenges

of Worldview Transformation: To Rethink and Refeel Our Origins and Destiny," in *Religion and Ecological Crisis: The Lyne White Thesis at Fifty*, eds.Todd LeVasseur and Anna Peterson (New York: Routledge / Taylor & Francis, 2017), 121–137; "An Ecological Imaginary: Evolution and Religion in an Ecological Era," in *Ecological Awareness: Exploring Religion, Ethics and Aesthetics*, eds. Sigurd Bergmann and Heather Eaton (Berlin: LIT Press, 2011), 7-23; "The Revolution of Evolution," *Worldviews: Environment, Culture, Religion*, Vol. 11 (1) (Spring, 2007): 6-31; and *The Intellectual Journey of Thomas Berry: Imagining the Earth Community* (Lanham: Lexington Press, 2014).

8. Klein and Edgar, *The Dawn of Human Culture*. There is little agreement on when, where, how, and which version of hominids began to manifest creative and symbolic thinking. See Celia Deane-Drummond and Agustin Fuentes, eds., *Theology and Evolutionary Anthropology: Dialogues in Wisdom, Humility, and Grace* (Abingdon: Routledge, 2020). See also John Noble Wilford, "When Humans Became Human," published in *The New York Times*, February 26, 2002. http://www.nytimes.com/2002/02/26/science/when-humans-became-human.html?pagewanted=all. Accessed September 8th, 2020.

9. Dixon, *Images of Truth*, 49.

10. Gaston Bachelard's stellar book, *The Poetics of Space*, describes in-depth how humans interact with spaces and matter via the imagination, symbolic consciousness, and interiority. Gaston Bachelard, *The Poetics of Space*, trans. M. Jolas (Boston: Beacon Press, 1964).

11. The web site http://www.journeyoftheuniverse.org/ provides excellent resources to appreciate the full range of the project.

12. Sam Mickey, Mary Evelyn Tucker, and John Grim, eds., *Living Earth Community: Multiple Ways of Being and Knowing* (Cambridge: Open Book Publishers, 2020).

13. My comments have nothing to do with intelligent design, linear progressions of life, or anthropocentrism of any kind. Life has regenerated following extinction periods, with an undeniable emergent complexity, including self and/or symbolic consciousness currently evident in many complex species.

Gut Theology: The Peril and Promise of Political Affect

Karen Bray

Since the first meeting, in 2016, of the American Academy of Religion's seminar on religion, new materialism, and planetary thinking, from which this volume materialized, a question has been nagging at me. It has wiggled its way into my brain, or my gut, or my gut-brain if we follow recent science, and it will not let me go. It is a question that has only grown louder as the Trump presidency unfolded and Trumpism hit its stride with the failed, but deadly, insurrection on January 6th. What if I was wrong? Not wrong in my horror or my politics, but wrong in my theology. I have been plagued by the thought that perhaps I was wrong to be so enamored with affect, with affect theory, with a poststructuralist resistance to capital "T" truth. I have been wondering about the risk of affect.

Of course, there is plenty of risk and guilt to go around, and the poststructuralist strawman certainly does not have the strength to wield the biggest of blows. And yet, I think many who gathered, first in person in our AAR seminar, and now in this volume, have had our roles to play: not just in (implicit) hesitancies to dismantle our own institutions' White supremacy, but also in an often unvoiced resonance between the rallying call expressed in Ann Cvetkovich's *Depression: A Public Feeling*, to embrace what Audre Lorde describes as "forms of truth that are felt rather than proven by evidence, the result of disciplined attention to the true meaning of 'it feels right to me'"[1] and to the birth of "alternative facts." I am horrified at the affect theory, the gut theology of resentment against those whose political depression my form of affect theology wanted to sensitively attune—women, queer folk, Black folk, transgender and nonbinary folk, the impoverished, the young, the disabled and debilitated, the colonized, the refugee, the immigrant (or as Sara Ahmed might put the list, the feminist killjoy, the melancholic migrant, the queer, the revolution-ary)[2]—embodied in such claims to factual alternatives. Are not "Make America Great Again," "Blood and Soil," and "You/Jews will not Replace Us," also ex-pressions of political depression? Or at the very least, are they not the moods of resentment bubbling for so long in the face of those other groups daring to kill the joy of White supremacist heteropatriarchy, but now unleashed in ways only really surprising to White liberals?

I do not wish to frame these affects as identical, nor am I eager to equate the feminist killjoy (a figure around whose prophetic nature much of my work is built) with the Nazi. And yet, my gut feeling, my worry, the grumblings that keep me up at night, are reminders that I have not done enough. These grumblings ask us to consider whether, in our exuberance about affect, new materialism, planetarity, and relationality, we have too quickly forgotten, or too lightly engaged, or too willfully ignored, or too hastily redeemed, the very *fact* that affect, material entanglement, and planetary relationality are not simply salvific. They are not merely the way out of this climate colonial mess. Rather, what such relational modes of knowing, or such attention to feeling should remind us of is that relations and affects are pharmakon-like, they are curse and cure.

Affect theorist Elizabeth Wilson reminds us, drawing on Jacques Derrida's concept of the pharmakon, that no remedy can ever be "just 'remedy' without also meaning 'poison' and 'philter.' The issue is not that the word *pharmakon* is incoherent, but that it 'partakes of both good and ill, of the agreeable and the disagreeable.'"[3] Wilson asks feminist affect theorists to be cautious of too quickly redeeming the "negative" affects we hold dear. She proposes that the very pharmakon nature of all politics and all affects cannot "mean that a harm is somehow, secretly, restorative (and thus not really a harm at all); rather, this is a claim that damage is a necessary condition of any endeavor to heal."[4] In other words, perhaps we need to recognize the damage done in our attempts to heal in the face of the traumas of Imperial truth-claims and modernist ideas of our fundamental autonomy and disentanglement from planetary relation. We might recognize the damage our very relationality produces. Or as ecstatic naturalist Robert Corrington has put it, we should not forget that the web is also a killing machine. That we can affect one another is also why my liberation can elicit male rage; it is why the-should-be-obvious assertion that Black lives matter, unlooses the White rage from some of the places it has lived—in state house-redistricting committees, and real estate redlining, and White flight into private and charter schools.[5]

And yet, just because the turn to affect, to the gut, has an even heightened sense of peril, does not mean we should overshadow its promise. Indeed, perhaps it is this too-easy slip from "it feels right to me" to "alternative facts," to which we *have* to attend. Perhaps in sensitive attention to the very affects arising from the different sides (and there are sides, no common call for a romanticized common good—thanks, but no thanks, Mr. Lilla—adequately attends to the histories of how the sides were created in the first place), a path toward trustier collective feelings, trustier methods of building justice, and the dream of a more ethical planetarity might be enacted.

Let us pause, let us attend exactly to what this affective turn is, what affect theory has been, and what it might do.

Turning to Affect

According to *The Affect Theory Reader,* a collection published in 2010 and edited by Melissa Gregg and Gregory J. Seigworth:

> Affect arises in the midst of *in-between-ness*: in the capacities to act
> and be acted upon . . . affect is found in those intensities that pass
> body to body (human, nonhuman, part-body, and otherwise), in those
> resonances that circulate about, between, and sometimes stick to bod-
> ies and worlds, *and* in the very passages or variations between these in-
> tensities and resonances themselves.[6]

Affect theory might be considered the critical exploration both of what types of acts, knowledge, bodies, and worlds are produced in this in-between space *and* of how we might better attend to affect's role in such a production. Think for example of the force of feeling produced when standing on the top of a mountain or in front of your favorite painting. Think of that first ineffable moment of terror that arises when you *feel* like something is off in your environment. Think of the spark, the tingle of expectation, before a first kiss. These pulsations, and all those precognitive pulsations that entangle us with one another (human, non-human, more-than-human), for which we do not have appropriate language; it is the study of them in which affect theorists engage.

However, affect theory is also the study of those feelings for which we have many names: rage, anger, madness, envy, anxiety, boredom, joy, happiness, optimism, pessimism, depression, and ecstasy. For Jasbir Puar, "Affect is at once an exchange or interchange between bodies and also an object of control."[7] Hence, the study of affect is also about how these feelings get coded within cultures or how they come to stick to certain types of bodies, objects, and choices. We can think, for instance, of which objects and subjects get coded as happy in the context of "the American Dream."

Here, a blonde, White, able-bodied spouse (of the "opposite" gender), a white picket fence, a suburban home, 2.5 kids, and a golden retriever all become shorthand for happiness. Happiness, in this sense, while not being inconsequent to those ineffable pulsations exchanged across human and other-than-human bodies, takes a very particular shape, one that gets narrowly defined and associated with particular human and "inhuman" figures. For instance, we might call to mind the figure of the *Happy Housewife* verses that of the *Angry Black Woman*. Affect theory in this sense can be considered the critical investigation into how others assume we should feel and how we are actually feeling.

There are multiple strains of affect theory one might take up in the study of affect, religiosity, and planetary thinking. According to Seigworth and Gregg, "There is no single, generalizable theory of affect: not yet, and (thankfully) there never will be. If anything, it is more tempting to imagine that there can only ever be infinitely multiple iterations of affect and theories of affect: theories as diverse and singularly delineated as their own highly particular encounters with bodies, affects, worlds."[8] For Gregg and Seigworth, affect inherently contains a multiplicity of forces, forces whose effects multiply within bloom spaces created by interactions with diverse and particular forms of bodies, other affects, and worlds. Hence, a generalizable or singular theory of affect cannot suffice; such a theory would indeed rob affect of the slipperiness of its own stickiness, or in other words, that part of affect that while sticking to certain bodies or worlds, and so threatening certain bodies and worlds, also contains the promise that such bodies and worlds might get unstuck.

While resisting a generalizable theory, Gregg and Seigworth still offer a preliminary topography of the field, which includes an overwhelming, blooming list of related fields and authors, including, but not limited to, feminist science studies, posthumanism and new materialism, cultural studies, and psychobiological approaches to affect.[9] I am drawn to such a comprehensive mapping and to such an attractive collection of thinkers. And yet, there are ways in which Gregg and Seigworth's genealogy feels at once too expansive, and simultaneously includes confusing cuts.

In *Religious Affects: Animality, Evolution, and Power* (arguably the first monograph on contemporary affect theory and religion), Donovan Schaefer takes a more pared-down, but no less complex, approach in his mapping of affect theory. Drawing on genealogies of affect theory, proffered by Gregg and Seigworth, Cvetkovich, and Puar, Schaefer identifies two primary currents in affect theory: the Deleuzian mode and the phenomenological mode.[10] This is a categorical division, supported by the mapping work proffered by Stephen Moore and Jennifer Koosed, editors of the 2014 special edition of the journal, *Biblical Interpretation*, on affect theory and biblical study.[11] While acknowledging that these modes do often converge, Schaefer's work marks the key divergences. According to Schaefer, "For some [Deleuzian] affect theorists such as Brian Massumi, Patricia Clough, and Erin Manning, the term affect rigidly excludes what are called emotions—felt experiences that are the pieces of your personhood. But others [working in the phenomenological mode] such as Silvan Tomkins, Eve Kosofsky Sedgwick, Sara Ahmed, Teresa Brennan, and Ann Cvetkovich, suggest that the consideration of emotions falls under the purview of affect theory."[12] The key distinction between the two modes hinges on whether or not affects are engaged as metaphysical or cultural phenomena. I am convinced by Schaefer's (and Moore and Koosed's) streamlined, yet complex, genealogy of affect. However, I find the division into only two modes insufficient for the purposes of political theology, and the current political mood. For one, such division risks reifying the false dualism between the material and the socially constructed, which this volume hopes to deconstruct. Such a risk is helpfully highlighted in Elana Jefferson-Tatum's chapter in this volume, "Africana Sacred Matters: Religious Materialities in Africa, the Caribbean, and the Americas," when she suggests that we divorce the metaphysical from the immanent and social, not only to our peril, but also at the expense of taking Africana lenses and African and Indigenous religions seriously. Additionally, I worry that settling on only two categories risks, despite best intentions and precautions against simplistic divisions, creating a binary insufficient for containing the shifting flows of convergence and divergence between theories of affect. For instance, while upholding the division between affect and emotion, Patricia Clough simultaneously writes, contra-Massumi, that "affect is not 'presocial' . . . There is a reflux back from conscious experience to affect, which is registered . . . as affect."[13] In other words, the social affects the metaphysical. Similarly, placing both the Tomkins-inflected psychobiological approaches to affect and those theories coming from within what Sara Ahmed has called "feminist cultural studies of emotion and affect"[14] under the phenomenological mode, risks eclipsing the key convergences of the psychobiological approach with the Deleuzian approach (including a certain resistance to the culturally discursive production of affect as a locus of investigation), *and* eclipsing the key divergences between psychobiological approaches to affect and those of queer, critical race, and feminist cultural theorists for whom the cultural production of emotion is crucial, outweighing any search for what affect is. While Schaefer avoids these risks through a slow and nuanced mapping of the complexities within and between his two modes, I have found a slightly modified genealogy more helpful for understanding the streams of affect theory, most significant for the work of political theology and planetary ethics.

My focus on a political theology of affect, one concerned with the ethical resistance to neoliberal capitalism and finely attuned to the political affects pulsating in our country, has led me to more fully divide (while recognizing key entanglements between, in particular, the interdisciplinary work of Eve Kosofsky Sedgwick) the psychobiological approaches to affect and the cultural studies approaches. Hence, in an attempt to strike a balance between Gregg and Seigworth's blooming list of approaches and Schaefer's streamlined modes, I frame affect theory through three interconnected and yet distinct lenses: the *psychobiological* lens, the *prepersonal* lens, and the *cultural* lens. I suggest, perhaps contra-Schaefer, that each of these strains has phenomenological inclinations; the key divergences, I find, stem from the interpretive schema with which they approach the phenomena engaged.

The psychobiological lens represented in the works of such thinkers as Sedgwick, Adam Frank, and Tomkins (Tomkins first coined the term "affect theory" in 1962), and more recently in Elizabeth Wilson's *Gut Feminism*, looks at how feelings are psychobiologically structured in ways that shape human (and sometimes intra-human/nonhuman) experience. Psychobiological approaches can be, but are not always, investigations of affects that cut across histories and cultures. The prepersonal lens, found in work that draws on the philosophies of Alfred North Whitehead and Gilles Deleuze (Massumi, Clough, Manning, Shaviro) takes affect as a force or intensity. Affects are felt data that affect us before we code them as emotions. It is important here to note that to understand affects as prepersonal is not to understand them as inconsequent to the social, or as unaffected by postpersonal emotion, but rather to understand affect as that which overflows the discursive production of emotional codes. And finally, how the cultural approach to affect, most readily found in the work of queer and feminist cultural studies and Black and ethnic studies, resists categorizing affects as presocial and focuses instead on how affects are produced through cultural and historical structures of power. Cultural theorists of affect look at which bodies, which choices, which feelings—stick.

Cultural theorists of affect, such as Ann Cvetkovich, Jasbir Puar, Sara Ahmed, and Lauren Berlant depathologize and deindividualize "negative" feelings. Instead of viewing these feelings as signs of sickness in the individual, they ask us to examine the diagnostic potential of such moods. How might envy, for instance, diagnose the mentality created in a society in which we are always striving, but failing to "keep up with the Joneses?" How might depression diagnose a society that asks us to be ever more efficient and productive, but cares little for the necessities of rest and reflection? How might rage diagnose what it feels like to have your life under threat or your authority under suspicion because of your race or gender? How might anxiety diagnose a society taught to be afraid of anyone who worships your God(s)? It is this strain of affect theory, the critical examination of culturally produced emotions, that for my purposes today is both most promising, and most perilous. For in all our talk of how it feels to be made depressed by capitalism, and White supremacy, and heteropatriarchy, I am afraid we have spent too little attention, or too few theoretical treatises, on what to do about the responsive anger and contesting truth that the claims of our depressive complaints continue to provoke.

Believing in History

Affect theorist Sara Ahmed writes, "How is it that we enter a room and pick up on some feelings and not others? I have implied that one enters not only *in* a mood, but *with* a history, which is how you come to lean this way or that. Attunement might itself be an affective history, of how subjects become attuned to others over and in time."[15] For Ahmed, attunement to the atmosphere of the room can mean learning to not bring up certain topics. In such understanding of attunement, we might find a call to better attend to all the moody histories that enter our rooms—histories of harassment that bubble up on social media, histories of exclusion and dehumanization that flow into the streets and topple monuments, histories unable to be written here because *I* have not yet been finely attuned enough. But we might also find, in the call to acknowledge moody histories, a reminder that Whiteness comes laden with its own moods— Whiteness is saturated with affect. These affects form obstacles for attunement to the moods of those others on whose effacement our Whiteness was built. We might begin this endeavor by acknowledging a resistance to such moody histories, our desire for the mood of innocence over guilt.

Perhaps we need to begin again, this time with belief. Or rather to the matter of believing; not so much how we believe, or in what we believe, but in whom we believe. Moody histories have been swirling around us and have become viscerally ambient in the six years since we first met in the seminar room to begin our work on new materialism, planetarity, and religion. It was in what felt, simultaneously, like a collective sigh of relief, an anxious yelp of pain, and a hurricane of rage, that the #MeToo movement swept across social media. It was as though the flood of stories coming out of women and nonbinary gendered people of assault and harassment was either going to lift all our boats or drown us once again in affects of frustration, dismissal, and denial. The collective scream of "enough!" spurred by the affects of rage and sorrow in the wake of George Floyd and Breonna Taylor's deaths, finally pushed thousands of White people into the streets in solidarity with Black life. And yet, the heightened visibility of such affects of grief, rage, and lament should also remind us of why so much flesh had to be violated and blood spilled before they could be felt and heard by the majority.

In her work on "complaint," worth quoting at length, Ahmed writes:

> The experience of a situation as something to be complained about is an experience of coming apart from a group . . . The violence of such utterances is what you are required not to notice in order to participate in the group. You have to laugh—and laugh convincingly—in order not to stand out. You can stand out by just experiencing violence as violence. And then the violence you fail not to experience as violence is redirected towards you; the violence that was already in the room is channeled in your direction. This is probably why some laugh; to avoid the channeling. Laughing could thus be considered a form of institutional passing; a way of avoiding standing out, of trying to slide by undetected. The problem of passing is that if someone fails to pass, those who have passed are still participating in what has left someone stranded. Being stranded is part of the experience of complaint; a sense that you have been cut off from a group that you had formerly understood yourself as part of; you come apart; things fall apart. Cutting yourself off can also be a judgement made about the complainer: as if you have caused your own alienation by not going along with something. This is how a complaint teaches us about culture; we learn what is required to participate in something. A complaint teaches us about we; how a bond becomes a bind. Those who complain are often judged as causing the problem they identify by failing to be part of a we.[16]

Guilty affects are not embodied by the original perpetrators of violence; rather guilt and shame stick to those doing the complaining. They fester in the souls of those who are now cut off from the community they once believed themselves to be a part of, or in which they were told they would one day have a place. They, no, we, could not attune to the cultural mood, so we are guilty. We are shamed. The room gets to keep its innocence. Inspired by Ahmed, I am left to wonder how much these recent #MeToo "complaints" will be a cure or a curse that damns us to relive, in perpetuity, our abuse. Where I lie, moment to moment, on the spectrum from curse to cure is largely related to response. How I feel minute to minute, how many of my colleagues and friends feel second to second, has depended on the response of the men in our lives. Some friends had their abusers apologize and say plainly that they were guilty; this was the beginning of catharsis. Others of us saw our harassers, the very men we wrote #MeToo about, claim their own innocence by calling the "other men," the "bad ones," to account. In these cursed cases mood went down—denial leading to depression.

In a 2007 interview given to Bill Moyers regarding the then upcoming book *The Cross and the Lynching Tree,* James Cone warned us that if America cannot get over its "innocence" we will never be able to build the beloved community.[17] It is, perhaps, in better attending to the affects of innocence and guilt where we might find a path toward those trustier modes of being related, of finding justice, of a planetarity loosed of its romanticism and open to its responsibility. So let us pause, attend to the guilt of White supremacist heteropatriarchy, and examine (just some of) the moody history that brought us to this place.

In *Stand Your Ground: Black Bodies and the Justice of God,* Kelly Brown Douglas explores the roots of what she calls our "Stand Your Ground" culture. Stand your ground culture marks Whiteness itself as sovereign property, such that George Zimmerman could kill Trayvon Martin with impunity, and such that we all know if it had been Martin who shot Zimmerman the same would not be true. Having no rights to even self-possession, self-sovereignty, means that in stand your ground culture, there is no property Black people are supposed to count as theirs; there is no ground on which to stand. Kelly Brown Douglas traces the development of this culture through American "founding fathers" who were enamored with the myth of Anglo-Saxon purity held over from Tacitus's *Germania,* the 98 AD treatise that marked ancient Germans as an exceptional race "possessing a peculiar respect for individual rights and an almost instinctive love for freedom."[18] Brown Douglas leads readers through the same founders' fears of the mixing of the blood of European colonists and African slaves, to a shift in the ability of non-Anglo Saxon European immigrants to become White because of their skin color. For instance, she notes cases of White domestic workers who refused to refer to their employers as "master," because only Black people could be servants and slaves. These low-paid workers preferred the word "help," and even borrowed the Dutch word "boss," to replace "master," even though boss in Dutch means master.[19] In other words, "Whatever the specific twists and turns on the path to constructing whiteness, the construction was done in opposition to Blackness. The 'new stock' immigrants constructed their white identities as 'not slaves' and as 'not blacks.'"[20] The mood of true Americanness, of exceptional Americanness, was bought through the ransoming of Blackness for Whiteness. The mood of innocence is today, bought by ransoming this mood history in favor of the moods of White resentment that twist history, decry critical race theory, and choose the righteousness of their false innocence over the healing potential of their repentance.

A mood of possession flows through this history and its contemporary de-nials. As Brown Douglas, quoting and building on the work of Cheryl Harris, puts it:

> "Whiteness and property share a common premise—a conceptual nucleus of a right to exclude." This right to exclude inexorably gives way to other fundamental rights—the right to claim land and the right to stake out space. These rights, Harris points out, were actually "rati-fied" at America's beginnings with "the conquest, removal, and exter-mination of Native American life and culture." From then on she says, "Possession and occupation of land was validated and therefore privi-leged" as a white property right.[21]

Brown Douglas further suggests that such White property rights do not just refer to the ownership of land, but more broadly the ownership of space itself, "[white space] travels with white people. It is the space that white people oc-cupy. This space is not to be intruded upon, hence the right of whiteness to exclude."[22] If it is my God given right to exclude, then there need not be a feeling of guilt or shame. I am innocent. The complainant, the trespasser, is guilty. Whiteness is part of the moody history I carry into this volume. If I deny this mood, if I refuse my guilt, if I run toward being the good planetary ethicist and poststructuralist anti-racist feminist I want to be, without attending to our inherited history, then affects of innocence curse the relationality we seek. The mood of individual pardon trumps us again.

Gutting the Guilt and Guilting the Gut

Perhaps this is obvious. Perhaps our guilt is clear. Perhaps it means little to point to the need to get over our innocence. It certainly cannot be a silver bullet that will get us out of the morass of affects and "truths" proffered today. I have not, necessarily, found an answer to the nagging questions with which we began. For indeed, guilt might lead us once again to be frozen in shame; knowing that there is nothing we do that can redeem the histories of violence endeavored for the construction of Whiteness and maleness and humanness as sovereign—as exception, as cherished property—we do nothing. And yet, to erase the guilt, to have forgiven our debts each time the bill comes due, has gotten us nowhere. Or rather, has trapped us in the violent cycles of complaint and denial, or worse, in affects of avoidance couched in moods of conciliation, of getting along, of romantic rationalities that elide difference in favor of a violent sameness. We can do better than that. We cannot pay all the debts, they are beyond payment, but we can better attend to what is owed. In that attending we might find ways of thinking planetarily that recognize our fundamental entanglements, as well as our responsibilities to the shifting terrains of encounters and the histories of power that led us to be tangled this way and not that, and which nurtured in each of us varying senses of the truth, or varying attachments to what feels right. Hence, in following "the right feeling" towards its moody history we might find ways of distinguishing trusty propositions versus reactionary, alternative facts.

For the concluding chapter of her *When Species Meet*, Donna Haraway provides the following two epigraphs, the first from Karen Barad and the second from Jacques Derrida:

> Knowing is a direct material engagement, a practice of intra-acting with the world as part of the world in its dynamic material configuring, its ongoing articulation. . . Ethics is about mattering, about taking account of the entangled materializations of which we are a part, including new configurations, new subjectivities, new possibilities— even the smallest cuts matter.

> KAREN BARAD, *Meeting the Universe Halfway*

> One never eats entirely on one's own: this constitutes the rule underlying the statement, "One must eat well." . . . I repeat, responsibility is excessive, or it is not a responsibility.

> JACQUES DERRIDA, "Eating Well, or the Calculation of the Subject"[23]

With these two injunctions, Haraway reminds us that any responsible affect theology or planetary ethic must include attending to even the smallest cut. She reminds us that such responsibility is excessive. We can never be innocent to what we owe to those we have consumed, to what and whom we have devoured on the way to our becoming. If the smallest cut matters, then I owe all cuts a fair hearing, let alone the cut of Blackness from Americanness. I am responsible for the cutting off of ugly, mooded histories to which I do not want to attend—those I do not want to claim. In other words:

> There is no way to eat and not to kill, no way to eat and not to become with other mortal beings to whom we are accountable, no way to pretend innocence and transcendence or a final peace. Because eating and killing cannot be hygienically separated does *not* mean that just any way of eating and killing is fine, merely a matter of taste and culture . . . The practice of regard and response has no preset limits, but giving up human exceptionalism has consequences that require one to

know more at the end of the day than at the beginning and to cast
oneself with some ways of life and not others in the never settled biopolitics of entangled species. Further, one must actively cast oneself
with some ways of life and not others *without* making any of three
tempting moves: (1) being self-certain; (2) relegating those who eat differently to a subclass of vermin, the underprivileged, or the unenlightened; and (3) giving up on knowing more, including scientifically, and
feeling more, including scientifically, about how to eat
well—together.[24]

There is no way to become without unbecoming, no way to construct without
destruction. To give up on Anglo-Saxon, and while we are at it Protestant
Christian and modern human, exceptionalism requires the same kind of excessive responsive attention. It requires us to be uncertain; to continue to attend
to (even as we responsibly contest) those we find distasteful, disgusting, vermin-
like; and to ever-desire to eat and kill well together.

To eat and kill well together does not mean we will not make cuts, just the
opposite. This is a call to decide whose feelings are trustier, with what moods/
modes of life we will cast our lot. But we have to make those decisions, those
cuts, with a finer attention to how we got to our lot in the first place, and with
a willingness to be undone by the others whom we have, frankly, killed. We
must admit that we remain entangled with those we have assumed were up
for consumption and disposal. We cannot get to this point of attention if we
believe we can check ourselves out of the regimes of killing. Our attention is
muddied if we find some self-satisfaction even here amongst our planetarily-
attuned kin; it is blurred when we assume that, if only we recognized our affective entanglements, we would all *feel* and do right. For, as Elizabeth Wilson
critiquing recent strains of feminist theory argues, "Politics is a broadly and
bitterly constituted activity; it is not a synonym for amelioration. The key question, then, is not one of choosing between harm or remedy, or adjudicating on
how much hostility or how much reform we are able to avoid or create. Rather,
feminist theory has the more engaging task of developing ways to exploit and
expand political terrains that are always pharmakological in character."[25] This
volume on planetary thinking, *is* a violence to non-planetary thinking. We risk
doing a violence to each of the modes of thought and ways of thinking religion
that we have not included as properly planetary. There is no innocence to be
had. Hence, let us own the guilt, let us sit with it, attend to it. Let us better feel
for when such violence has not meant killing and eating well, but rather de-
vouring for ill. There may even be hope in recognizing that we cannot always
know which feelings to trust, because such unknowing is the reminder that
we should strive to know better by the end of the day than we did at its start.

Such uncertainty, the uncertainty carried by affect, allows for our undoings, for a better sensitivity to the promise and peril of our relations. As Puar notes, "Affect is precisely the body's hopeful opening, a speculative opening not wedded to the dialectic of hope and hopelessness but rather a porous affirmation of what could or might be."[26] Or, as I have offered in my own work, we might find ways to gravely attend to how we are or are not killing well. "Grave attending" is the act of being brought down by the gravity of what is and a witnessing to the graves of that which we wish would stay buried, a listening to the ghosts of what might have been (all those irredeemable subject positions and moody complaints we tried to closet away). To gravely attend to the violence of our own affect-laden politics is to look to the moody past so that we might find trustier ways of traversing our planetary webs in the present.

It will not be easy. It will take time. But I know in my gut we can attend better, because lately, nothing feels right to me.

Notes

1. Ann Cvetkovich, *Depression: A Public Feeling* (Durham: Duke University Press, 2012), 77.

2. Sara Ahmed, *The Promise of Happiness* (Durham: Duke University Press, 2010).

3. Elizabeth Wilson, *Gut Feminism* (Durham: Duke University Press, 2015), 99.

4. Ibid., 143.

5. For an excellent mapping of the history of American White rage, see Carol Anderson, *White Rage: The Unspoken Truth of Our Racial Divide* (New York: Bloomsbury, 2016).

6. Melissa Gregg and Gregory J. Seigworth, eds., *The Affect Theory Reader* (Durham: Duke University Press, 2010), xi.

7. Jasbir Puar, *The Right to Maim: Debility, Capacity, Disability* (Durham: Duke University Press, 2017), 1.

8. Gregg and Seigworth, *The Affect Theory Reader*, 3–4.

9. Gregg and Seigworth's map includes eight approaches to affect theory, a summary of which follows: (1) phenomenologies and post-phenomenologies of "sometimes archaic and often occulted practices of human/nonhuman [interaction]"; (2) theories of assemblage that engage the ontological entanglement of the human/machine/inorganic, which include, for Gregg and Seigworth, cybernetics, neurosciences, and bio-informatics / bio-engineering; (3) nonhumanist philosophies centered on "linking the movements of matter with a processual incorporeality (Spinozism)," particularly in critical stances that seek to move beyond "various cultural limitations" in philosophy through: feminist (Rosi Braidotti, Elizabeth Grosz, Genevieve Lloyd, and Moira Gatens); Italian autonomism (Paolo Virno or Maurizio Lazzaratto); cultural studies (Lawrence Grossberg, Meaghan Morris, and Brian Massumi); and political philosophy (Giorgio Agamben, Michael Hardt, and Antonio Negri); (4) psychological and psychoanalytic inquiry (early Sigmund Freud,

Silvan Tomkins, Daniel Stern, and Mikkel Borch-Jacobsen); (5) politically engaged critiques of the normative power of affect, which view affects more as collective than as individual (often undertaken by queer theorists, subaltern peoples, feminists, and disability theorists); (6) critical, often humanist, turns away from the "linguistic turn" in order to explore non-discursive and ethico-aesthetic forces of feeling (Raymond Williams, Frantz Fanon, Walter Benjamin, Susanne Langer, and John Dewey); (7) engagement with affect to interrogate subject or self-based philosophies (often comes from postcolonial, hybridized, and migrant voices); (8) science studies, often drawing on the work of Alfred North Whitehead, that embrace a pluralistic approach to materialism and ontology (Stengers). Gregg and Seigworth, *Affect Theory Reader*, 6–9. Beyond this map, Gregg and Seigworth list others that could have been included, but whose work, according to them, is not definitive of the field: Donna Haraway, Erin Manning, William Connolly, J.K. Gibson-Graham, Lisa Blackman, John Protevi, Sianne Ngai, Ghassan Hage, Jane Bennett, Paul Gilroy, Karen Barad, Steven Shaviro, Elizabeth Wilson, Alphonso Lingis, and Michael Taussig. Gregg and Seigworth, *Affect Theory Reader*, 9.

10. Donovan Schaefer, *Religious Affects: Animality, Evolution and Power* (Durham: Duke University Press, 2015). Kindle Location: 542.

11. Jennifer L. Koosed and Stephen D. Moore, "Introduction: From Affect to Exegesis," *Biblical Interpretation*, Vol. 22 (Leiden: Brill, 2014): 381–387.

12. Schaefer, *Religious Affects*, Kindle location: 571–574.

13. Patricia Ticineto Clough, "Introduction," in *The Affective Turn*, eds. Patricia Tincineto Clough and Jean Halley (Durham: Duke University Press, 2007), 2.

14. Ahmed, *Promise of Happiness*, 13.

15. Sara Ahmed, "Not in the Mood," *New Formations*, Vol. 82 (January 2014): 18.

16. Sara Ahmed, "Cutting Yourself Off," feministkilljoys blog, November 3, 2017, accessed November 11, 2017. https://feministkilljoys.com/2017/11/03/cutting-yourself -off/.

17. Bill Moyers, Interview with James Cone, *Bill Moyers The Journal*, November 23, 2007, accessed November 11, 2017, http://billmoyers.com/content/james-cone-on -the-cross-and-the-lynching-tree/.

18. Kelly Brown Douglas, *Stand Your Ground: Black Bodies and the Justice of God* (New York: Orbis Books, 2015), 5.

19. Ibid., 36.

20. Ibid.

21. Ibid., 41–42.

22. Ibid., 42.

23. Donna Haraway, *When Species Meet* (Minneapolis: University of Minnesota Press, 2008), 285.

24. Ibid., 295.

25. Wilson, *Gut Feminism*, 166–167.

26. Puar, *The Right to Maim*, 19.

The Entangled Relations of Our Ecological Crisis: Religion, Capitalism's Logics, and New Forms of Planetary Thinking

Matthew R. Hartman

A Cinematic Meditation

In 2018, a Canadian documentary generated some buzz on the independent film festival circuit for its striking footage of the effects of climate change. The film, *Anthropocene: The Human Epoch*, follows the work of a group of international climate scientists, layering voiceovers describing their research with captivating and alarming footage of the Earth.[1] The official description of the film on the documentary's website calls it "a cinematic meditation on humanity's massive reengineering of the planet," effectively blending artistic, spiritual, and scientific language to characterize the project. The *Anthropocene* trailer features a series of provocative images that quickly transition from one sensational scene to the next. "Humans go from being participants in the whole Earth to being a dominant feature," remarks a voiceover to a zoomed-out shot of a broad cityscape that engulfs the entire screen; "dominating the oceans, the landscape, agriculture, animals," says another, this time over video of trees being destroyed through explosive images with apocalyptic resonance. The footage is spectacular and shocking as the film showcases the impact of humans on the environment, serving a performative function to induce a sense of alarm in the viewer.

Through dramatic music, gripping scenes, and dire warnings about the future, the film advances a particular narrative of the Anthropocene to contextualize and explain humanity's current environmental situation. Originally a term used in the field of geology, the Anthropocene has become increasingly prevalent in the wider culture—from visual media like the BBC's *Planet Earth* series, to literary works in climate fiction or "cli-fi," to popular music such as Canadian indie-pop star Grimes's 2020 album "Miss Anthropocene," and podcasts like young-adult novelist John Green's "Anthropocene Reviewed," which purports to review different aspects of a human-centered planet on a five-star scale. The name of the documentary itself—*Anthropocene: The Human Epoch*—participates in the ongoing construction of this narrative, further establishing the Anthropocene in the cultural imaginary as a kind of artistic and technological tour de force that shapes how we think about the historical time in which we live. The Anthropocene is increasingly understood in the scientific community and more popular spaces as denoting a new era in which humans are irrevocably changing the contours of the planet, its origins often located sometime around the Industrial Revolution and the rapid increase in human technological advancement.

But who constitutes the "Anthropos" of the Anthropocene? If the increased adoption of the term Anthropocene by the wider culture advances a particular narrative of a universal humanity presumed to be at the center of the universe, what might such a narrative mean for questions of justice, sustainability, and planetary thinking? Perhaps even more importantly, does the language and narrative of such a "human epoch" move us toward a meaningful response to climate chaos and ecological devastation? While perhaps useful in certain fields such as geology and the biosciences, I am skeptical of certain cultural assumptions advanced by an Anthropocene narrative that considers humanity the preeminent subjects of history. I argue that this narrative follows reductive binaries that separate environment/human and nature/culture, thus obscuring the intricate relationships between the human and other-than-human. Rather than advancing narratives that locate technological foundations rooted in Western hegemony, I am, instead, more interested in addressing the underlying relational logics of the contemporary ecological crisis in pursuit of what Karen Bray calls "a planetarity loosed of its romanticism and open to its responsibility" in her chapter from this collection[2] (see, "Gut Theology: The Peril and Promise of Political Affect").

Beginning with a critique of subjectivity in the Anthropocene, I challenge the inherent assumption of the "Anthropos" as an undifferentiated whole, set apart from nature. I primarily draw on the work of sociologist Jason Moore, who has proposed the concept of the "Capitalocene,"[3] which centers the logics of capitalism, and its world-ecology relations, as an alternative to the Anthropocene.[4] After contesting the epistemic and methodological legacy of the human/nature binary, I then challenge the historical narrative of the Anthropocene that traces modernity's formation in the Enlightenment and locates the origins of the environmental crisis in early industrializing Europe. Instead, I focus on the period of European conquest and the coconstitutive relations of colonialism, theology, and capitalism as constructing the West and engendering the ecological crisis. I emphasize the importance of theology and Christian discovery in this formation, highlighting entanglements less prevalent in Moore's analysis. By locating the origins of ecological degradation, not in industrial Europe, but in the period of colonial expansion and capitalist formation that preceded it—an inherently theological project—the environmental crisis becomes less a crisis of technology and is instead formed by relations of power.

Anthropocene or Capitalocene? The Politics of Subjectivity

In May of 2000, the International Geosphere-Biosphere Programme published a short essay in their *IGBP Newsletter* by chemist Paul Crutzen and biologist Eugene Stoermer. Titled "The 'Anthropocene,'" the essay marks one of the first uses of the term to describe "the current geological epoch" and "emphasize the central role of mankind in geology and ecology."[5] Crutzen and Stoermer propose the latter part of the eighteenth century as the beginnings of the Anthropocene, noting dramatic growth in greenhouse gases beginning in that period, as well as the correlation with the invention of the steam engine often accredited to James Watt in the late 1700s. The essay concludes with a warning about climate change and a challenge for scientists and innovators to put their engineering and research skills together to address the crisis.

While proposed as a geological term in an international science newsletter, the Anthropocene has gained popular traction and is used with increased frequency in the wider culture. Crutzen and Stoermer broadly attribute atmospheric changes related to climate change to the actions of "mankind," clearly locating the roots of the climate crisis in industrial Europe. While the Anthropocene might be useful for a particular scientific or geological narrative of climate change, it is rather monolithic in its categorization of "mankind" and seems to further a human/environment binary that, according to Whitney Bauman in his chapter in this collection (see, "Developing a Critical Romantic Religiosity for a Planetary Community"), characterizes much of the language of the natural sciences since World War II.[6] Such modern reductive categories do not account for the intricate relations between humans and their environments. This begs the question, is there another term or narrative that might better elucidate the complex formations of ecological crisis?

Jason Moore engages this critique in his concept of the Capitalocene, which focuses on entangled relationships as opposed to the more universal language of the Anthropocene. The Capitalocene refers to "a system of power, profit, and re/production in the web of life" that locates the formation of capitalism's logics in the period of European colonial expansion rather than the mere technical advancements and innovations of later industrial Europe.[7] Three areas of emphasis stand out in Moore's understanding of the Capitalocene and critique of the Anthropocene narrative: first, the Anthropocene follows a problematic logic that creates a human/nature binary and converts them into abstractions; second, the history of capitalist origins is also the origin of environmental crisis; and finally, capitalism is understood as a "world-ecology of power" that is fundamentally relational.[8]

In locating the ecological crisis in post-Enlightenment, industrial Europe, there is an assumption of Enlightenment-era logics framing the narrative. Europe becomes the subject of history and all else is relegated to "other." The European subject, then, is abstracted over and against nature, constructing a binary rooted in Cartesian dualism. Moore refers to this kind of analytical reduction of Enlightenment thought as "Green Arithmetic" which "offers a Human/Nature binary that can proceed only by converting the living, multi-species connections of humanity-in-nature and the web of life into dead abstractions—abstractions that connect to each other as cascades of consequences rather than constitutive relations."[9]

However, the categories of the human/nature binary are never absolute or static. They are, instead, part of an identity-making project in which groups of people are classified as nature, and thus transformed into abstractions. This transformation is fundamental to the formation of the logics of capitalism, as primitive accumulation becomes not only the accumulation of objects and capital, but an active, ecological process of abstraction that creates new sites of accumulation. Moore calls the praxis of primitive accumulation "one of accumulating and organizing not only human bodies, but of assigning their value through the Humanity/Nature binary."[10] The identity-making project of classifying Europeans as the subjects of history establishes a hierarchy that divides non-Europeans through a process of racialized assemblages and gendered construction.[11]

Capitalism's Logics and Racialized Geographies

Viewed through a lens of bifurcation, there are two primary logics of capitalism: that people are separated from their land and homes via primitive accumulation, and that humans are separated from other humans through racism, sexism, and other logics of domination and separation. Thus, the separation of accumulation is not a one-time event—an original sin of political economy, if you will, suspended in historical time—but an ongoing process encapsulated in a logic of domination.[12] Moore refers to the creation of these abstractions as "Cheap Natures," a concept which:

> . . . embodies a logic of *cheapening* in an ethico-political sense, relocating many—at times the majority of—humans into Nature, the better to render their work unpaid, devalued, invisibilized. Early primitive accumulation's epochal achievement went far beyond the expulsion of the direct producers from the land. It turned equally on the expulsion of women, indigenous peoples, Africans and many others from Humanity.[13]

There is a kind of totalizing process which reduces human and environmental others to abstractions that are "ready to be accumulated" by the European subject. This reduction is the *cheapening* referred to by Moore, which is a process of the logics of capitalism.

The question of subjectivity in the Anthropocene is a crucial one, as the dominant narrative seems to presume a White, male subject of European descent. This is precisely Kathryn Yusoff's critique of the geological history, embedded in the narrative of the Anthropocene, that not only is the human subject set over and against an object of nature, but that humans are further separated through racialized hierarchies. In her critical work, A *Billion Black Anthropocenes or None*, Yusoff specifically connects biopolitical organizations of the *inhuman as matter* to the *inhuman as race* as foundational to modern colonial geographies.[14] In this way "histories of the Anthropocene ubiquitously begin with meditations on the great white men of industry and innovation to reinforce imperial genealogies," thus advancing narratives that maintain societal organization along hierarchical lines of race and class.[15] Nature is deemed object, as other humans are designated as part of this notion of nature.

Such histories of racial hierarchies and intricate power relations that shape capitalist logics are absent in the dominant Anthropocene narrative that privileges the industrious European of the Industrial Revolution. However, accumulation is ongoing and envelops humans and the environment in a web that is constructed socially and materially. Moore asks, "Are we really living in the *Anthropocene*—the 'age of man'—with its Eurocentric and techno-determinist vistas?"[16] The Anthropocene understands the human subject as a monolithic, "undifferentiated whole," that obscures more than it illuminates.[17] However, a focus on the underlying relational logics of capitalism in Moore's Capitalocene might just shift attention to material realities, better emphasizing human-ecological entanglements that are environment-making.

Theology, Colonialism, and the Origins of Climate Crisis

As science and technology become the center of modernity's project, other forces—such as capitalist relations, gender constructions, racialized assemblages, and theological formations—are cast to the periphery. This has profound implications for addressing the ecological crisis: if environmental degradation is fundamentally a problem of technological advancement rooted in industrial Europe, then following the same trajectory, solutions will also be technological. This is the narrative arc of the Anthropocene where industrialization acts as a kind of *deus ex machina* both in the origins and presumptive solutions for ecological crisis, effectively ignoring complex relationships of power.[18]

In Lynn White Jr.'s classic 1967 essay "The Historical Roots of Our Ecological Crisis," he critiques the influence of what he calls "Judeo-Christian teleology" on environmentally destructive practices.[19] Calling Christianity "the most anthropocentric religion the world has seen," White makes a strong link between Jewish and Christian religious teachings and what he sees as the medieval origins of ecological crisis.[20] Though widely critiqued for a very narrow view of a particularly European understanding of Christianity, White's argument is important in that it connects religion as an underlying factor of the environmental crisis—even if he does understand religion as monolithic. A key problem I see in White's argument is that he is ultimately concerned with the role of technology and the tools that make environmental destruction so effective, and religion becomes a kind of singular force in justifying this project. Industry mixed with European Christianity become the harbingers of history, organizing the rest of the world in their wake.

Neither technology nor religion alone offers a sufficient enough explanation for the origins of the ecological crisis. It is on this point that we can return to Moore and his view of capitalism as a key formation for this discussion. Moore offers a particular logic that is both connected to and expands on technological advancements and religious developments. Challenging the dominant historical narrative that dates to the Industrial Revolution, Moore describes what he calls the rise of the "Age of Capital" between the years 1450 and 1750, writing, "Alongside new technologies, there was a new *technics*—a new repertoire of science, power and machinery—that aimed at 'discovering' and appropriating new Cheap Natures."[21] By distinguishing between "technology" and "technics," Moore is making a discursive move to focus on the logics and relations that undergirded and informed the development of new technologies— what he refers to as a "new repertoire"—rather than the abstract technologies themselves.

Colonial discovery is a formational *technic*, which is an inherently theological project and predates the Industrial Revolution of the eighteenth century. Challenging the typical narrative of the Enlightenment and industrialization of Europe as the significant historical events that have primarily shaped the modern era, Enrique Dussel has argued that the West was in fact made through discovery and colonialism—which in turn laid the foundations for the Enlightenment and Scientific Revolution.[22] In a passage important to Catherine Keller in her work *Cloud of the Impossible,* Dussel points out that the conquering spirit of colonialism comes before the enlightened thinking of European philosophy and science: the *ego conquero* of empire precedes the *ego cogito* of Descartes.[23] Similarly, Moore compares the figures of Descartes and Cortés, noting that the logic of Enlightenment (Descartes) does not exist without the conquering of Crusade (Cortés). In this way, the mind/body dualism of Descartes—which leads to what Moore refers to as the abstractions of the nature/society binary—is not only a Cartesian construction, but a *Cortesian* one as well.[24]

Colonialism and the conquest of the world by Europe is central to the history and construction of modernity. Joerg Rieger notes that while colonialism and empire revolve around money and power, enacted in an outward appropriation of land and resources, they also fundamentally "shape us all the way to the core of our being."[25] More than just the technological means of conquest or the resources stolen, the project of imperialism is transformational on an individual and collective level—a project that envelops economy, politics, and religion, and their various intersections. By locating the origins of ecological crisis in the industrialization of Europe, the Anthropocene narrative neglects the logics of power in the colonial period. The narrative is more concerned with abstract, technological processes instead of the all-encompassing project of European colonialism, undergirded by theological formations.

However, if the history of modernity, and thus the origins of ecological crisis, are located in the coconstitutive projects of colonialism, Christian discovery, the Scientific Revolution, and the early formations of capitalism, then the narrative trajectory changes. A more complex, entangled history that recognizes relations of power requires an analysis of environment-making solutions versus mere abstract, technical fixes. Similarly, if theological formations are understood as constitutive of the logics of the West, then religion becomes an important force in considering new materialist and imaginative responses.

Capitalist Formations, New Materialisms, and World-Ecological Relationships

The Anthropocene narrative that separates humanity from the rest of nature ignores the material realities of our entangled existence. We are intimately entwined with and shaped by the environments we inhabit. This is not to say that humans do not impact and transform nature, but that the process is more relational than monolithic. Humans are not the dominant subjects acting on a passive nature-as-object, but rather are part of a more interactive process that entangles us with what Stacy Alaimo calls the "stuff of the world."[26] It is in this sense that capitalism's logic is environment-making, a process which is a dialectic between project and process.[27]

Under the logic of the Anthropocene, there is an assumption that, as new ways of appropriating resources are articulated, or new technologies invented, the world is then transformed. But the project of environment-making precedes the effects of technology. There is a drive to discover, to invent, to develop new, Cheap Natures that instills a kind of creative element in capitalism.[28] Capitalism does not only produce tools of technology, but it engenders a particular ethic that envelops. While the historical manifestations of capitalism may have shifted as new discoveries, inventions, and resource exploitation took place under the guise of modernity and scientific enlightenment, capitalism's underlying logic remained and adapted over time.

In this way, "capitalism becomes something more-than-human," writes Moore, "It becomes a world-ecology of power, capital, and nature."[29] This "more-than-humanness" is a characteristic similar to what Timothy Morton has called a "hyperobject," which effectively shapes time and space, while simultaneously transgressing those bounds.[30] There is both an immanence and a physicality to the hyperobject of capitalism. Given capitalism's early formations in Christian colonial conquest, we can add theology to this co-constitutive web as well. Through capitalism's appropriation of nature and human labor, its logic transcends the monolithic binary assumed in the logic of the Anthropocene. While the effects of capitalism may appear as abstraction, organizing the world into categories that follow a nature/society binary, the logic of capitalism is more enveloping and functions as a process that converts people and environment into Cheap Natures. Through this system, different categories of humans are essentially categorized as "nature," to be subsumed and appropriated under capitalism's logic.

New materialist discourse can illuminate these concepts of environment-making and world-ecology relationships as transcending a human/nature dualism. Alaimo refers to the concept of "trans-corporeality" which draws on feminist and materialist discourses of the body, emphasizing that "the substance of what was once called 'nature,' acts, interacts, and even intra-acts within, through, and around human bodies and practices."[31] This is similar to Catherine Keller's assertion that we not only exist on the Earth, but are gathered within it—that we, in effect, exist *as* it.[32] This emphasis collapses the human/nature binary to focus on the relational realities of existence.

The entangling logic of the Capitalocene is the inverse of the abstracting logic of the Anthropocene. We are intricately bound to one another through the inter- and intra-planetary forces that shape life on Earth. To update the old adage: we are not only products of our own environments; we are part of them as well. As Timothy Morton would say, flipping another old saying, the sum of the parts is actually greater than the whole.[33] This material reality shapes and is shaped by our systems of power.[34] We must recognize the environment-making processes and world-ecology relations of which we are a part, participating in systems of power that constantly shape and are shaped by ecological entanglements.

Imagining New Relationships

At the end of his 2016 book *The Great Derangement*, novelist Amitav Ghosh addresses the role of literature and culture in confronting climate change, before making an interesting turn in his argument. Having spent the majority of the book making the case that the climate crisis is not only a crisis of science and politics, but "also a crisis of culture, and thus of the imagination," Ghosh concludes by juxtaposing the 2016 Paris Climate Agreement with Pope Francis's 2015 encyclical *Laudato Si'*.[35] For Ghosh, the technical and neoliberal frame-work of the Agreement ultimately rings hollow as "the current paradigm of perpetual growth is enshrined at the core of the text."[36] In the same way the film *Anthropocene: The Human Epoch*, referenced at the outset of this essay, offers "a cinematic meditation" on the Anthropocene and how it is character-ized by modern human technological advancements, the Agreement, for Ghosh, offers a kind of political or economic meditation on, to use the documentary's language, "humanity's massive reengineering of the planet."[37] In contrast to the opaqueness of the Agreement, Ghosh finds a kind of openness and imag-inative language at the heart of the encyclical where human/nature relationships are recognized in the pursuit of justice.

Such an imaginative turn demonstrates an attention to environment-making processes and our entangled existence with one another and our environments. Recognizing these world-ecology relationships moves away from the abstract, universal logics of the Anthropocene's historical narrative, to focus on complex material relations of power. As is evidenced by the Capitalocene, the role of theology through discovery and conquest is central to the logics of capitalism and the formations of ecological degradation. Any response that seriously ad-dresses the ecological crisis must take the various theological assemblages and relations of power into account.

Theologians and environmentalists are increasingly drawing attention to the existing, hollow assessments and, in turn, entangled relations of our ecological crisis. We see this in Pope Francis's notion of "integral ecology" that Ghosh cites, which recognizes the entangled histories and goals of social and environmental justice.[38] Similarly, the "global village" that Sallie McFague imagines in her work is a kind of planetary theology that recognizes intricate relationships of all beings.[39] Joerg Rieger articulates what he calls the "common good" as an alternative to the abstracting logics of a market born out of empire, arguing that attention must be paid to power relations and that all have access to a "productive participation in life."[40] These offerings from leading theological scholars, among others, are but a few examples of the ways in which the ecological crisis must be understood as a complex series of power relationships rather than abstractions. As Reiger argues in his chapter in this volume (see, "Which Materialism, Whose Planetary Thinking?"), new materialisms can effectively offer alternative religious modes of engaging environments that recognize and confront existing power dynamics.[41] These power dynamics not only include the continued accumulation of capital, but the ways in which the Anthropocene advances colonial narratives that are both racialized and ecological. Perhaps religion can offer a generative approach to address the environmental crisis—understood both as a forming logic and as a relational response.

Notes

1. Jennifer Baichwal, *Anthropocene: The Human Epoch*. Documentary directed by Jennifer Baichwal, Nicholas de Pencier, and Edward Burtynsky (Toronto: Mercury Filmworks, Inc., 2018).

2. Karen Bray, "Gut Theology: The Peril and Promise of Political Affect," in *Earthly Things: Immanence, New Materialisms, and Planetary Thinking*, eds. Karen Bray, Heather Eaton, and Whitney Bauman (New York: Fordham University Press, 2023).

3. See Jason W. Moore, "The Capitalocene, Part I: On the Nature and Origins of Our Ecological Crisis," *The Journal of Peasant Studies*, Vol. 44 (3) (2017): 594–630 and "The Capitalocene, Part II: Accumulation by Appropriation and the Centrality of Unpaid Work/Energy," *The Journal of Peasant Studies*, Vol. 45 (2) (2018): 237–279.

4. Moore calls global warming "capital's crowning achievement." See Moore, "The Capitalocene, Part II," 237.

5. Paul J. Crutzen and Eugene F. Stoermer, "The 'Anthropocene,'" *IGBP Newsletter*, Vol. 41 (May 2000): 17.

6. Whitney Bauman, "Developing a Critical Romantic Religiosity for a Planetary Community," in *Earthly Things: Immanence, New Materialisms, and Planetary*

Thinking, eds. Karen Bray, Heather Eaton, and Whitney Bauman (New York: Fordham University Press, 2023).

7. Moore, "The Capitalocene, Part I," 594.

8. Ibid., 595.

9. Ibid., 598.

10. Ibid., 600.

11. For a helpful discussion of "racializing assemblages" and the socio-political construction of racialized bodies, see Alexander G. Weheliye, *Habeas Viscus: Racializing Assemblages, Biopolitics, and Black Feminist Theories of the Human* (Durham: Duke University Press, 2014). And for an important look at the relationship between gender and nature and the subjugation of both due to the power relations of patriarchy, see the classic work by Carolyn Merchant: *The Death of Nature: Women, Ecology, and the Scientific Revolution* (New York: Harper Collins, 1980).

12. This is a reference to Marx's line in *Capital, Volume I* that "primitive accumulation plays approximately the same role in political economy as original sin does in theology." See Karl Marx, *Capital: A Critique of Political Economy, Volume I* (1867), trans. Ben Fowkes (New York: Random House, 1977), 873.

13. Moore, "The Capitalocene, Part II," 242.

14. Kathryn Yusoff, *A Billion Black Anthropocenes or None* (Minneapolis: University of Minnesota Press, 2018), 5.

15. Ibid., 15.

16. Moore, "The Capitalocene, Part I," 596.

17. Ibid., 595.

18. Ibid., 608.

19. Lynn White Jr., "The Historical Roots of Our Ecological Crisis," *Science*, Vol. 155 (3767) (March 1967): 1203–1207.

20. Ibid., 1205–1206.

21. Moore, "The Capitalocene, Part I," 610.

22. Enrique Dussel, "Europe, Modernity, and Eurocentrism," *Nepantla: Views from South*, Vol. 1 (3) (2000): 470.

23. Enrique Dussel, *The Invention of the Americas: Eclipse of "the Other" and the Myth of Modernity*, trans. Michael D. Barber (New York: Continuum, 1995), 43. Also see Catherine Keller's discussion of Dussel's comparison of *ego cogito* and *ego conquero* in *Cloud of the Impossible: Negative Theology and Planetary Entanglement* (New York: Columbia University Press, 2015), 257; see especially the entirety of chapter 8, "Crusade, Capital, and Cosmopolis: Ambiguous Entanglements."

24. Moore references work by Bikrum Gill, who seems to be the first to propose this dual language of "Cartesian" and "Cortesian" in a PhD dissertation. See Bikrum Gill, "Race, Nature, and Accumulation: A Decolonial World-Ecological Analysis of Indian Land Grabbing in the Gambella Province of Ethiopia," PhD dissertation (Toronto: York University, 2016). Also see Moore, "The Capitalocene, Part II," 244.

25. Joerg Rieger, *No Rising Tide: Theology, Economics, and the Future* (Minneapolis: Fortress Press, 2009), 19.

26. Stacy Alaimo, *Exposed: Environmental Politics and Pleasures in Posthuman Times* (Minneapolis: University of Minnesota Press, 2016).

27. Moore, "The Capitalocene, Part II," 256–258.

28. Moore, "The Capitalocene, Part I," 621.

29. Moore, "The Capitalocene, Part II," 239.

30. Timothy Morton, *Hyperobjects: Philosophy and Ecology after the End of the World* (Minneapolis: University of Minnesota Press, 2013), 2.

31. Stacy Alaimo, *Bodily Natures: Science, Environment, and the Material Self* (Bloomington: Indiana University Press, 2010), 1–3.

32. Catherine Keller, *Political Theology of the Earth: Our Planetary Emergency and the Struggle for a New Public* (New York: Columbia University Press, 2018), 5–6.

33. Morton calls this *subscendence*. That a person is much more than just some kind of abstract label of "human." He articulates this concept in a number of places, but perhaps most concisely in his recent book *Humankind*. See Timothy Morton, *Humankind: Solidarity with Non-Human People* (New York & London: Verso, 2017).

34. Moore speaks of an "epistemic rift" at the center of a logic of dualism that neglects relations of power. "The heart of the problem," writes Moore, "is that Nature/Society dualism not only poses analytical barriers but reproduces 'real world' systems of domination, exploitation, and appropriates." See Moore, "The Capitalocene, Part I," 601.

35. Amitav Ghosh, *The Great Derangement: Climate Change and the Unthinkable* (Chicago: The University of Chicago Press, 2016), 9.

36. Ibid., 154.

37. Baichwal, *Anthropocene: The Human Epoch*, documentary.

38. Pope Francis, *Laudato Si': On Care for Our Common Home*, encyclical letter (Huntington: Our Sunday Visitor, 2015); see especially chapter four, "Integral Ecology."

39. Sallie McFague, *Life Abundant: Rethinking Theology and Economy for a Planet in Peril* (Minneapolis: Fortress Press, 2001); see especially chapter five, "The Ecological Economic Model and Worldview."

40. Rieger, *No Rising Tide*, 157.

41. Joerg Rieger, "Which Materialism, Whose Planetary Thinking?" in *Earthly Things: Immanence, New Materialisms, and Planetary Thinking*, eds. Karen Bray, Heather Eaton, and Whitney Bauman (New York: Fordham University Press, 2023), this volume, 148–160.

Solidarity with Nonhumans: Being Ecological with Object-Oriented Ontology

Sam Mickey

New materialism is among a variety of contemporary schools of thought that seek to regenerate human contact with the real world, including object-oriented ontology (OOO). OOO is technically not a kind of materialism. Indeed, Graham Harman, the philosopher who first coined "object-oriented philosophy" in the late 1990s, has described the position as "immaterialism."[1] In some sense, OOO and new materialism are diametrically opposed to one another, yet they also share many theoretical and practical commitments, which can be broadly categorized as ecological. What follows is an account of the ecological rendering of OOO in the work of Timothy Morton, including their notion of *hyperobjects*—entities that are massively distributed relative to humans like: global warming; global capitalism; the COVID-19 pandemic; the Internet; and the Earth. Indicating how Morton's ecological iteration of OOO intersects with their understanding of Buddhist notions of emptiness, *karma*, and compassion, I present OOO as an ally with new materialisms, religious perspectives on ecology, and other modes of planetary thinking in solidarity with nonhumans.

The Spectral Plain

Ecology is about interconnectedness, what Morton calls *the mesh*, but it is also about the entities that are interconnected—*strange strangers*.[2] While interconnectedness is fundamental to ecological thought, the entities that are interconnected are not, thereby, reducible to the mesh. Strange strangers are irreducible to their relations or to their constituent parts and processes. Furthermore, strange strangers are not simply others. They are also beings with which one is most familiar and most intimate, including oneself.

Morton inherits their view in part from Jacques Derrida, whose deconstruction (following Martin Heidegger's *Destruktion*) is one of Morton's "favorite philosophical regions."[3] Morton follows the deconstruction of the "metaphysics of presence," which, simply stated, is the metaphysical idea that something must be present for it to exist. (BE, 67) The term "strange stranger" is Morton's way of translating Derrida's *arrivant* ("newcomer," "guest"), which is someone whose arrival always remains "to come" (*à venir*) in the future (*l'avenir*). (ET, 143n, 72) The guest comes from an unknown past, and thus the arrival of the guest is also the haunting return of a specter, a ghost (*revenant*), demanding justice. Ontology is "hauntology" for Derrida, such that, to exist is to be an undecidable flickering of past and future, a kind of appearing that is never quite present.[4] In Morton's words, ". . . what I mistakenly call 'present' is a kind of relative motion between two sliding trains of past and future. I call it *nowness* to differentiate it from a reified atomic 'present' that actually I don't think truly exists." (BE, 82) It is worth noting that Karen Barad considers quantum entanglement in hauntological terms as well.[5] OOO, new materialism, and spirits all converge in hauntology.

The deconstruction of presence shows up in Heidegger's tool analysis in *Being and Time*, where he demonstrates the way in which being present-at-hand (*vorhanden*) depends on being ready-to-hand (*zuhanden*). A tool can become present, for instance, when it breaks and has to be fixed, but when everything is working, it must withdraw (*zurückzuziehen*) from any direct theoretical or practical concern to stay ready.[6] Morton puts it this way:

> Things are present to us when they stick out, when they are malfunctioning. You're running through the supermarket hell bent on finishing your shopping trip, when you slip on a slick part of the floor (someone used too much polish). As you slip embarrassingly toward the ground, you notice the floor for the first time, the color, the patterns, the material composition—even though it was supporting you the whole time you were on your grocery mission. Being present is secondary to just sort of happening, which means, argues Heidegger, that *being isn't present*, which is why he calls his philosophy deconstruction or destructuring. What he is destructuring is the metaphysics of presence. (BE, 7)

The distinction between withdrawal and presence in the tool analysis is a crucial component of OOO, which Morton adapts from Harman.[7] For Harman, every object is a fourfold that is split along one axis into the withdrawal of the real and the relational interactions of the sensuous, and along another axis between objects and their qualities, thus rendering a quadruple object (real object, real qualities, sensuous object, sensuous qualities). (BE, 80, 150–52) The terminology of OOO can cause confusion. To be sure, objects are *not* opposed to subjects. They include any entity whatsoever, whether simple, complex, natural, artificial, self-aware, insentient, etc. There is no subject/object dualism for OOO. Accordingly, OOO is against any kind of objectification that would treat an object as lacking agency, intrinsic value, or interpretive capacity. Historically, only a few privileged (e.g., wealthy, White, cisgendered, heterosexual, able-bodied, male) humans tend to get the status of full subject, and that status comes at the expense of the subjectivity of all other human and nonhuman objects.

Preserving the irreducibility of objects, OOO avoids two kinds of reductionism, "undermining" and "overmining," which can combine into "duomining." (BE, 41–52) Undermining reduces things to their constituent parts or to an underlying field, like reducing things to matter or energy. Overmining reduces them to their effects or relations with overarching systems, like reducing objects to products of language and social construction, or to momentary occasions within networks of interactions. An example of duomining could involve reducing things to undifferentiated singularities (undermining) that only obtain specific qualities through the interpretive activity of humans (overmining).

Another reductionism rejected by OOO is anthropocentrism. Against the notion that reality is only what is accessible to human interpretation, OOO holds that every object has some interpretive openness, and thus every object can have some limited access to the appearances of other objects, some way of touching other entities. This is what Harman calls "vicarious causation," which is an immanent version of occasionalism—a concept that Harman borrows from the Ash'arite school of Islamic theology, including thinkers like the eleventh-century theologian Al-Ghazali, for whom all entities are withdrawn from direct contact and relate indirectly through the one and only causal power, God. Approximately five centuries later Nicolas Malebranche formulated this famous thesis of occasionalism, declaring that he seeks ". . . to prove in few words that there is only one true cause because there is only one true God; that the nature or power of each thing is nothing but the will of God; that all natural causes are not *true* causes but only occasional causes.[8]

For Harman, it is not God, but the interpretive openness of things themselves, that provides mediated access. Objects touch through translation, interpreting one another's qualities while otherness remains withdrawn—objects *"touch without touching."* (BE, 150) When fire burns paper, the fire does not touch the paper itself, but translates the paper's flammable qualities into the fire's terms. Like humans anthropomorphize nonhumans, fire pyromorphizes whatever it burns. Causality happens in translations between sensual objects (the mesh), while real objects (strange strangers) are withdrawn. Causality is thus an aesthetic process of objects interpreting each other's sensuous appearances. *"The aesthetic dimension is the causal dimension,"* as Morton puts it.[9] This sense of causality is expressed in Mahāyāna Buddhism, specifically in an example where Nāgārjuna discusses fuel and fire, applying a figure common to Indian logic: the tetralemma (affirmation, negation, both affirmation and negation, neither affirmation nor negation):[10]

> Nagarjuna, the great philosopher of Buddhist emptiness (*shunyatā*), argued that a flame never really touches its fuel—nor does it fail to touch! . . . If it did so, then the fuel would be the flame or vice versa, and no causality could occur. Yet if they were totally separate, no burning could take place. Nagarjuna argues that if something were to arise from itself, then nothing would happen. Yet if something were to arise from something else that was not itself, then nothing can happen either. A mixture of these views (both–and and neither–nor) is also possible, since such a mixture would be subject to the defects of each one combined. For instance, on this view, the idea that things arise neither from themselves nor from something else is what Nagarjuna calls nihilism, on which basis anything at all can happen. The logic of causal explanations, he argues, is circular. Emptiness is not the absence of something, but the nonconceptuality of reality: the real is beyond concept, because it is real. (RM, 73–4)

Emptiness is not a lack. It is not an absence hidden behind the magical display of the mesh. It is the nonconceptual reality of the display itself, happening all by itself, withdrawn and groundless.

Sharing the non-theistic standpoint of Buddhism, Morton explains vicarious causation, not through a mediating God, but through emptiness, which is simply the openness of appearances, the groundlessness whereby a flame is free to flicker in a relation without relation that leaves fuel consumed:

> There is no "causation" as such—that's a superficial illusion, a presence-at-hand as Harman would say. Like Al-Ghazali, for whom God provides the causal links between unlinkable objects, a kind of magic happens (without God) and we see flames emerging out of candlewicks and billiard balls smacking one another. There is nothing underneath this display. And the display happens whether "we" observe it or not. (RM, 74)

OOO is like new materialism, adhering to a reality that happens whether humans observe it or not. Undoing the privileged ontological status of humans, OOO and new materialism share a "flat ontology."[11] It flattens the hierarchies that separate the irreducibly strange differences of objects into categorically distinct kinds, like mind/body, or worlding/worldless. Such categories are used to justify reducing humans to their appearances (e.g., color, gender, and ability) and excluding them from ethical concern—treating humans inhumanely, as in racism, sexism, and ableism. That also justifies the exclusion of nonhumans from ethical concerns. Flat ontology does not bring all beings up to the level of humans but down to the level of strange strangers. The flat field of OOO is a field of specters: "the *Spectral Plain*." (BE, 126–30) "There are some basic rules of politeness on the Spectral Plain, and these have to do with the idea of '*hospitality to strangers*'." (BE, 128) This ethics of hospitality is found in Derrida, for whom hospitality involves welcoming the stranger—the guest, the ghost— without appropriating or assimilating the stranger's otherness.[12] Welcoming the other is ultimately an impossible task. Some hostility always infects hospi- tality, hence Derrida's portmanteau, "hostipitality" (*hostipitalité*). (Bray, "Gut Theology," 419) As Karen Bray argues in "Gut Theology: The Peril and Promise of Political Affect" in this volume, this interweaving of hostility and hospitable responsibility is never simple or innocent, but always involves ethical culpabil- ities and political complicities.

Hyperobjects

If planetary thinking is a kind of holism, Morton specifies that it must be a holism for which the whole is *smaller than the sum of its parts*. Every object is a whole composed of parts, which are also composed of parts, potentially into an infinite regress of objects inside of objects, which means that an object is "literally out scaled by its parts. It is bigger on the inside," and thus "however absurd and amazing it sounds, we need to say 'the whole is always smaller than the sum of its parts.'"[13] Wholes might be spatially or temporally larger than their parts, but however "physically huge," wholes are "ontologically tiny," meaning that the existence of wholes is insubstantial, hollowed out by con- tingencies and constraints. (HK, 102) This resonates with Terrence Deacon's theory of emergence, for which "the whole is less than the sum of its parts."[14] For Deacon and Morton, material components do not make a whole what it is. Rather, emptiness does, as indicated by Deacon's ambiguous spelling of wholes as "(w)holes," in a chapter that opens with a Daoist saying from Laozi, "Thirty spokes converge a the wheel's hub, to a hole that allows it to turn. . . . Though we can only work with what is there, use comes from what is not there." (HK, 18)

Like OOO, Deacon's theory is not a materialism, since it is not the material of something that makes it what it is. Deacon calls it "absentialism." (HK, 42) It can also be described in terms of naturalism.[15] Moreover, it very nearly resembles emergence theory, which Kevin Schilbrack elucidates upon in his chapter in this volume, "Emergence Theory and the New Materialisms." Deacon's naturalism is based in scientific and philosophical inquiry. It is not based primarily or exclusively on Daoism. Nonetheless, this naturalism contains its own religious dimension, as in religious naturalism, the plea of which Carol Wayne White articulates in her essay, "Planetary Thinking, Agency, and Relationality: Religious Naturalism's Plea." This approach to naturalism avoids appeals to otherworldly transcendence, but it does not simply favor immanent matter. Rather, it focuses on immaterial dynamics of emergence. This is similar to what Morton calls "subsendence." In contrast to the idea that a whole transcends its parts toward greater presence, a whole for OOO is withdrawn "because it *subscends* its appearance in a way that is not constantly present." (HK, 106) Unlike the transcendence of "explosive holism," for which "the parts are reducible to the whole," subscendence involves an "implosive holism" in which all objects are "spectral beings" that can be seen both as wholes that are less than their parts, and as *"partial objects"* ("parts of a whole that they exceed"). (HK, 70)

This is not an individualist claim that wholes do not exist. Wholes and parts are equally real, and every part is itself a whole composed of parts. When you look for any whole object, all you find is its enmeshed parts, which interrelate to make up the whole; the whole remains withdrawn. I look for my hand, and I find skin, fingers, nails, bones, veins, etc. I look for *Homo sapiens*, and I find humans. I look for climate, and I find weather. "Climate is ontologically smaller than weather. Weather is a symptom of climate, but there is so much more to weather other than simply being a symptom of climate. A shower of rain is a bath for this bird. It's a spawning pond for these toads. It's this soft delicate pattering on my arm. It's this thing I wrote some sentences about." (HK, 103) Moreover, saying that the climate subscends weather does not mean that it is in some world behind the scenes, "empirically shrunken back or moving behind; it means—and this is why I now sometimes say 'open' instead of 'withdrawn'—*so in your face that you can't see it.*" (HK, 37)

Explosive holism disempowers objects from playing their parts, as can be seen in utilitarian, correlationalist, and Gaian holisms. (HK, 121–22) Utilitarianism is an explosive holism of the greater good, which is a sum of the goods of all sentient individuals, making it justifiable to kill one person in order to save ten. The correlationist holism is Kantian, where the real is correlated to a subject, who becomes "the Decider" for all moral value, whether that value is centered in the autonomy of humans, the subjectivity of animals, or as biocentrism has it, the self-organization (autopoiesis) of organisms. Gaian holism sets the transcendent bar for value at the whole biosphere, making every entity a replaceable component of Gaia. Implosive holism lowers the bar to welcome all objects.

OOO avoids all of the mega-categories of explosive holism, hence Morton's "ecology without nature," which is also without world, matter, and the present.[16] Where there was a background or a horizon or a container there is now only a proliferation of entities that are massively distributed in spacetime relative to human scales; *hyperobjects*. (H, 1) Hyperobjects are viscous; "they 'stick' to beings that are involved with them," such that there is no "away," only the unbearable intimacy of the mesh. (H, 1, 31) They are sticky yet nonlocal, seemingly everywhere and nowhere, in our face all the time but too strange to simply locate. Global warming is massively distributed across fracking wells, rising sea levels, industrial agriculture, hamburgers, extreme weather events, and more, and yet none of those parts renders global warming present.

Hyperobjects involve multiple, imbricated scales of time and space. (H, 65) Global warming is happening where I am, here and now, yet it is also happening over the course of centuries and across Earth's surface, radiating innumerable spatiotemporal ripples; political administrations, polar ice caps, research grants, a mass extinction event, seasons, information technologies, hurricanes, the hydrologic cycle, a conference presentation, and a drop of rain during a rainstorm. Hyperobjects emit multiple, spatiotemporal scales as well as their own causality. Morton describes this as "interobjectivity." That is, the translation process whereby strangers touch without touching; "the magic of real objects." (H, 89) I feel a rain drop, which is part of a storm, which is part of the local meteorological system, which is connected to global weather patterns, which display the current state of the climate. "When it rains on my head, climate is raining." (H, 76)

Ethics is itself a hyperobject. Hospitality usually refers to a human subject welcoming another human, or in more ecological readings it refers back to a human subject who is welcoming human and nonhuman others. Along with those options (human-to-human and human-to-nonhuman), OOO adds nonhuman-to-nonhuman relations to ethics. OOO hears compelling demands issuing to and from every entity, thus entangling human ethics within a massive mesh of ethical injunctions of hospitality to strangers.

Spectral Attunement

Welcoming hyperobjects requires a hospitable atmosphere. "Mood," "atmosphere," and "attunement," are all translations of the German noun *Stimmung*, like the "tuning" of a guitar or the "voice" (*Stimme*) of a singer. Kant describes aesthetic experience as a *Stimmung*.[17] For Heidegger's existential phenomenology,

all experience involves *Stimmung*, such that mood is an ontologically primordial condition through which humans find themselves in the world. Like an
existential affect or feeling, *Stimmung* is what *"makes it possible first of all to
direct oneself toward something."* (BT, 176) Undoing the anthropocentrism of
Kant and Heidegger, OOO finds attunement happening in all beings. Every
object has its own tune or timbre. (BE, 94) In the "sticky mesh of viscosity"
that entangles humans with a hyperobject object like consumerism or the
climate, "I find myself tuned by the object." (H, 30)

Tuning is how all things relate, touching each other vicariously by resonating
with each other's qualities; "attunement is the mode in which causality happens." (BE, 90) To put it another way, *karma* is the mode in which causality
happens. The atmospheric vibes of *Stimmung* can be thought of in terms of
karma—the Sanskrit word for "action," denoting a cause-effect principle found
in Hinduism, Jainism, and Buddhism. According to Morton, the causal dimension where all things are enmeshed, such that every single thing is in-
between the other things, can be described in terms of what Tibetan Buddhists
call a *bardo*. Karma is the mode of causality in a *bardo*. *Bardos* are different
types of in-between spaces, including the *bardo* that crosses between life and
death. In a *bardo*, one is said to be blown about by "the winds of karma," winds
that circulate the patterns of action accumulated in one's life. (RM, 179)

The winds of *karma* compose the atmospherics of coexistence—the interconnected tunings, vibes, or energies of things. Consider these remarks from
the Tibetan Buddhist meditation master, Chögyam Trungpa Rinpoche:

> All the processes that take place in the universe are dependent on the
> environmental situation of karma. It is rather like the atmosphere that
> the planet requires in order to function, in order for things to grow.
> When we talk about the karmic situation, we are speaking about the
> sense of individual relationship to the given situation, whatever it is.
> Any given situation is bounded by cause and effect, dependent on
> some cause and effect . . . So, altogether when we discuss karma, we
> are discussing energy.[18]

The aim of religious practice is not to completely harmonize with one's
karmic situation. For Buddhists, all life involves some kind of suffering or unsatisfactory incompleteness (*dukkha*). Care or compassion (*karuna*) requires an
acceptance, not an erasure, of the existential turbulence in the karmic
atmosphere.

Unlike an opera singer's pitch, ethics can never be perfect, since the injunc-
tion to welcome strange strangers is impossible to fulfill: a welcomed stranger
is not a stranger. The imperfection of ethics is exacerbated by hyperobjects.

Impossible to handle, the Anthropocene has to be handled, nonetheless. *"The time of hyperobjects is a time of hypocrisy."* (H, 6) The choice to care about the planet is hypocritical since an individual cannot be hospitable to a stranger at that scale. Moreover, the cynical choice to give up and avoid handling planetary issues is still a choice. Global warming sticks to the cynic just as much as it does to everyone else. The cynic tries to avoid hypocrisy but ends up being hypocritical about hypocrisy. (H, 148) Cynical reason corresponds with a phase of ethical development that Hegel calls "the beautiful soul:" someone who seeks harmony while avoiding anything in the external world that would contaminate the purity of that ideal.[19] "Beautiful soul syndrome," as Morton terms it, names the position of the cynic in the Anthropocene: "Beautiful me over here, corrupt world over there." (H, 154) Much to the beautiful soul's chagrin, that view *is* the corruption it sees in the world.

Whereas the beautiful soul seeks unambiguous and harmonious decisions, *hyposubjects*, who accept the hypocritical conditions of decisions, seek a stranger mood, an attunement more fit for specters, a "floating of decision" in "spectral attunement." (HK, 82) Along with hypocrisy, this spectral attunement is characterized by weakness and lameness, *"weakness* from the gap between phenomenon and thing," and *"lameness* from the fact that all entities are fragile." (H, 2) The gap between a thing and its appearance is the space of attunement, which involves weakness insofar as nothing can make direct contact with anything. I cannot do anything directly to Earth's oceans, and yet everything I do indirectly impacts them. The fragility of things is their subscendence, whereby things are inconsistent, torn into pieces, failing to coincide with themselves, in the way that rain is and is not climate, and I am and am not *Homo sapiens*. (H, 195–96)

The weakness, lameness, and hypocrisy of spectral attunement means that everything I do for the sake of planetary community is inadequate, even self-defeating. That is why Morton calls their ecological thought *dark* ecology. That darkness is depressing, but fighting that depression only causes more problems. "Don't fight it," advises Morton; instead, try to "tunnel down" so that the darkness becomes more mysterious than depressing, and finally becomes "dark and sweet like chocolate." (DE, 117) Morton discusses a series of tunings that tunnel through dark ecology, beginning with guilt and shame. (DE, 131–35) For those who are cynical about their catastrophic complicity, feelings of guilt and shame arise. Accepting that complicity is depressing, leaving you feeling the unbearable imprints of all the beings enmeshed in this massive problem.

The more you accept the melancholy of this situation, the more horrifying it becomes. The horror starts to appear ridiculous after a while, as it becomes obvious that your attempt to find an anthropocentric escape from ecological horror *is* ecological horror. The attempt of humans to disengage from the ecological mesh is "the Severing." (HK, 13–18) It produces the anthropocentric attitudes that began operating in civilization during the Neolithic development of agriculture, were subsequently exacerbated by the hierarchical and transcendent agendas of religions in the Axial Age, and then further exacerbated by modern risk society and global capitalism. The Severing causes ecological problems to which humans respond by trying to sever their connections to ecological problems, thus producing more ecological problems and more severing, and so on unto mass extinction.

Moreover, Morton's diagnosis of the Severing in religions after the Axial Age does not amount to a totalizing rejection of religions. They practice (and have written about) Vajrayana Buddhism, and their notions of the mesh and strange stranger reflect the interdependence and emptiness of things. They are affirmative of "the VIP lounges of agricultural-age religions," which is to say, the esoteric practices and lineages that have maintained intimacy with non-humans. (HK, 26) They encourage us to talk, like Yoda does, about "the Force," an energy field not unlike the "animal magnetism" of Franz Anton Mesmer. (BE, 80) Morton is not against the essentialism of religions, only the metaphysics of presence that considers essence directly knowable. They avoid gnostic positions that fall for transcendence, as well as agnostic positions that fall for the cynical reasoning of much critical theory; they propose an "ecognosis:"

> Ecognosis is like knowing, but more like letting be known. It is something like coexisting. It is like becoming accustomed to something strange, yet it is also becoming accustomed to strangeness that doesn't become less strange through acclimation. Ecognosis is like a knowing that knows itself. Knowing in a loop—a weird knowing. (DE, 5)

Tendencies toward anthropocentrism, androcentrism (patriarchy), and transcendence in Axial Age religions are indicative of an aversion toward this loopy way of knowing. Morton coins a word for this kind of fear—Buddhaphobia.[20] When you realize that you are the ecological loop you are trying to escape, then tragedy starts feeling like comedy; the border separating humans from the mesh is shown to be ridiculous, and the horror of planetary destruction now appears as a more ambiguous or absurd strangeness. The laughter that comes with ridicule can give way to fascination. "Fascinated, I begin to laugh with nonhumans, rather than at them (horror, and ridicule), or at and with my fellow humans about them (shame and guilt)." (DE, 147) Fascination opens onto a deeper sadness, the sadness of beauty, the sadness of attuning to that which you can never grasp. Within that sadness is a longing, which subtends the "*basic anxiety*" manifest in guilt, shame, horror, and ridicule. Coined by Buddhist teacher Chögyam Trungpa Rinpoche, this "basic anxiety" relates to Heidegger's notion of anxiety (*Angst*) as an attunement that discloses the openness of human existence. (RM, 203)

"We have anxiety because we care." (DE, 152) The withdrawal of things is what makes it possible for humans to care about them, but it also renders care impossible. When I try deleting that anxiety, attempting to care in the perfectly right way, my attunement becomes something like guilt, shame, sadness, horror, or ridicule; but if I accept that anxiety, the ongoing frustration of longing can feel weirdly playful and fun. Care becomes a little careless or carefree. It is a *"playful care,"* "care with the care/less halo," like commitment without attachment to outcome—a spectral care that indicates, not a lack of seriousness, but a sort of open-heartedness or effortlessness, a *"playful seriousness"* open to the ambiguous exigencies of planetary thinking. (BE, 131) Can planetary ethics become something you enjoy? Living during a climate emergency and a global pandemic, such a question could sound flippant. While it is playful, it is deadly serious about promoting the kind of emotional resilience required for sustained engagement in care for others and for oneself.

Joy still involves pain, mourning, and unrequited longing, hence the darkness of this joy. Joy can energize ongoing engagements in the tediously local, yet massively distributed, tasks of coexisting peacefully in the Anthropocene. Where explosive holisms assimilate the plurality of objects, joy is a mood that lets itself be tuned by strangers. As nonhumans are always already tuning you, joy is always already happening, whether you pay attention or not. It is a "basic effervescence" accompanying the basic anxiety of coexistence. (DE, 155) Joy is what anxiety feels like when you let it well up and do not try to erase it. It feels like a hospitable atmosphere, like a strange solidarity with nonhumans, where "solidarity is the default affective environment of the top layers of Earth's crust." (HK, 14) Solidarity is the default atmosphere of the symbiotic real. Since you are always already symbiotic, finding solidarity is a challenge, not because it is lacking or too difficult, but because of the exorbitant excess of available connections, making it too easy. (BE, 157)

One area of human experience excels in inspiring joy amidst the anxiety of never-ending longing: consumerism. It gets boring to chase after new products and trends all the time, yet boredom with the treadmill of consumer pleasure is part of what people enjoy. This ennui is the main ingredient in consumerist experience, "stimulated by the boredom of being constantly stimulated." (HK, 65) Ennui is the atmosphere with which humans can attune to Earth's atmosphere. "Ennui is the correct ecological attunement!" (HK, 66) Consider the example of consumerist experience, which is coupled with the capitalist economy as a driving force of today's planetary crisis. Consumerism seems antithetical to planetary community, especially considering the beautiful-soul syndrome of scholars and activists who are disgusted by consumerism. But the ennui consumerism inspires is, strangely, the best mood for humans to find joy and solidarity in their responses to global, ecological issues. The problem is that *"capitalism is not spectral enough."* (HK, 63) Consumers are entranced by specters that take on lives of their own, like apps, clothes, cars, shoes, iPhones, and so many other products, but there are countless more specters, with more loops for more longing, like rivers, chimpanzees, the biosphere, the atmosphere, neighborhoods, compost heaps, and electrical grids. The slightly disgusted enjoyment of a bored consumer is careless and carefree, unburdened by cries that seek relief from anxiety, like the classic ethical cry, "What are we going to do?" (BE, xxvii) Looking for something to do to ensure a safe escape from ecological crisis *is* the karmic situation composing the crisis. *Karma* includes wise and compassionate action, such as the activity of the Buddha, as well as the action of one caught in *samsara*—the cycle of confusion and suffering. The difference is duality. In the samsaric condition, according to Trungpa, *karma* is "energy that moves from here to there and then bounces back," which is "the definition of duality." More specifically, it is "duality in the sense of the neurosis of dualistic fixation."[21] Enlightened energy undoes the dualistic fixation. This does not mean that we should try to escape from our tendency to escape. That would be more of the same problem—the same neurotic fixation.

A compassionate response to the planetary crisis does not mean that you have to worry yourself with obsessively questioning, "What should I do?" It means letting things be, letting yourself be in unbearably intimate relations with nonhumans, trusting the process of *karma* and letting samsaric energy mutate. It means trusting solidarity with nonhumans and accepting imperfection. The courage to be compassionate constitutes a mutation of the *karma* driving the climate crisis. Trungpa puts it simply, "Part of compassion is trust. If something positive is happening, you don't have to check up on it all the time. The more you check up, the more possibilities there are of interrupting the growth. It requires fearlessness to let things be."[22]

Notes

1. Graham Harman, *Immaterialism: Objects and Social Theory* (Cambridge: Polity Press, 2016).

2. Timothy Morton, *The Ecological Thought* (Cambridge: Harvard University Press, 2010), 15. [Abbreviated **ET** throughout this chapter]

3. Timothy Morton, *Being Ecological* (Cambridge: MIT Press, 2018), 70. [Abbreviated **BE** throughout this chapter]

4. Jacques Derrida, *Specters of Marx: The State of the Debt, the Work of Mourning, and the New International*, trans. Peggy Kamuf (New York: Routledge, 1994), 10, 161, 196.

5. Karen Barad, "Quantum Entanglements and Hauntological Relations of Inheritance: Dis/continuities, SpaceTime Enfoldings, and Justice-to-Come," *Derrida Today, Vol.* 3 (2) (2010): 240–68.

6. Martin Heidegger, *Being and Time*, trans. John Macquarrie and Edward Robinson (New York: Harper & Row, 1962), 99. [Abbreviated **BT** throughout this chapter]

7. A key difference between Harman and Morton is that the former draws little from Derrida and is highly critical of Derrida's anti-realism. Graham Harman, *Object-Oriented Ontology: A New Theory of Everything* (London: Pelican, 2018), 199.

8. Nicolas Malebranche, *The Search for Truth and Elucidations of the Search for Truth*, trans. Thomas M. Lennon and Paul J. Olscamp (Cambridge: Cambridge University Press, 1997), 448.

9. Timothy Morton, *Realist Magic: Objects, Ontology, Causality* (Ann Arbor: Open Humanities Press, 2013), 20. [Abbreviated **RM** throughout this chapter]

10. Nāgārjuna, *The Fundamental Wisdom of the Middle Way*, translated with commentary by Jay L. Garfield (Oxford: Oxford University Press, 1995), 28–30.

11. Harman, *Object-Oriented Ontology*, 54.

12. For Derrida, "deconstruction is hospitality to the other." Jacques Derrida, "Hostipitality," in *Acts of Religion*, ed. Gil Anidjar (New York: Routledge, 2002), 364.

13. Timothy Morton, *Humankind: Solidarity with Nonhuman People* (New York & London: Verso, 2017), 106. [Abbreviated **HK** throughout this chapter]

14. Terrence Deacon, *Incomplete Nature: How Mind Emerged from Matter* (New York: W.W. Norton, 2012), 43.

15. Jeremy Sherman, *Neither Ghost nor Machine: The Emergence and Nature of Selves* (Columbia: Columbia University Press, 2017), 107.

16. Timothy Morton, *Ecology without Nature: Rethinking Environmental Aesthetics* (Cambridge: Harvard University Press, 2007), and *Hyperobjects: Philosophy and Ecology after the End of the World* (Minneapolis: University of Minnesota Press, 2013), 92. [Abbreviated **H** throughout this chapter] Morton is not aiming for any transcendence of these terms. The point is not to take away anybody's favorite symbols. "I balk at saying *without* in the sense of 'utterly without' or 'beyond'"; a "formula like that tries to progress once and for all like the modernity of which it is sick." Timothy Morton, *Dark Ecology: Toward a Logic of Future Coexistence* (New York: Columbia University Press, 2016), 83. [Abbreviated **DE** throughout this chapter]

17. Immanuel Kant, *Critique of Judgment*, trans. Werner S. Pluhar (Indianapolis: Hackett, 1987), 445–46.

18. Chögyam Trungpa, *The Future Is Open: Good Karma, Bad Karma, and Beyond Karma* (Boulder: Shambhala Publications, 2018), 3.

19. Georg Wilhelm Friedrich Hegel, *Phenomenology of Spirit*, trans. A. V. Miller (Oxford: Oxford University Press, 1977), 406.

20. Timothy Morton, "Buddhaphobia: Nothingness and the Fear of Things," in *Nothing: Three Inquiries in Buddhism*, eds. Marcus Boon, Eric Cazdyn, and Timothy Morton (Chicago: University of Chicago Press, 2015), 187.

21. Trungpa, *The Future Is Open*, 3–4.

22. Chögyam Trungpa, *Mindfulness in Action: Making Friends with Yourself through Meditation and Everyday Awareness* (Boston: Shambhala Publications, 2015), 56.

Developing a Critical Romantic Religiosity for a Planetary Community

Whitney A. Bauman

There are many diverse sources for thinking "immanently" about our place within the rest of the natural world. They can be found within the "immanent" customs of what we Moderns think of as the world's religious traditions. Animisms, pantheisms, panentheisms, cosmovisions, and other ways of thinking ideas and matter onto a single plane, all beg the question of what is "new" about the so-called "new materialisms."[1] These "new" ways of thinking often lack intentional grounding in the ideas that precede them. Affect theory, new materialisms, process thought, political ecology/theology, neo-animisms, and emergence theories have surfaced, in part, to react against the methodological reductionism in the natural sciences, and the persistence of idealism and dualism in some religious and philosophical thought.[2] In this chapter, I draw from both new and old immanent ways of thinking in order to develop what I am calling a *Critical Planetary Romanticism* (CPR) for the Earth. In doing so, I argue that the productive and reductive model of science is an anomaly within the sciences that dominates only from approximately WWII until, roughly, the publication of Rachel Carson's *Silent Spring in 1962*. I follow the work of a German romantic scientist, Ernst Haeckel, to highlight the contested, polydox beginnings of the reductive and productive model of the natural sciences and the short (though not uncontested) reign of said methodology. Haeckel and other romantic scientists' thinking about the nature of nature ranged from idealisms, dualisms, and reductionisms to a type of pantheism. Specifically, I will explore the non-reductive, open triune Monism found in Haeckel's work. I argue that he employed a critical, romantic approach to "nature" because he understood that placing humans within an evolutionary and ecological perspective meant that science must involve hermeneutics. Humans no longer stand outside of nature, looking down from some sort of objective space, but are interpreting nature (including humans) from within. Sadly, Haeckel and many other romantic scientists of his era hitched romanticism to localism and nationalism and did not have a concept of the planetary. This was all prior to the "great acceleration" after WWII, which led to unimaginable globalization and regular engagement with multiple perspectives on reality and multiple ways of being embodied in the world. It was also prior to the process of climate change, which also dethrones the modern human from its perceived exceptional, managerial space above the rest of the natural world and helps (re)place Moderns within a planetary community. I argue that we can reread critical romanticism from this space of a globalized planet whose climate is changing, and rethink romanticism beyond any local or national boundaries, from within a planetary perspective. This is not a naïve romanticism or return to some sort of direct access to pure nature: we are not thinking like mountains nor are we the universe thinking itself. Rather, it is what we might call a CPR. Planetary, here, builds on the work of Gayatri Spivak, and suggests that we are first and foremost planetary creatures among other creatures.[3] *Critical*, following theories of the Frankfurt School variety, but also including queer theories, critical race theories, Marxist theories, affect theories, disability studies, and other critical hermeneutical theories which recognize that we have no direct access to reality (within or without, so to speak). *Romantic*, because unlike many critical theories, we ought always to include our entangled realities, embedded within and among many other types of planetary systems and creatures.

The development of a CPR for the Earth is also an attempt to escape the pre/post Modern and colonial, that keep us locked into a linear understanding of time that privileges "the West," and which has little-to-no regard for the rest of the planetary community, except in as much as it continues to fuel this Western "tunnel of time."[4] The breaking open of this tunnel of time and the "logocentrism" associated with it, in the end, offers up a religiosity beyond salvation and more akin to "attunement."[5] I conclude this chapter with a possible definition of "religiosity" from within a CPR; but first, in order to break open the tunnel of time and allow for planetary realities to flourish, I begin with a brief undoing of the modern mechanization of nature.

Incomplete Mechanization:
Making Newton's Cosmology Immanent

Early attempts at describing the laws of "biology" drew from Newtonian and other types of physics, which subsequently drew from theological metaphors: of a hierarchical order made up of distinct entities and elements created as such "in the beginning." Attempts to mechanize the cosmos, set in motion by an intelligent being in such a way that laws would ensure its continuation, were paralleled in the emerging sciences of geology and biology. Just as God was pushed further and further from questions of cosmology, so too would God be pushed further and further from intervening in an emerging picture of geology and biology that was increasingly mechanical. This worldview went hand-in-hand with the Industrial Revolution: it was a worldview that understood most of the non-human world (and some of the human world—slaves, Indigenous peoples, and laborers) in terms of what resources they might provide for the projects of human cultivation and civilization. By the eighteenth and nineteenth centuries, many scientists, influenced by romanticism and the problems that arose from industrialization, realized that this reductive and productive model of nature did not quite work, nor did the theological model of creation which many regarded as dualistic. There was not yet a naturalistic worldview that encompassed all the sciences into a single vision: and which methodology would be best for such a worldview was far from agreed upon. The German scientist, philosopher, and artist Ernst Haeckel was among the philosophers/scientists trying to articulate such a worldview and foundation for all the sciences (human and natural).

For Haeckel, a reductive model of the world could not be possible because it recapitulated the dualism assumed by dogmatic theology, in particular the separation of humans from the rest of the natural world. Descartes' "ghost in the machine" was one example, as was Kant's *ding an sich* and *a priori* knowledge—all of which Haeckel argued against.[6] These types of dualism were just repeating years of wrong-headed, dogmatic theology. Attempts to reduce everything to either material *or* ideals/spirit, did not (thought Haeckel) account for the reality of the world we experience.[7] For these reasons, Haeckel proposed a triune monism of energy, matter, and "sensing" or "feeling." These three elements were present in life "all the way down" and this meant that the entire world could not be reduced and divided in the sort of mechanistic, hierarchical way that the Industrial Revolution demanded.

Haeckel's Non-Reductive Naturalistic Worldview

In bringing physics, chemistry, biology, embryology, ecology, and cosmology together into a single scientific narrative, Haeckel also had to account for linguistics, anthropology, and eventually theology (or at least religiosity) as part of this same narrative. Whereas Newton had *largely* (though not intentionally or personally) done away with the need for God in understanding the heavens through his mechanization of the cosmos, and Darwin had *largely* (intellectually though not necessarily personally) done away with the need for God and soul with his "descent of man," Haeckel wanted to ensure that we could do away with God-talk when discussing meanings, values, truth, and beauty in the world. These were not properties of a soul endowed by God or any other ethereal realm, but emergent from the processes of evolution.

Whatever was meant by "soul" (and consciousness), it could not survive the process of death and new life, because individuals emerged through the process of reproduction within species, and from common ancestors over time. Haeckel argues:

> Our human body has been built up slowly and by degrees from a long series of vertebrate ancestors, and this is also true of our soul; as a function of our brain it has gradually been developed in reciprocal action and re-action with this its bodily organ. What we briefly designate as the "human soul," is only the sum of our feeling, willing, and thinking— the sum of those physiological functions whose elementary organs are constituted by the microscopic ganglion-cells of our brain.[8]

One of Haeckel's main critiques of the Christian projection was its anthropocentrism. Again, developments in physics, chemistry, embryology, zoology, ecology, botany, and cosmology would help to correct that "anthropomorphic" projection of a God in human form by showing that humans were evolutionary products of the same processes as all other life. However, for Haeckel and other romantic scientists the answer was not reductionism: if humans were a part of this process, then a reductive model would also reduce humans to mere resources. Instead, the answer was to understand the primary substance as triune, "all the way down." He writes:

> The monistic idea of God, which alone is compatible with our present knowledge of nature, recognises the divine spirit in all things. It can never recognise in God a "personal being," or, in other words, an individual of limited extension in space, or even of human form. God is everywhere. As Giordano Bruno has it: "There is one spirit in all things, and nobody is so small that it does not contain a part of the divine substance whereby it is animated." Every atom is thus animated.[9]

I would argue that, far from being a reductionist, Haeckel might, today, be grouped with emergent theorists. His understanding of monism had qualities of emergent newness and was not what we would call reductionist. I offer a different interpretation of his triune monism than the one found in Mary-Jane Rubenstein's (and William James's) work.[10] I contend that the structure of his monism is plural in its triune nature, and this resists a reductive monism. The idea that ecologies (including all things human) drive emergent evolutionary changes also means a plurality of things emerge from the evolutionary process.

Haeckel's appreciation of plurality is evident in his thousands of paintings and drawings of various "art forms" in nature, and in his insistence that education should be reformed so that students spend more time studying nature rather than texts. For Haeckel, the study of nature was about wonder rather than production. Rubenstein and others (e.g., Stephen Gould and Daniel Gasman) are correct in calling out his Eurocentric, racist, and antisemitic orderings of life (which, of course, places Germans at the top of the human hierarchy). I posit that this is partly a function of his turn toward nationalism after WWI, rather than a result of his monism per se. In other words, he hitches his romantic and evolutionary worldview to "the nation" rather than to an understanding of "the planetary." But more on this later. For now, I turn to a discussion of Haeckel's reading of Darwin.

Although Haeckel is often thought of (when not being equated with Nazi science) as merely a popularizer of Darwin's theory, he built on the science with his own study of embryology, geology, and zoology. For Haeckel, the environment was the biggest factor in determining how a specific individual developed. He understood recapitulation as taking place across all species: ontogeny recapitulates phylogeny.[11] This reasoning states that each individual organism, depending on where it is in the evolutionary tree, undergoes the developmental processes of all those species "below" it on the tree. Differences between, and within, organisms emerge through interactions with the ecosystems in which organisms find themselves:

> Haeckel also required that unpredictable environmental influences
> and other historical contingencies intervene continually and let the
> embryo break free from its past in order to vary and adapt in novel
> ways. It served his ideological purposes and polemics against religion
> to have nature create forms that one could not have predicted from
> knowledge of the Creation and the laws of development and heredity.[12]

For Haeckel, then, changes in organisms are primarily a result of interactions with the surrounding environment, and when these environments shift, different organisms will thrive. All human cultures could be explained by these geological/environmental differences. For Haeckel (among other monists and materialists at the time):

> The dynamism was not limited to the individual, for the individual
> was embedded in the larger systems of its genealogical lineage, its

species, and the collective of interacting species. Human society, too, was a dynamic system, moving along a developmental trajectory (*Entwicklungsbahn*).[13]

Haeckel's view of evolution, and the relationships between organisms that make up an ecological community, is similar to what Donna Haraway and others name as *sympoiesis* rather than merely *autopoiesis*.[14] Sympoiesis recognizes that individual organisms are a poor metric for understanding the variety of bodies and things in the world; rather organisms must be thought of in their coevolving lives together. This is just as true at the atomic level as it is at the biological and social level. Haeckel's understanding of "culture" also demands that we look at the ways in which cultures evolve with one another and with the rest of the natural world. While for him, such differences were ranked in dangerous and damaging ways, he nonetheless understood these multiple and mixed, evolving relationships. He thought that religion and philosophy, based upon anything other than the scientific observations of and sensory data gathered from the natural world, were less developed and needed reform. Unfortunately, as a scholar of his day, (as seen in many of his now infamous drawings of evolutionary trees) this also helped fuel depictions of European humans at the top of the evolutionary development *vis a vis* other peoples: ideas which helped fuel Nazi ideology in the early twentieth century.[15] His evolutionary theory was a bit of a Rorschach test for what people wanted to project onto nature: Marxists/Materialists, Fascists, Theists, and Idealists all rejected him for depicting nature as opposing their views in one way or another, and Haeckel rejected them (or would have in the case of the fascists as he was dead before their rise in power) for imposing ideology onto nature.

It is important to note that Haeckel's triune monism (energy, matter, and feeling) was a full-blown type of immanent, religious naturalism: he wanted to explain everything based upon his own understanding of nature through science. He believed that the sciences would, indeed, provide a better route to understanding everything—including philosophy, culture, psychology, and religion—more fully than anything the Earth had previously seen. However, included in his vision of the sciences were both the *natur-* and *geisteswissen-schaften:* the human, social, and natural sciences were all on a single plane, but had to make sense according to our sensory observations of the natural world, rather than our ideas about ultimate, unseen reality. He was concerned with how the sciences could aesthetically and reasonably provide us with a better picture of the rest of the natural world than any religious idea of reality:

> Monistic investigation of nature as knowledge of the true, monistic
> ethic as training for the good, monistic aesthetic as pursuit of the

beautiful—these are the three great departments of our monism: by
the harmonious and consistent cultivation of these we effect at last
the truly beatific union of religion and science, so painfully longed
after by so many today. The True, the Beautiful, and the Good, these
are the three august Divine Ones before which we bow the knee in
adoration.[16]

Through his travels, Haeckel's monism led him to a huge appreciation of the
immense diversity of the natural world, as seen in all of his nature drawings.
He was mesmerized by the sheer diversity of the world and the way that everything fit together. Because his evolutionary views promoted diversity in all life,
he also saw this in the human world. He even supported Magnus Hirschfield's
Institute for the Study of Sexuality in Berlin during the late nineteenth and
early twentieth centuries.[17]

Haeckel tried to make a meaningful whole out of the world he observed
and experienced, using all the best science of the time, and these experiences
were limited by his own cultural/historical location. In this way, Haeckel
provides us with an early version of religious naturalism that thinks with and
within the natural world, and not with abstractions from the world. However,
we ought to be aware of the ways in which he read his culturally located
chauvinism and racism back into the natural world as "natural." This ought
to be a reminder to always think *critically* about what we are reading into
"nature" as good, bad, progress, better, worse, etc.[18] Hence the need for a
critical planetary romanticism.

The Industrial Revolution, the Romantics, and WWII

The Industrial Revolution had, by Haeckel's time, co-opted many of the natural
sciences into a productionist, reductionist model, and it was Haeckel's view
that the true goal of the natural sciences was more about instilling a sense of
wonder about the world and our place therein than it was about technology
transfer.[19] Matter was not, for Haeckel, just dead stuff that was somehow fodder
for human ends, but was a living, evolving entanglement that included human
beings and human cultures.

The sciences had not yet fully settled on a reductive and productive method
as the universal basis for the natural sciences, even as WWII began. Wilhelm
Bölche, Gustav Fechner, Haeckel, and other German scientists had been articulating foundations that were non-reductive. The German romantic scientists
clashed with the British empiricists, who were largely reductive and productive
in their approach to the sciences. During WWII, the race to create the atomic

bomb and develop other weapons (chemical, biological, and mechanical) in order to win the war, led to a whole generation of scientists—chemists, biologists, and physicists especially—being sucked into a reductive and productive model of the sciences.[20] Of course, not all sciences and not all scientists were swept into the reductive/productive model—non-reductive models have always persisted,[21] but what I am arguing is that the reductive and productive model became the dominant model during and after WWII. The technology transfer of the sciences during that period was so successful, in fact, that after the war these methods were applied to industrializing agriculture around the world with the "green revolution," and toward improving communication, transportation, and production technologies in general. This reductive and productive model was not widely challenged from within the sciences again until Rachel Carson published *Silent Spring*. She, and others, helped to point out the problem with the reductive and productive models of science. We can now critically revisit German romantic scientists as resources for a new basis of the human and natural sciences. Although Haeckel's thinking became nationalistic and his romanticism and interpretation of evolution were tied together in some very dark ways, perhaps we can pick up a variation of his non-reductive materialism, a type of new materialism, by couching critical theory and re-attunements to nature within a planetary perspective?

Toward a Critical Planetary Romanticism (CPR) for the Earth

I want to start reflecting on what a CPR for the Earth might look like, through a definition of what "religiosity" might mean if humans and all things human (ideas, reason, thought, religions, technologies, etc.) are part of the ongoing evolutionary process, and if we humans are but one of many agents/actants in the planetary community.[22] A CPR depends upon "religiosity" because, unlike religion, which is too often associated with a noun or a thing, religiosity, like curiosity, is an orientation toward experiencing the world. It draws attention to the moment-by-moment process of an evolving planetary community and our need to "keep up" with those changes (as a change in one is a change in all). I devote the rest of this chapter to an immanent understanding of religiosity that encompasses the re-attunement of the evolving contexts in which we live in order to cocreate worlds that listen to, and care deeply for, the multiple matterings that make up the planetary community:

> Considering its constitutive attunement to interconstitution, pantheism
> in its pluralistic and even monistic forms combats the willful ignorance
> and ethical quietism of our astonishingly undead political theisms.[23]

The concept of attunement can be found in many of the essays in this volume that deal with immanence in "world religions," especially the more meditative and contemplative activities of these traditions. In her essay, "The Animist, Almost Feminist, Quite Nearly Pantheist Old Materialism of Giordano Bruno," and in her recent book, *Pantheologies: Gods, Worlds, and Monsters*, Mary-Jane Rubenstein talks about the need for attuning. This attunement is also articulated by Bruno Latour's understanding of religion which deals with the "close up and near,"[24] and by Sam Mickey who speaks of the long history of *stimmung* and the attunement between evolving objects in his chapter in this volume, "Solidarity with Nonhumans: Being Ecological with Object-Oriented Ontology."[25] It can be found in Kevin Minister's understanding that religious ways of being are "constituted through conversions—patterns of turning in space that give the sense that the community is there."[26] (see "Interreligious Approaches to Sustainability Without a Future: Two New Materialist Proposals for Religion and Ecology.") Karen Bray calls for such attunement and "staying with the trouble" in her essay "Gut Theology: The Peril and Promise of Political Affect," and in her most recent book, *Grave Attending*.

All of these authors point to the idea of religiosity (reinterpreting the worlds in which we live and binding them back together) as a form of re-attunement. Precisely because we are living in a context that is immanent, multiple, and agential, all our desires, ideas, and emotions are wrapped up in our evolving planetary contexts. This means that there is never an arrival at a final point in which we have a smooth space of objectivity from where we can view the whole thing. This would be stasis, or equilibrium, which are both tantamount to death from within a biological, ecological, and even cosmological perspective. If our planetary contexts are always changing, our relationships (internal and external) are always changing, then we must "re-attune" to these ever-emergent contexts. To re-attune in this context is to be in better touch with these changes and how they affect bodies differently. It is to recognize that human ideas and concepts can never fully capture the newly emerging ways of being/becoming in the world. In the end, it suggests that religiosity is a form of knowing that is more about attunement to ethics and aesthetics for a planetary community. In what ways do our current orderings affect different "earth bodies"[27] differently and what types of worlds do we want to help cocreate that consider the needs of all these different Earth bodies?

Many extant religious traditions started because of this type of attunement (or what John Cobb calls a "secularizing moment").[28] Jesus and the Roman Empire, Moses and Exodus, Muhammad and the Quraysh, and Buddha and Lord Mahavir vis-à-vis the Vedic background in India: these are all examples of re-attuning in ways that pay deep attention to how world orderings affect bodies differently and in particular, call attention to injustices. Other religiosities understand that re-attunement is necessary from time to time because the worlds in which we live become ossified into human concepts of the ways we think reality is. Shamanic experiences, meditation, yoga, and other practices help us to deconstruct our worlds so that we might re-attune to our connectivity with planetary others.

Other peoples and places have practiced this re-attunement because of apocalypses brought on by colonization (and now climate change); such as in Indonesia, India, African countries, and Latin American countries. In many areas of Indonesia, for instance, a type of hybrid religiosity is practiced that combines local traditions (Javanese, Balinese, etc.), waves of colonization (Indic, Muslim, and Christian), and the unique ecology and landscapes of the given island. In Indian, African, Afro-Caribbean, and Central and South American traditions, one can also find such hybrid religiosities. Each time there is an apocalypse (slavery, direct colonization, economic colonization, and intensive resource extraction that goes along with these economic and political systems of domination), people find ways to live through the end of the world by a re-attunement that combines the old and new worlds in which they are living.[29] Post-colonial and decolonial scholarship, queer theories, critical race theories, affect theories, feminist theories, and many other critical types of theories all help us to re-attune to new worlds and to how these worlds are oppressive in different ways to different embodiments.

Finally, in places like China, the former USSR nations, and South America, one cannot ignore the Marxian re-attunements: think of liberation theologies and the pedagogy of the oppressed. I am reminded of Joerg Rieger's call for religion to resist capitalism's efforts to gloss over the contributions and costs of labor, and (following Asasi-Díaz) to focus on the everyday: a re-attunement to the realities of class struggles and the struggles of daily life.[30] Post-Marxist scholarship suggests that Marxism needs significant re-attuning to recognize that we are in a planetary setting (not just human) and that globalization and climate change mean that we cannot be isolationists. Re-attuning means to take all of these theories, coming out of various oppressions, seriously. Such an intersectional approach will help us pay attention to the various worlds and how these worlds are affecting bodies differently.

Similar to Kevin Minister's call, to "orient the study of religion to a present sustainability without hope for a future,"[31] this understanding of religiosity understands time as an ever-expanding and evolving present. In other places, I have argued (along with Kevin O'Brien) that thinking about the future from the present will only project more of the same onto the future.[32] Instead, a hope-filled way of understanding the future is necessarily an agnostic one. It is not yet there, so we can only hope forward by looking back (within the planetary matrix) and thinking, feeling, and acting toward newly emergent possibilities that result from re-attuning. All our ideas and actions will have unintended consequences, but a critical romanticism of constant re-attunement will help ground us in the evolving, planetary community in ways that keep us firmly within the multiple worlds that make up the planetary at any given time. This is why retrieval of alternative narratives from within the modern West and the lifting up of voices from outside the modern West, in the ways this volume attempts to do, is so important. Why look at Haeckel, Bruno, Cusa, or immanent thinking within extant "religious traditions" if not to disrupt and challenge colonizing narratives?

It is important to note here, following Walter Mignolo and his analysis of the Zapatista movement, that worlds must remain plural.[33] The point is not to create a common world, as that always ends in the colonial "tunnel of time." A tunnel of time which is exacerbated by the petrocultures that have fueled "globalatinization," which Terra Rowe addresses in her contribution to this volume, "Oily Animations: On Protestantism and Petroleum." A fossil-fueled tunnel of time prevents us from attuning to the realities of the present as we are propelled into a future of "more of the same." What if, instead of this fast-paced, linear time, which is literally outstripping the capacities of the planet, and which has created mass economic inequities and environmental and social injustices: we (from whatever worlds we find ourselves in) imagined ourselves on the periphery of the expanding planetary community (which at any time is made up of multiple worlds) looking back toward and listening to the various "earth bodies" (from within our own worlds) in the hope that by attuning to this present—pregnant with multiple perspectives and multiple possibilities—we might release new possibilities for connections and, thus, new possibilities for becoming—as planetary creatures among other creatures?

> Immanentist modalities of ontological relationalism, ethical animacy, and pluralist cosmopolitics here undo any disembodied transcendence. Instead, there unfolds an intercarnational rhizome pressing beyond the capitalist/climate apocalypse toward a new atmosphere and earth.[34]

Now, at the end of this chapter, I fear that I have taken David Byrne's command to "Stop Making Sense" a bit too seriously. But, to some degree, we need to practice the unknowing that Catherine Keller has written so much about.[35] We need to also practice the *Queer Art of Failure*, as Halberstam has suggested, (and I embellish). In a system which is fundamentally unjust and which is destroying planetary systems, we must fail.[36] We need a new Monkey Wrench gang and perhaps a new form of Dada-ism, in order to articulate these "new materialisms" that: see humans as actants among other actants (Latour); see all matter as agential and alive (animisms, neo-animisms, pantheisms, panentheisms); understand all of life (ideal and material, cultural, and natural) as on a single plane of existence (Deleuze and Guattari); seek to attune to the sympoietic emergence of "monstrous" embodiments (Haraway); and attempt to welcome the planetary "strange strangers" (Morton), in an effort to coconstruct worlds that are more attuned to the evolving needs, desires, hopes, and dreams of these multiple Earth bodies and systems.

This is one possibility for a CPR for the Earth. An ongoing process of re-attuning is needed, and it is not just re-attuning to human voices but to the "voices" of the rest of the natural world. As Elana Jefferson-Tatum suggests in her work on Afro-Caribbean traditions, we are possessed by the gods that we cocreate. Quoting an Ewe Gorovodu priest, she writes: "We Ewe are not like Christians, who are created by their gods. We Ewe create our gods, and we create only the gods that we want to possess us, not any others."[37] From this perspective we might say that we have created the gods of globalization and neo-liberal capitalism (see Matthew Hartman's essay in this volume), or the gods of fossil fuel (see Terra Schwerin Rowe's essay in this volume) or maybe both; and climate weirding is those gods coming back to possess us, to tell us something.[38] Fires in California, bleaching of reefs, salinization of ground water, droughts, rising temperatures, melting glaciers, and mass extinctions, are the Earth's way of talking to us and telling us to exercise these gods of planetary destruction. These are planetary citizens and systems telling us that we must re-attune to our creaturely, embodied, contexts.

These re-attunements will also require something like "planetary public spaces" in order to make the connections between worlds that make up the planetary community at any given time. We need these spaces for articulating what types of worlds we want to live in and to assess how those worlds affect others within the planetary community, in better and worse ways. If there is one thing that the academics in this collection can do, especially those of us who are fortunate enough to be tenured, it is to work across disciplines and across institutions, across the university/public divide in order to be midwives for these "planetary public spaces."

Notes

1. Mary Evelyn Tucker, "Confucianism as a Form of Immanental Naturalism," in *Earthly Things: Immanence, New Materialisms, and Planetary Thinking*, eds. Karen Bray, Heather Eaton, and Whitney Bauman (New York: Fordham University Press, 2023); Graham Harvey, "We Have Always Been Animists . . . ," in *Earthly*

Things: Immanence, New Materialisms, and Planetary Thinking, eds. Karen Bray, Heather Eaton, and Whitney Bauman (New York: Fordham University Press, 2023); and John Grim, "Indigenous Cosmovisions and a Human Perspective on Materialism," in *Earthly Things: Immanence, New Materialisms, and Planetary Thinking*, eds. Karen Bray, Heather Eaton, and Whitney Bauman (New York: Fordham University Press, 2023).

2. Karen Bray, "Gut Theology: The Peril and Promise of Political Affect," in *Earthly Things: Immanence, New Materialisms, and Planetary Thinking*, eds. Karen Bray, Heather Eaton, and Whitney Bauman (New York: Fordham University Press, 2023) and Kevin Shilbrack, "Emergence Theory and the New Materialisms," in *Earthly Things: Immanence, New Materialisms, and Planetary Thinking*, eds. Karen Bray, Heather Eaton, and Whitney Bauman (New York: Fordham University Press, 2023).

3. Gayatri Spivak, *Death of A Discipline* (New York: Columbia University Press, 2003), 71ff.

4. Walter Mignolo, *The Darker Side of Western Modernity: Global Futures, Decolonial Options* (Durham: Duke University Press, 2011) and Teresa Brennan, *Globalization and Its Terrors: Daily Life in the West* (New York: Routledge, 2003).

5. Mary-Jane Rubenstein, "The Animist, Almost Feminist, Quite Nearly Pantheist Old Materialism of Giordano Bruno." in *Earthly Things: Immanence, New Materialisms, and Planetary Thinking*, eds. Karen Bray, Heather Eaton, and Whitney Bauman (New York: Fordham University Press, 2023).

6. Ernst Haeckel, *The Wonders of Life: A Popular Study of Biological Perspective* (New York: HarperCollins, 1905), 10–11.

7. Elizabeth Grosz provides a historical analysis of Western attempts to think beyond the ideal-material dualism, within immanent frameworks in *The Incorporeal: Ontology, Ethics, and the Limits of Materialism* (New York, Columbia University Press, 2017).

8. Ernst Haeckel, *Monism as Connecting Religion and Science: The Confession of Faith of a Man of Science* (London: Adam and Charles Black, 1895), 15.

9. Ibid., 18–19. Haeckel draws upon Bruno as he does in many places. This links up nicely with some of the work Mary-Jane Rubenstein has done on Bruno, and Catherine Keller has done with Cusa.

10. Mary-Jane Rubenstein, *Pantheologies: Gods, Worlds, and Monsters* (New York: Columbia University Press, 2018), 22–33.

11. Robert J. Richards describes Haeckel's biogenetic law well: "A chief feature of Haeckel's evolutionary doctrine that supposedly distinguishes his views from those of Darwin is the principle of recapitulation. Haeckel put the principle thusly: The organic individual . . . repeats during the quick and short course of its individual development the most important of those changes in form that its ancestors had gone through during the slow and long course of their paleontological development according to the laws of inheritance and adaptation," in *The Tragic Sense of Life: Ernst Haeckel and the Struggle over Evolutionary Thought* (Chicago: University of Chicago Press, 2008), 148.

12. Sander Gilboff, *HG Bronn, Ernst Haeckel, and the Origins of German Darwinism* (Cambridge: MIT Press, 2008), 23.

13. Ibid., 40.

14. Donna Haraway, *Staying with the Trouble: Making Kin the Chthulucene* (Durham: Duke University Press, 2016), 58.

15. Richards, *Tragic Sense of Life*, 489–512. At the same time, Haeckel's writings were eventually banned early on by the Nazi regime because many believed they had "Marxist" implications and normalized diversity a bit too much.

16. Haeckel, *Monism as Connecting Religion and Science*, FN 20.

17. See, e.g.: "The feminist Helene Stocker found Haeckel's viewpoint attractive for providing a secular, scientific approach to ethics that denied original sin and could counter conservative Christian objections to women's emancipation. For the sexual reformer Magnus Hirschfeld, Haeckel's authority brought human love and sex under the purview of science and made homosexuality a biological condition instead of a form of moral depravity." Gilboff, *H.G. Bronn*, 19.

18. On a good day, Haeckel would warn against assuming that we had the capacity to interpret "bare facts" in nature: "No science of any kind whatever consists solely in the description of observed facts," in Ernst Haeckel, *The Wonders of Life*, 5–6.

19. Ibid.

20. David Kaiser, *How the Hippies Saved Physics: Science, Counterculture, and the Quantum Revival* (New York, W.W. Norton, 2011).

21. Kate Rigby, *Reclaiming Romanticism: Towards an Ecopoetics of Decolonization* (London: Bloomsbury, 2021) and Kocku von Stuckrad, *A Cultural History of the Soul: Europe and North America from 1870 to the Present* (New York: Columbia University Press, 2022).

22. Whitney A. Bauman, "Critical Planetary Romanticism: Ecology, Evolution and Erotic Thinking," *Journal for the Study of Religion, Nature, and Culture*, Vol. 15 (1) (2021): 33–52.

23. Rubenstein, *Pantheologies*, 175.

24. Bruno Latour, "Thou Shalt Not Freeze-Frame" in *Science, Religion, and the Human Experience*, ed. James Proctor (Oxford: Oxford University Press, 2005).

25. Sam Mickey, "Solidarity with Nonhumans: Being Ecological with Object-Oriented Ontology," in *Earthly Things: Immanence, New Materialisms, and Planetary Thinking*, eds. Karen Bray, Heather Eaton, and Whitney Bauman (New York: Fordham University Press, 2023).

26. Kevin Minister, "Interreligious Approaches to Sustainability Without a Future: Two New Materialist Proposals for Religion and Ecology." in *Earthly Things: Immanence, New Materialisms, and Planetary Thinking*, eds. Karen Bray, Heather Eaton, and Whitney Bauman (New York: Fordham University Press, 2023).

27. Glen Mazis, *Earth Bodies: Rediscovering our Planetary Senses* (Albany: SUNY Press, 2002).

28. John Cobb, *Spiritual Bankruptcy: A Prophetic Call to Action* (Nashville: Abingdon, 2010).

29. Kyle Powys White, "Indigenous Science (fiction) for the Anthropocene: Ancestral Dystopias and Fantasies of Climate Change Crises" *Environment and Planning: Nature and Space*, Vols. 1–2 (2018): 224–242.

30. Jorge Rieger, "Which Materialism, Whose Planetary Thinking?" in *Earthly Things: Immanence, New Materialisms, and Planetary Thinking*, eds. Karen Bray, Heather Eaton, and Whitney Bauman (New York: Fordham University Press, 2023).

31. Minister, "Interreligious Approaches."

32. Whitney A. Bauman and Kevin O'Brien, *Environmental Ethics and Uncertainty: Wrestling with Wicked Problems* (New York: Routledge, 2019).

33. Mignolo, *The Darker Side of Modernity*.

34. Catherine Keller, "Amorous Entanglements: The Matter of Christian Panentheism," in *Earthly Things: Immanence, New Materialisms, and Planetary Thinking*, eds. Karen Bray, Heather Eaton, and Whitney Bauman (New York: Fordham University Press, 2023).

35. Catherine Keller, *The Face of the Deep: A Theology of Becoming* (New York: Routledge, 2003) and *Cloud of the Impossible: Negative Theology and Planetary Entanglement* (New York: Columbia University Press, 2015).

36. Jack Halberstam, *Queer Art of Failure* (Durham: Duke University Press, 2011).

37. Elana Jefferson-Tatum, "Africana Sacred Matters: Religious Materialities in Africa, the Caribbean, and the Americas," in *Earthly Things: Immanence, New Materialisms, and Planetary Thinking*, eds. Karen Bray, Heather Eaton, and Whitney Bauman (New York: Fordham University Press, 2023).

38. Mary Keller, in her work with Afro-Caribbean Indigenous traditions has also written about Climate Change as possession. We are not in control but are possessed by these collective spirits (or gods) we have cocreated over time. See, Mary Keller, "The Spirit of Climate Change," conference paper given at the meeting of the International Society for the Study of Religion, Nature, and Culture (Cork, Ireland: Cork University, June 14, 2019).

Matter Values: Ethics and Politics for a Planet in Crisis

Philip Clayton

Changing Actions, Changing Metaphysics

For humanity to face the most devastating crisis in the history of our species, planetary thinking must be returned to its rightful place. It names our home ground, whether the topic is science, religion, value, ethics, or politics. Planetarity is the heart of "the new possible."[1]

Having spent some decades constructing theories in order to derive ethical conclusions from them, I am pleased now to join these coauthors in turning that approach on its head. Some will label our approach an ethic in search of a metaphysic. More accurately, it is a planetary ethic mating with the many different forms of language that support it: narratives, metaphors, myths, heuristics, biographies, and evocative analogies. Naturally, some will desire, instead, a theoretical framework that will "ground" a planetary ethic and politics, in good old academic style. After all, the belief that one is standing on solid ground tends to make people stay awhile, and staying awhile is necessary for sustained and effective activism. But, together with many of my coauthors, I deny that such metaphysical grounding is necessary for coherence, much less for action. That is fortunate, since solid grounds are hard to come by these days—especially non-polluted grounds.

New Materialism is an odd partnership for religion; it unsettles comfortable dichotomies both left and right.[2] Although Crockett and Robbins still use the term "theology,"[3] I have chosen not to make it part of this closing planetary appeal. It is also important to question some core assumptions of the "religion and science" world, in which I have lived for several decades, in the interest of bonding science and value in a rather different way. By proceeding without metaphysical grounding, my thought experiment becomes "unjustified," as one says—except by the parched ground, dead soil, and undrinkable water that grounds radical action, philosophical grounding be damned.

Another way to set out on this journey is to practice the "natural piety" of allowing the material world to unveil itself on its own terms, without requiring a full translation (reduction) of the phenomena to scientific or philosophical laws. Authors in this collection have described a vast array of cultures, religions, and philosophies without subjecting them to master narratives, allowing the foreignness of the phenomena to manifest without domestication. One feature of the foregoing chapters that I admire is their weaving together of otherwise incompatible discourses.

New Materialism and Mainline Religion-and-Science

What happens when one contrasts New Materialism with the mainline discussion of Religion and Science (R&S) over the last fifty years? Is this volume a new chapter in R&S, or a new book altogether?

Looking back, one can identify five phases of the R&S discussion, divided (roughly) into decades:

> 1960s: *The collapse of positivism and the rediscovery of the questions.*
> The 1950s were still dominated by positivist thinkers in the philosophy of science such as Hans Reichenbach and Carl Hempel. But by the mid-1960s, a series of challenges to the traditional model (Stephen Toulmin, N.R. Hanson, and W.V.O. Quine), popularized in Thomas Kuhn's *Structure of Scientific Revolutions* (1962), had begun to change the paradigm. Changes in the understanding of science opened the door to a reengagement of religion and science, and an explosion of books in this genre began to appear.
>
> 1970s: *The search for a method.* Ian Barbour and other scholars began to outline the parameters for the disciplined study of the relations between the sciences and religion. It is not that early works in this genre—Ian Barbour's *Myths Models and Paradigms*; T.F. Torrance's *Theological Science*; Wolfhart Pannenberg's *Theology and the Philosophy of Science*; and soon thereafter the early works of Arthur Peacocke and John Polkinghorne—were all in agreement. But each of these authors shared the drive to establish a disciplined mode of scholarship in this field.
>
> Late '80s and early '90s: *Why (and how) it is rational.* By this point, the early methodological proposals had given rise to a series of works on

methodology, epistemology, and rationality. Nancey Murphy and I offered competing models for the emerging field, historians of religion and science provided the first detailed expositions of the history of the R&S debate, and numerous scholars such as Wentzel van Huyssteen advanced "critical realism" as a framework for the field as a whole.

1990s: *Templeton funding and the rapid expansion of the field.* A decade of ambitious and well-funded projects followed. The Center for Theology and the Natural Sciences hosted almost $30 million of projects, including the CTNS/Vatican Observatory series of conferences on divine action, Science and the Spiritual Quest, and the Science and Religion Course Program. Other centers sprang up with similarly ambitious projects. Scientists and theologians began to appear publicly within major science departments; courses were established at hundreds of universities; dozens of conferences took place each year; and the discussions began to spread beyond Christian theology and beyond the English-speaking nations.

By *September 11th, 2001: Entrenchment.* The debate between Intelligent Design and the New Atheists had commenced by the end of the 1990s, but it exploded into public attention with the collapse of the Twin Towers. Scientists felt that the very possibility of science was under attack from the religious right, and an increasing number of religious people became convinced that science was out to destroy their faith. With the major cultural shift after 9/11, the period of sustained support for bold exercises in bridging across the science/religion divide began to decline. It is most accurate to construe the decade that followed primarily in terms of warring factions, combined with a sense of incommensurability between facts, values, and religions.[4]

During most of these fifty years of academic R&S in the West, the primary focus was on Christianity, with religious pluralism only entering the discussion more recently. The failure of mainline R&S to achieve a productive pluralism offers several important lessons.

(1) It turns out to be hard to pluralize academic fields when they are defined by just two opposing terms, such as "religion" and "science." Failing to overcome the dualisms, scholars obscure their failures by hacking the big terms into ever tinier pieces, fighting ever more vehemently for smaller and smaller victories. Consider the history of R&S: as "religion" becomes problematic, advocates focus on theism. If there are distinctively theistic insights into R&S,

then clearly there are also specifically Christian answers; if Christian, then also evangelical; if evangelical, then also Baptist—you get the point. On the other hand, if there is science, then there is biology . . . neurology . . . cognitive science . . . the cognitive science of religion (CSR) . . . *eliminativist* CSR . . . the Pascal Boyer program of eliminativist CSR inspired by the New Atheism . . . and so forth.

(2) The multiplication of special interest groups gradually hardened into oppositions. As the camps became more and more narrow, the rhetoric grew harsher. To return to the previous example, Baptists on the one side and New Atheist eliminativists on the other, each began to develop more vehement apologetics for their own views. The biocultural, anti-theist program of LeRon Shults[5] pitted itself against evangelical divine action theories, and they against his atheism. What were once dialogues tended to devolve into combatants behind their fortifications, lobbing shells at one another. Many have become disillusioned with dualistic formulations—as we have seen repeatedly in the earlier chapters of this book—and are looking for new mediating categories, such as religious naturalism or process-relational thought. Others are turning to traditions other than classical Christian theology, finding new openings for connection (see section one above), or are calling for completely "rebooting science and religion."[6]

New Materialism (NM) has much to learn from this brief history. If we now *also* set up dichotomies and dualisms—new vs. old materialism, NM vs. pan-psychism, immanent religiosities vs. transcendent religions—they will bear similar fruits. If academic language and its cascading, ever more abstract distinctions, and "-isms" become NM's motor, they are certain to drive us away from activists . . . and them from us. Academic Marxism, preoccupied with its internecine battles, lost touch with factory workers and the unemployed (tragically, in my view); will we do the same? NM, too, faces temptations to police its boundaries, enforcing pure immanence, challenging any transcending language, or keeping itself narrowly and highly academic.

But, as the introduction to this volume shows, there is hope. To eschew dichotomies, and to celebrate values all the way down, is built into NM's very genes. As Whitney Bauman writes, religion and science *can* be a way to bring together ideas and matter.[7] Note that Jane Bennett's "vibrant matter" breaches the realm of the religious, while immanent religiosities are pervasively material, embodied fully and without remnant.

Fortunately, "value-free" is no longer an option. Climate change produces eco-injustices that grow more pressing daily. Black and Brown bodies suffer from White fumes. The rich consume carbon at vastly higher rates than the poor; yet it is only the rich who can afford to protect themselves from scorching summers and freezing winters, from floods and famines, and from COVID-19 and the economic devastation that it has wrought. The need to be advocates for the Earth and its ecosystems, against the forces that are destroying her, has become a grounding and all-encompassing value.

Emergent Complexity

Emergent Complexity is the scientific and philosophical program that seeks to understand complex organisms, their mental states, and their cultural products, in terms of the evolutionary increase in complexity.[8]

The Emergent Complexity (EC) project fits smoothly within the New Materialism research program because it also brackets out metaphysical questions. It does not need to decide between an ultimate theism and an ultimate atheism, panpsychism, or anti-psychism. Like any scientific inquiry, it represents a more focused and empirical research program.

Note that the EC project contrasts sharply with old-style materialism.[9] Traditional materialists made the (metaphysical) claim that material particles, and the compound bodies constructed out of them, are all that exist; all other types of reality are eliminated. Some wondered whether gravity could even qualify as a force; they called it "spooky action at a distance," since it was so different from the billiard-ball type of interaction that Newton's three laws describe. And if gravity was suspect, consciousness far more so.

As a further contrast to old-style materialism, NM does not require one to be value-free. Not to be distracted by intractable (and epistemically problematic) metaphysical claims on the one hand, and to be able to develop powerful ethical and political critiques and responses on the other, puts us in the place we need to be if we are to face the environmental and social crises that are engulfing our planet.

Finally, classic materialism demanded what NM does not: full reduction to lower-level laws. Its failure to comprehend cultural artifacts on their own terms eventually spelled its demise. Since cultural artifacts and values are emergent natural realities, they fail the law-based requirement, which thus relegates them to the status of mere constructs and illusions. Personhood suffered a similar fate. For NM, by contrast, the "integrated physicality"[10] of persons is an emergent reality. But not for the old-style materialist Dennett, following Nietzsche:

> "Body am I, and soul"—thus speaks the child. And why should one
> not speak like children? But the awakened and knowing say: body am
> I entirely, and nothing else; and soul is only a word for something

about the body. The body is a great reason, a plurality with one sense, a war and a peace, a herd and a shepherd. An instrument of your body is also your little reason, my brother, which you call "spirit"—a little instrument and toy of your great reason . . . Behind your thought and feelings, my brother, there stands a mighty ruler, an unknown sage— whose name is self. In your body he dwells; he is your body. There is more reason in your body than in your best wisdom.[11]

One rejection begets another. Dennett's black-and-white negations give rise to equal and opposite reactions, a dynamic we also saw between the New Atheism and Intelligent Design. The dualists line up against Dennett, defending strong dualism, fighting for pure spirit, and reintroducing God as foundation. Dennett and his allies shoot them down with equal glee.

Emergent Complexity, by contrast, does not perpetuate the ultimate claims of either side. It is a research program aimed at understanding natural emergence as far as it will take us. When one runs out of material emergents (not likely prior to the heat death of the universe!), EC as a research program ends, and folks can turn their attention to other fields. In the meantime, EC supports continuing attempts to integrate understanding across all natural systems, cultural artifacts, and emerging ideas.

Admittedly, I am envisioning an expansive NM here, one that includes panpsychists and theologians of immanence along with the usual allies. If our attempt to sidestep the well-known stalemates is successful, attention can shift to the study of natural systems as valuable from the start, recognizing that humans are as embodied as any other part of nature, and no less when we are being religious. I doubt that NM is a permanent resting place (what theoretical discourse is?), but for the moment it is a place from which to understand, criticize, and reconstruct—in ways that more traditional religion-and-science discourse has not yet accomplished.[12]

Toward an Ethics of Matter

Emergent Complexity goes hand in hand with New Materialism in the exploration of living matter. Driven by an ethical (and ultimately a political) quest, we recognize values across the biosphere, understood as a single, interconnected system of value and valuing.

"Matter values" in two senses: living things are material agents acting in webs of value, and the resulting material systems matter for the fate of the planet. Emergent New Materialism gives rise to a planetary ethic, and that ethic in turn gives rise to a politics for a planet in crisis. As Holmes Rolston writes:

Value is a frequently encountered term in evolutionary biology and ecosystem science . . . "An ability to ascribe value to events in the world, a product of evolutionary selective processes, is evident across phylogeny. Value in this sense refers to an organism's facility to sense whether events in its environment are more or less desirable" (Dolan 2002: 1191). Adaptive value, survival value, is the basic matrix of Darwinian theory. An organism is the loci of values defended; life is otherwise unthinkable. Such organismic values are individually defended; but, ecologists insist, organisms occupy niches and are networked into biotic communities. At this point ethicists wonder whether there may be goods (values) in nature which humans ought to consider and care for. Animals, plants, and species, integrated into ecosystems, may embody values that, though nonmoral, count morally when moral agents encounter these.[13]

The biosphere was packed with living, interpreting systems well before human beings came onto the stage, agents who already possessed many of the properties that are manifested today in higher organisms. Every organism, every living thing, is an agent composed of communities of living parts. They sense or perceive their environment, process data, and make appropriate responses. The lifeworld is agent-centered; it is an ontology of living agents.[14]

Humans know that the primary agents in their lives—partners, children, friends, extended family—are of value, which means that these others have a claim on us to also be *treated as* valuable. The task for our species today is to follow this same ethical practice of valuing outwards until it includes *all* agents.[15] Every living thing is an agent, we have seen. All agents value; all are of value; all are treated as valuable. (My more orthodox Whiteheadian friends extend matter/value to every bit of matter, as the classical panpsychists did, whereas Jesper Hoffmeyer and others in biosemiotics extend them to every living being. We need to be a movement that lives with both groups, rather than ostracizing either.)

Philosophers like to add metaphysical compliments: rights language, created-by-God language, deontologically and not merely utilitarianly valuable, or valuable thanks to a general theory of "axiology." But here I reverse the order: the raw ethical claim comes first, and the abstractions and speculations then play their supporting roles after the fact.

Valuing matter extends further. Some biologists affirm that *the biosphere itself* is an agent.[16] The biosphere is also a whole that is greater than the sum of its parts. Like other organisms, it adjusts and adapts so that the living parts of which it consists can continue to emerge and flourish. There is value in this self-regulating living system of living systems that is not merely derived from its parts. Lovelock explains that:

> The entire range of living matter on Earth, from whales to viruses, and from oaks to algae, could be regarded as constituting a single living entity, capable of manipulating the Earth's atmosphere to suit its over-all needs and endowed with faculties and powers far beyond those of its constituent parts.[17]

At another point he adds, "planetary self-regulation is purposeful, or involves foresight or planning by the biota."[18] Scientifically, a living organism is a system of systems that function together in an interdependent fashion, thereby creating a new emergent system that then becomes an agent in its own right. We are not responsible only to isolated individuals, but also to that communion of bodies that together make up the system of all living things. The sum total of life on this planet is thus a locus of intrinsic value, and its value calls forth our lived valuing.[19] Elisabeth Sahtouris calls this the *EarthDance*[20]:

> All evolution—of the great cosmos and of our own planet within it—is an endless dance of wholes that separate themselves into parts and parts that join into mutually consistent new wholes. We can see it as a repeating, sequentially spiraling pattern: unity→individuation →tension/conflict→negotiation→resolution →cooperation→new levels of unity, and so on.[21]

To consider the whole network of life as perceiving, acting, and valuing suggests a life-teleology that includes and extends beyond human flourishing. "We . . . are living beings within a larger being."[22]

Marrying matter and value changes one's view of the world and,[23] more importantly, one's action commitments. Stephan Harding puts it beautifully, "Thus the great archetypes of Gaia and *anima mundi* that figure so importantly in the human soul could well be prefigured in some mysterious way not in some abstract realm far from this world, but in the very molecules and atoms that constitute our palpable, sensing bodies."[24]

Not surprisingly, this ethics of matter and the Earth brings discomfort to traditional theologies, and to many traditional models of biology as well. A metaphysics of dead matter and isolated atoms blinds one to one's primary rootedness within an agential, interdependent, sacred world. If a metaphysics should grow out of these ethical roots, far better that it be one of *participation, reciprocity, and mutual indwelling*; and if a theology, then one akin to Jay McDaniel's: as we journey into the wilderness, "we journey into the very God whose body is that wilderness. As our senses join soil, air, water, and wilderness, they join God."[25]

Toward a Politics of Matter in Crisis

Ours is an ethic of verbs. *Matter values, both* because all material agents act within webs of value, *and* because the fate of our planet hinges on the active systems of matter that we call bodies.

Emergent NM—one that incorporates Emergent Complexity—gives rise to a planetary ethic. That ethic in turn gives rise to a politics for a planet in crisis.

One species has become dominant on this planet. The Age of the Anthropocene threatens nature as the womb that nurtures the actions of all living things. The Anthropocene could also mean the death of a large portion of the biosphere by human hands.

> A few years ago an iceberg twice the size of Luxembourg broke off from the Larsen C ice shelf in Antarctica; soon, the trillion-ton, 5,800 km² iceberg was threatening wildlife on the South Georgia Island off Argentina.[26] On March 15, 2022 the Conger Ice Shelf in East Antarctica collapsed completely, following a period of extreme heat in the region.[27]
>
> Carbon dioxide in the atmosphere settles into the oceans as C^{14}. As C^{14} bonds with ocean water and sinks deeper, it kills off phytoplankton, which are the foundation of the ocean's food chain.
>
> Consider the huge island of plastic in the Pacific, sometimes stretching to the size of Mexico. As the sun breaks down the plastic, toxic chemicals such as bisphenol, PCBs, and polystyrene are released into the food chain. Through a process known as "trophic transfer," microplastics move upward along the food chain. Albatrosses that imbibe sufficient amounts suffer much higher infant mortality, with up to one-third of their offspring dying.[28] On March 24, 2022, the first study was published showing microplastics in human blood.[29]
>
> Climate models show that Africa, which contributes just 7% of the total greenhouse gases and has the fewest resources to prepare for climate change, will face the most catastrophic effects of climate disruption.[30]

Bodies are being destroyed. But bodies matter—first to themselves, then to their ecosystems, and then, finally, to the perpetrators who, knowingly or not, are dependent on their existence. One does not need metaethics or metaphysics to detect this, to realize the danger, and to hear the call to action.

The Anthropocene was begotten by, and begets, anthropocentrism. For anthropomorphs, the wrong of species extinction is derivative, deduced from its impact on this one species alone. But an ethics of bodies is *not* derivative in this sense, any more than the suffering of an oil-soaked sea gull can be measured by the degree of suffering of the CEO of Exxon. Ethics is not born with Exxon (though it may well die there).[31]

Ethical judgments are statements about bodies, the suffering and abuse of bodies, the disappearance of bodies. Often, these disappearances can be traced back to the *polis*, as in the thirty thousand "disappearances" in Argentina in the late 1970s. Today, humans are building *polis* writ large, a global polis or, as poli-ticians like to say, a *cosmo*-polis. Polis-ing at the global level magnifies the disappearances a millionfold. The calving of glaciers, the disruption of phyto-plankton, the death of albatross chicks—these are poli-tical acts, acts against bodies. The fact that the bare hands of Exxon CEO, Darren Woods, do not destroy bodies does not change the ethical/political reality any more than the fact that Eichmann did not use a rifle does.[32]

To cease *global*izing is to begin to re*planet*ize; it means thinking and living as what we always were: vulnerable, interdependent members of the composite community of Earth's communities. To cease *transcendent*izing is to give up the myth that we are the Great Exception to the circle of life;[33] it means heed-ing the beautiful music of life and death that sounds in our bodies, dancing with the rhythm of the biosphere.[34] In embracing immanent religiosities, whether we draw from the New Materialisms, or from other theories of imma-nence, we proclaim the sacredness of living dust. We may forsake (the dream of) fellowship with the angels, but we regain, in return, the belonging to the single planetary community, which is our rightful heritage.

Notes

1. Philip Clayton et al., eds. *The New Possible: Visions of our World beyond Crisis* (Eugene: Cascade, 2021).

2. Lisa Stenmark and Whitney Bauman, eds., *Unsettling Science and Religion: Contributions and Questions from Queer Studies* (Lanham: Lexington Books, 2018), especially chapter 1, Kirianna Florez and Philip Clayton, "Both/And: Science, Religion, and the Fluidity of Identity."

3. Clayton Crockett, *Religion, Politics, and the Earth: The New Materialism* (New York: Palgrave Macmillan, 2012); and *Derrida after the End of Writing: Political Theology and New Materialism* (New York: Fordham University Press, 2017).

4. Philip Clayton, "The Fruits of Pluralism: A Vision for the Next Seven Years in Religion/Science," *Zygon: Journal of Religion and Science*, Vol. 49 (2014): 1–13.

5. F. LeRon Shults, *Iconoclastic Theology: Gilles Deleuze and the Secretion of Atheism* (Edinburgh: Edinburgh University Press, 2014), especially chapters 5 and 6.

6. This was the title of a large session at the 2019 AAR in San Diego. In the midst of difficulties, it is encouraging to see the Templeton Foundation beginning to

concentrate on what "human flourishing" might mean in a genuinely planetary context, redefined against the frightening backdrop of a single species becoming the dominant geological force and greatest single threat to the Earth's biosphere as a whole.

7. Whitney Bauman, personal email.

8. Philip Clayton, *Mind and Emergence: From Quantum to Consciousness* (Oxford: Oxford University Press, 2004); *In Quest of Freedom: The Emergence of Spirit in the Natural World* (Göttingen: Vandenhoeck & Ruprecht, 2009); *The Re-emergence of Emergence: The Emergentist Hypothesis from Science to Religion*, co-edited with Paul Davies (Oxford: Oxford University Press, 2006); "On the Plurality of Complexity-Producing Mechanisms," in *Complexity and the Arrow of Time. eds.* Charles Lineweaver, Paul C.W. Davies, and Michael Ruse (New York: Cambridge University Press, 2013), 332–51.

9. Philip Clayton, "Unsolved Dilemmas: The Concept of Matter in the History of Philosophy and in Contemporary Physics," in *Information and the Nature of Reality, eds.* Paul Davies and Niels Henrik Gregersen (Cambridge: Cambridge University Press, 2011), 32–52.

10. Sarah Lane Ritchie, "Integrated Physicality and the Absence of God: Spiritual Technologies in Theological Context," *Modern Theology* (open access), first published February 11, 2021; https://doi.org/10.1111/moth.12684.

11. Friedrich Nietzsche, "On the Despisers of the Body," from *Thus Spake Zarathustra*, quoted by Daniel Dennett in *Kinds of Minds: Toward an Understanding of Consciousness* (New York: Basic Books, 1996).

12. Even the John Templeton Foundation, which made the explosion of R&S work possible starting in the 1980s, now engages these topics with different concepts, programs, and allies than it did during the heyday of the science-and-religion movement.

13. Holmes Rolston, "Science and Religion in the Face of the Environmental Crisis," in *The Oxford Handbook of Religion and Ecology*, ed. Roger S. Gottlieb (New York: Oxford University Press, 2006), 376–397, quote p. 378.

14. Elisabeth Sahtouris, *Earth Dance: Living Systems in Evolution* (Lincoln: University Press, 2000), 12.

15. "Following . . . outwards," would have to be a paper in its own right.

16. This is the "Gaia hypothesis," proposed by James Lovelock and further developed by Lynn Margulis. In the Lovelock/Margulis hypothesis, the biosphere as a whole is living entity that regulates itself in specific ways in order to support the continuation of life. If there are interactions that occur only at the level of the biosphere as a whole and not merely as an aggregate of its parts, as appears to be the case, then Gaia also has features of agency.

17. James Lovelock, *Gaia: A New Look at Life on Earth* (Oxford: Oxford University Press, 1987), 9.

18. James Lovelock, "Hands up for the Gaia hypothesis," *Nature*, Vol. 344 (6262) (1990): 100–2.

19. It is unfortunate that humans seem able to conceive the neutral gender only in mechanistic, non-agential ways, rather than organically. It is not, in principle, necessary to gender an entity in order to make it alive; asexual reproduction is widespread across the biosphere (e.g., bacteria).

20. Sahtouris, *EarthDance*, 4.

21. Sahtouris, *EarthDance*, 24. See also Elizabeth Singleton and Philip Clayton, "Agents Matter and Matter Agents: Agency, Interpretation, and Value from Cell to Gaia," in *Entangled Worlds: Religion, Science, and New Materialisms*, eds. Catherine Keller and Mary-Jane Rubenstein (New York: Routledge, 2016).

22. It is also an ontology of continuous interactions between part and whole. The relationships between organisms, species, and ecosystems are ubiquitous As Elizabeth Sahtouris writes, "No being in nature, outside our own species, is ever confronted with" the choice between individual and whole. She adds, "if we consult nature, the reason is obvious. The choice makes no sense, for neither alternative can work. No being in nature can ever be completely independent, although independence calls to every living being, whether it is a cell, a creature, a society, a species, or a whole ecosystem. See Sahtouris, *EarthDance*, 2.

23. John Cobb writes, "Many people affirm immanence, but few explain it. Immanence requires what Substance thinking views as impossible, namely, that two things occupy the same space at the same time. More or less by definition, this cannot be true of substances" (unpublished lecture). See also the organic worldview defended by Brian Swimme and Thomas Berry, *The Universe Story* (San Francisco: HarperSanFrancisco, 1994).

24. Stephan Harding, *Animate Earth: Science, Intuition and Gaia* (White River Junction: Chelsea Green Pub. Co., 2006), 88.

25. Jay B. McDaniel, *Of God and Pelicans: A Theology of Reverence for Life* (Louisville: Westminster/John Knox Press, 1989), 91. In this onto-ethic we find ourselves with a living, sacred, interdependent, yet fully natural living planet on the one hand, and a radically immanent divine on the other.

26. https://www.theguardian.com/world/2017/jul/12/giant-antarctic-iceberg-breaks -free-of-larsen-c-ice-shelf, viewed Feb. 25, 2021.

27. https://www.reuters.com/business/environment/thinning-antarctic-ice-shelf -finally-crumbles-after-heatwave-2022-03-25/.

28. See Chris Jordan's article in the *New York Review of Books* (November 11, 2009), "Midway: Message from the Gyre," http://www.nybooks.com/daily/2009/11/11 /chris-jordan/, viewed Feb. 25, 2021.

29. Heather A. Leslie et al., "Discovery and Quantification of Plastic Particle Pollution in Human Blood," *Environment International* (2022).

30. https://www.downtoearth.org.in/news/climate-change/africa-the-least -responsible-but-most-vulnerable-to-climate-change-60669, viewed Feb. 25, 2021.

31. Body-ethics is not defined by the impact of mass-death and mass-extinction on humanity.

32. I owe this argument to Emil L. Fackenheim; see e.g. A *Political Philosophy for the State of Israel* (Jerusalem: Jerusalem Center for Public Affairs, 1988).

33. Brilliantly debunked by Catherine Keller in *Political Theology of the Earth: Our Planetary Emergency and the Struggle for a New Public* (New York: Columbia University Press, 2018).

34. See Kimerer LaMothe, "Dancing Immanence: A Philosophy of Bodily Becoming," in *Earthly Things: Immanence, New Materialisms, and Planetary Thinking*, eds. Karen Bray, Heather Eaton, and Whitney Bauman (New York: Fordham University Press, 2023).

Acknowledgments

The editors would like to thank all of the contributors to this volume and to the five-year seminar out of which this volume emerged, "New Materialism, Religion and Planetary Thinking." From 2016 to 2020, the contributors to this seminar, hosted by the American Academy of Religion, met every year to think together toward what appears in this volume. We believe the process has produced an excellent outcome. We would also like to thank the American Academy of Religion for having enough faith in this project to give us space on the schedule of the annual meeting for four years in a row. In addition, we would like to thank all the habitats that make the thinking of the co-editors possible including: the Transdisciplinary Theological Colloquia of Drew University, the Forum on Religion and Ecology at Yale, and the International Society for the Study of Religion, Nature, and Culture. Without such habitats, among many others not named here, this type of scholarship and thinking together is not possible. Thank you to our publishers at Fordham University Press, especially John Garza, for taking on this large volume, and Lis Pearson for the thorough editing job. Finally, thank you to all those colleagues, students, and friends who are working to resist and put an end to current systems that promote heteropatriarchy, racism, and ecological destruction, and instead co-create new spaces for planetary futures that are more just and sustainable in ways that we may not even yet imagine. It is to all of you that this work is dedicated.

Bibliography

"A Visit to Aida Refugee Camp." *The Israeli Committee Against House Demolition*, 2016. https://icahd.org/2016/06/09/a-visit-to-aida-refugee-camp/.

Abímbólá, Kólá. *Yorùbá Culture: A Philosophical Account.* Birmingham: Iròkò Academic Publishers, 2006.

Abram, David. *The Spell of the Sensuous: Perception and Language in a More-than-Human World.* New York: Vintage, 1996.

Ahmed, Sara. "Orientations Matter." In *New Materialisms: Ontology, Agency, Politics*, edited by Diana Coole and Samantha Frost. Durham: Duke University Press, 2010.

———. *The Promise of Happiness.* Durham: Duke University Press, 2010.

———. "Not in the Mood." *New Formations*, Vol. 82 (18).

———. "Cutting Yourself Off." *feministkilljoys blog* (November 3, 2017). https://feministkilljoys.com/2017/11/03/cutting-yourself-off/.

Akwesasne Notes, ed. "Basic Call to Consciousness." *Native Voices* (2005).

Alaimo, Stacy and Susan Hekman, eds. *Material Feminisms.* Bloomington: Indiana University Press, 2008.

Alaimo, Stacy. *Bodily Natures: Science, Environment, and the Material Self.* Bloomington: Indiana University Press, 2010.

———. *Exposed: Environmental Politics and Pleasures in Posthuman Times.* Minneapolis: University of Minnesota Press, 2016.

Albert, Bruce and Davi Kopenawa. *Yanomami, o espírito da floresta.* Rio de Janeiro: Centro Cultural Banco do Brasil/Fondation Cartier, 2004.

Alexander, Samuel. *Space, Time, and Deity.* Volumes 1–2. London: MacMillan, 1920.

Ames, Roger and David Hall. *Thinking From the Han: Self, Truth and Transcendence in Chinese and Western Culture.* Albany: SUNY Press, 1998.

———. *Thinking Through Confucius.* Albany: SUNY Press, 1987.

Ames, Roger. "Religiousness in Classical Confucianism: A Comparative Analysis."
    Asian Culture Quarterly, Vol. 12 (2) (1984).
Anderson, Carol. *White Rage: The Unspoken Truth of Our Radical Divide.* New
    York: Bloomsbury, 2016.
Antonaccio, Maria. "De-moralizing and Re-moralizing the Anthropocene." In
    Religion in the Anthropocene, edited by Celia Deane-Drummond, Sigurd
    Bergmann, and Markus Vogt. Eugene: Cascade Books, 2017.
Anzaldúa, Gloria E. *Borderlands/La Frontera: The New Mestiza.* San Francisco:
    Aunt Lute Books, 1987.
Apess, William. *On Our Own Ground: The Complete Writings of William Apess,*
    A Pequot, edited by Barry O'Connell. Amherst: University of Massachusetts
    Press, 1992.
Aquinas, Thomas. *Summa Theologiae,* 5 volumes. Translated by Fathers of the English
    Dominican Province. Allen: Christian Classics, 1981.
Aristotle. "Metaphysics." In *The Complete Works of Aristotle: The Revised Oxford*
    Translation, translated by W. D. Ross, edited by Jonathan Barnes. Princeton:
    Princeton University Press, 1971.
————. "Physics." In *The Complete Works of Aristotle,* translated by R. P. Hardie
    and R. K. Gaye, edited by Jonathan Barnes. Princeton: Princeton University
    Press, 1984.
Armstrong, Jeanette. *The Native Creative Process.* Penticton, BC: Theytus Books
    Ltd., 1991.
Arthur, Mathew. "Writing Affect and Theology in Indigenous Futures." In *Religion,*
    Emotion, and Sensation: Affect Theories and Theologies, edited by Karen Bray
    and Stephen Moore. New York: Fordham University Press, 2020.
Astor-Aguilera, Miguel and Graham Harvey. "Introduction: We have never
    been individuals." In *Rethinking Relations and Animism: Personhood and*
    Materiality, edited by Miguel Astor-Aguilera and Graham Harvey. New York:
    Routledge, 2018.
Aveni, Anthony F. *World Archaeoastronomy.* Cambridge: Cambridge University
    Press, 1988. Selected papers from the Second Oxford International Conference
    on Archaeoastronomy, Merida, Mexico, 1986.
Bachelard, Gaston. *The Poetics of Space.* Translated by M. Jolas. Boston: Beacon
    Press, 1964.
Baichwal, Jennifer. *Anthropocene: The Human Epoch.* Documentary. Directed by
    Jennifer Baichwal, Nicholas de Pencier, and Edward Burtynsky. Toronto: Mercury
    Filmworks, Inc., 2018.
Barad, Karen. "Quantum Entanglements and Hauntological Relations of Inheritance:
    Dis/continuities, SpaceTime Enfoldings, and Justice-to-Come." *Derrida Today,*
    Vol. 3 (2) (2010): 240–68.
————. "What Flashes Up: Theological-Political-Scientific Fragments." In *Entangled*
    Worlds: Religion, Science, and New Materialisms, edited by Catherine Keller and
    Mary-Jane Rubenstein. New York: Fordham University Press, 2017.

————. *Meeting the Universe Halfway: Quantum Physics and the Entanglement of Matter and Meaning*. Durham: Duke University Press, 2007.

Barnes, R. H., Andrew Gray, and Benedict Kingsbury, eds. *Indigenous Peoples of Asia*. Ann Arbor: Association for Asian Studies, 1995. Originated from a 1989 conference.

Basso, Keith. *Wisdom Sits in Places: Landscape and Language Among the Western Apache*. Albuquerque: University of New Mexico Press, 1996.

Battiste, Marie and James Henderson. *Protecting Indigenous Knowledge and Heritage*. Saskatoon, SK: Purich Publishing, 2000.

Bauman, Whitney and Kevin O'Brien. *Environmental Ethics and Uncertainty: Wrestling with Wicked Problems*. New York: Routledge, 2019.

Bauman, Whitney. "Developing a Critical Romantic Religiosity for a Planetary Community." In *Earthly Things: Immanence, New Materialisms, and Planetary Thinking*, edited by Karen Bray, Heather Eaton, and Whitney Bauman. New York: Fordham University Press, 2023.

————. "Critical Planetary Romanticism: Ecology, Evolution, and Erotic Thinking." *Journal for the Study of Religion, Nature, and Culture*, Vol. 15 (1) (2021): 33–52.

————. *Religion and Ecology: Developing a Planetary Ethic*. New York: Columbia University Press, 2014.

Bellegarde-Smith, Patrice and Claudine Michel. "Danbala/Ayida as Cosmic Prism: The Lwa as Trope for Understanding Metaphysics in Haitian Vodou and Beyond." *Journal of African Religions*, Vol. 1 (4) (2013): 458–487.

Benjamin, Walter. "Theses on the Philosophy of History." In *Illuminations: Essays and Reflections*, translated by Harry Zohn, edited and intro by Hannah Arendt. Boston: Mariner Books, 2019.

————. "On the Concept of History." In *Selected Writings*, Volume 4 (1996–2003), edited by Marcus Bollock and Michael W. Jennings. Cambridge: Harvard University Press, 2004.

Bennett, Jane. *Vibrant Matter: A Political Ecology of Things*. Durham: Duke University Press, 2010.

Benson, Melinda Harm and Robin Kundis Craig. *The End of Sustainability: Resilience and the Future of Environmental Governance in the Anthropocene*. Lawrence: The University Press of Kansas, 2017.

Berkes, Fikret. *Sacred Ecology: Traditional Ecological Knowledge and Resource Management*. Philadelphia: Taylor & Francis, 1999.

Berkovits, Eliezer. *Faith after the Holocaust*. Newark: Ktav Publishing, 1973.

Berry, Thomas. *Religions of India*. New York: Columbia University Press, 1992.

Berthrong, John. *All Under Heaven*. Albany: SUNY Press, 1994.

Betcher, Sharon V. "Crypt/ography: Disability Theology in the Ruins of God." http://www.jcrt.org/archives/15.2/betcher.pdf.

————. "Of Disability and the Garden State." *Religious Studies News* (March 2013).

————. *Spirit and the Obligation of Social Flesh: A Secular Theology for the Global City*. New York: Fordham University Press, 2014.

Bettelheim, Judith. "Palo Monte Mayombe and Its Influence on Cuban
    Contemporary Art." *African Arts, Vol.* 34 (2) (Summer 2001): 36–96.
Bhambra, Gurminder K. *Rethinking Modernity: Postcolonialism and the Sociological
    Imagination.* Basingstoke: Palgrave Macmillan, 2007.
Bielefeldt, Carl. "The Mountain Spirit: Dōgen, Gary Snyder, and Critical Buddhism."
    Zen Quarterly, Vol. 11 (1) (1999).
———. "Buddhism." In *Mountains and Waters without End* (unpublished).
———. "Dōgen's Shōbōgenzō Sansuikyō." In *The Mountain Spirit,* edited by Michael
    Charles Tobia and Harold Drasdo. Woodstock: The Overlook Press, 1979.
———. *Dōgen's Manuals of Zen Meditation.* Berkeley: University of California Press,
    1990.
Biernacki, Loriliai and Philip Clayton, eds. *Panentheism Across the World's Traditions.*
    New York: Oxford University Press, 2014.
Bird-David, Nurit and Danny Naveh. "Relational Epistemology, Immediacy, and
    Conservation: Or, What Do the Nayaka Try to Conserve?" *Journal for the Study
    of Religion, Nature, and Culture, Vol.* 2 (1) (2008): 55–73.
Bird-David, Nurit. "'Animism' Revisited: Personhood, Environment, and Relational
    Epistemology." *Current Anthropology, Vol.* 40 (Supplement 1999): S67–S91.
——— "Persons or Relatives? Animistic Scales of Practice and Imagination." In
    Rethinking Relations and Animism: Personhood and Materiality, edited by Miguel
    Astor-Aguilera and Graham Harvey. New York: Routledge, 2018.
Bishara, Amahl and Lajee youth. *The Boy and the Wall.* 2005. https://www.shop
    palestine.org/product-p/book_kids_boy_wall.htm.
Black, Mary B. "Ojibwa Power Belief System." In *The Anthropology of Power,* edited
    by Raymond D. Fogelson and Richard N. Adams. New York: Academic, 1977.
Blakeslee, Sandra and Matthew Blakeslee. *The Body Has a Mind of its Own: How
    Body Maps in Your Brain Help You Do (Almost) Everything Better.* New York:
    Random House, 2007.
Bodhi, Bhikku, trans. *In the Buddha's Words: An Anthology of Discourses from the
    Pali Canon.* Boston: Wisdom Publications, 2005.
Booth, Newell S. "An Approach to African Religions." In *African Religions: A
    Symposium,* edited by Newell S. Booth. New York: Nok Publishers Ltd. 1977.
Botterweck, Johannes G., Helmer Ringgren, and Heinz-Josef Fabry, eds. "Tiqwâ."
    In *Theological Dictionary of the Old Testament.* Grand Rapids: Eerdmans
    Publishing, 2006.
Braidotti, Rosi. "The Politics of 'Life Itself' and New Ways of Dying." In *New
    Materialisms: Ontology, Agency, Politics,* edited by Diane Coole and Samantha
    Frost. Durham: Duke University Press, 2010.
Brassier, Ray. "Develeving: Against 'Flat Ontologies.'" https://uberty.org/wpcontent
    /uploads/2015/05/RayBrassierDevelevingAgainstFlatOntologies.pdf.
Bray, Karen. "Gut Theology: The Peril and Promise of Political Affect." In *Earthly
    Things: Immanence, New Materialisms, and Planetary Thinking,* edited by Karen

Bray, Heather Eaton, and Whitney Bauman. New York: Fordham University Press, 2023.

Brennan, Teresa. *Globalization and Its Terrors: Daily Life in the West*. New York: Routledge, 2002.

Brown Douglas, Kelly. *Stand Your Ground: Black Bodies and the Justice of God*. New York: Orbis Books, 2015.

Bruno, Giordano. "Cause, Principle and Unity." In *Cause, Principle and Unity and Essays on Magic*, translated by Richard J. Blackwell, edited by Richard J. Blackwell and Robert de Lucca. Cambridge Texts in the History of Philosophy. Cambridge: Cambridge University Press, 1998.

———. *The Ash Wednesday Supper*. Translated by Edward A. Gosselin and Lawrence S. Lerner. Renaissance Society of America Reprint Texts. Toronto: University of Toronto Press, 1995.

Bühlmann, Vera, Felicity Colman, and Iris van der Tuin. "Introduction to New Materialist Genealogies: New Materialisms, Novel Mentalities, Quantum Theory." *Minnesota Review*, Vol. 88 (2017).

Butler, Jenny. "Druidry in Contemporary Ireland." In *Modern Paganism in World Cultures: Comparative Perspectives*, edited by Michael F. Strmiska. Santa Barbara: ABC-Clio, 2005.

Cajete, Gregory. *Native Science: Natural Laws of Interdependence*. Santa Fe: Clear Light, 1999.

Calcagno, Antonio. *Giordano Bruno and the Logic of Coincidence: Unity and Multiplicity in the Philosophical Thought of Giordano Bruno*. Translated by Eckhard Bernstein. Renaissance and Baroque Studies and Texts. New York: Peter Lang, 1998.

Caputo, John D. *The Insistence of God: A Theology of Perhaps*. Bloomington: Indiana University Press, 2013.

Caterine, Darryl. "The Haunted Grid: Nature, Electricity, and Indian Spirits in the American Metaphysical Tradition." *Journal of the American Academy of Religion*, Vol. 82 (2014): 371–397.

Chan, Wing-tsit. *A Source Book in Chinese Philosophy*. Princeton: Princeton University Press, 1963.

Chapple, Christopher Key. "Immanence in Hinduism and Jainism: New Planetary Thinking?" In *Earthly Things: Immanence, New Materialisms, and Planetary Thinking*, edited by Karen Bray, Heather Eaton, and Whitney Bauman. New York: Fordham University Press, 2023.

———. ed. *Jainism and Ecology: Nonviolence in the Web of Life*. Cambridge: Harvard University Press, 2002.

Chen, Mel Y. *Animacies: Biopolitics, Racial Mattering, and Queer Affect*. Durham: Duke University Press, 2012.

Chidester, David. "Classify and Conquer: Friedrich Max Müller, Indigenous Traditions, and Imperial Comparative Religion." In *Beyond Primitivism:*

Indigenous Religious Traditions and Modernity, edited by Jacob K. Olupona. New York: Routledge, 2004.

———. *Empire of Religion: Imperialism and Comparative Religion.* Chicago: The University of Chicago Press, 2014.

———. *Savage Systems: Colonialism and Comparative Religion in Southern Africa.* Charlottesville: University of Virginia Press, 1996.

Choi, Jin Young and Joerg Rieger. *Faith, Class, and Labor: Intersectional Approaches in a Global Context.* Edited by Jin Young Choi and Joerg Rieger. Eugene: Pickwick, 2020.

Clayton, Philip and Elizabeth Singleton. "Agents Matter and Matter Agents: Interpretation and Value from Cell to Gaia." In *Entangled Worlds: Religion, Science, and New Materialisms,* edited by Catherine Keller and Mary-Jane Rubenstein. New York: Fordham University Press, 2017.

Clayton, Philip and Paul Davies, eds. *The Re-Emergence of Emergence: The Emergentist Hypothesis from Science to Religion.* Oxford: Oxford University Press, 2006.

Clayton, Philip, Kelli M. Archie, Jonah Sachs, and Evan Steiner, eds. *The New Possible: Visions of Our World beyond Crisis.* Eugene: Cascade Books, 2021.

Clayton, Philip. "Matter Values: Ethics and Politics for a Planet in Crisis." In *Earthly Matters: Immanence, New Materialisms, and Planetary Thinking,* edited by Karen Bray, Heather Eaton, and Whitney Bauman. New York: Fordham University Press, 2023.

———. "On the Plurality of Complexity-Producing Mechanisms." In *Complexity and the Arrow of Time,* edited by Charles H. Lineweaver, Paul C. W. Davies, and Michael Ruse. Cambridge: Cambridge University Press, 2013.

———. "The Fruits of Pluralism: A Vision for the Next Seven Years in Religion/Science." Zygon: *Journal of Religion and Science,* Vol. 49 (2) (June 2014): 430–42. https://doi.org/10.1111/zygo.12092.

———. "Unsolved Dilemmas: The Concept of Matter in the History of Philosophy and in Contemporary Physics." In *Information and the Nature of Reality: From Physics to Metaphysics,* edited by Paul Davies and Niels Henrik Gregersen. Cambridge: Cambridge University Press, 2011.

———. In *Quest of Freedom: The Emergence of Spirit in the Natural World,* edited by Michael G. Parker and Thomas M. Schmidt. Göttingen: Vandenhoeck & Ruprecht, 2009.

———. *Mind and Emergence: From Quantum to Consciousness.* Oxford: Oxford University Press, 2004.

Clough, Patricia Ticineto. "Introduction." In *The Affective Turn,* edited by Patricia Tincineto Clough and Jean Halley. Durham: Duke University Press, 2007.

Coates, Ken S. A *Global History of Indigenous Peoples: Struggle and Survival.* Basingstoke, UK: Palgrave Macmillan, 2004.

Cobb, John B., Jr. *Beyond Dialogue: Toward a Mutual Transformation of Christianity and Buddhism.* Philadelphia: Fortress Press, 1982.

———. *Spiritual Bankruptcy: A Prophetic Call to Action*. Nashville: Abigdon, 2010.

Connolly, William. *Climate Machines, Fascist Drives, and Truth*. Durham: Duke University Press, 2019.

———. *A World of Becoming*. Durham: Duke University Press, 2010.

———. *The Fragility of Things: Self-Organizing Processes, Neoliberal Fantasies, and Democratic Activism*. Durham: Duke University Press, 2013.

Cook, Francis H. "Dōgen's View of Authentic Selfhood and Its Socio-ethical Implications." In *Dōgen Studies*, edited by William R. LaFleur. Honolulu: University of Hawaii Press, 1985.

———. *Sounds of Valley Streams: Enlightenment in Dōgen's Zen*. Albany: SUNY Press, 1988.

Coole, Diana and Samantha Frost. "Introducing the New Materialisms." In *New Materialisms: Ontology, Agency, Politics*, edited by Diana Coole and Samantha Frost. Durham: Duke University Press, 2011.

Cornell. Laura. "Green Yoga: Contemporary Activism and Ancient Practice; A Model for Eight Paths to Green Yoga." In *Yoga and Ecology: Dharma for the Earth*, edited by Christopher Key Chapple. Hampton: Deepak Heritage Books, 2009.

Crockett, Clayton and Jeffrey Robbins. *Religion, Politics, and the Earth: The New Materialism*. New York: Palgrave Macmillan, 2012.

Crockett, Clayton. *Derrida after the End of Writing: Political Theology and New Materialism*. New York: Fordham University Press, 2018.

Crosby, Donald. "Matter, Mind, and Meaning." In *The Routledge Handbook of Religious Naturalism*, edited by Jerome Stone and Donald Crosby. New York: Routledge, 2018.

———. *Living with Ambiguity: Religious Naturalism and the Menace of Evil*. Albany: SUNY Press, 2008.

———. *The Thou of Nature: Religious Naturalism and Reverence for Sentient Life*. Albany: SUNY Press, 2013.

Crosby, Donald, and Jerome Stone, ed. *The Routledge Handbook of Religious Naturalism*. London & New York: Routledge, 2018.

Cruikshank, Julie. *Do Glaciers Listen: Local Knowledge, Colonial Encounters, and Social Imagination*. Vancouver and Toronto: University of British Columbia Press, 2005.

Crutzen, Paul J. and Eugene F. Stoermer. "The 'Anthropocene.'" *IGBP Newsletter*, Issue 41 (2000): 17–18.

Curd, Patricia. "Anaxagoras." In *Stanford Encyclopedia of Philosophy*. https://plato.stanford.edu/entries/anaxagoras/ (October 1, 2015).

Cvetkovich, Ann. *Depression: A Public Feeling*. Durham: Duke University Press, 2012.

Dagron, Tristan. "David of Dinant–Sur Le Fragment [Hyle, Mens, Deus] Des Quaternuli." *Revue de Métaphysique et de Morale*, Vol. 40 (2003): 419–436.

Daniel, Yvonne. *Dancing Wisdom: Embodied Knowledge in Haitian Vodou, Cuban Yoruba, and Bahian Condomble*. Urbana: University of Illinois Press, 2005.

Dark Mountain Project. "The Devil's Door." Accessed August 13, 2020. https://dark
     -mountain.net/the-devils-door-a-call-for-contributions-to-issue-12/.Document27.
Davis, Bret. "The Presencing of Truth: Dōgen's Genjōkōan. In *Buddhist Philosophy:
     Essential Readings*, edited by Willian Edelglass and Jay Garfield. New York:
     Oxford University Press, 2009.
Davis, Edward B. "Myth 13: That Isaac Newton's Mechanistic Cosmology
     Eliminated the Need for God." In *Galileo Goes to Jail and Other Myths About
     Science and Religion*, edited by Ronald L. Numbers. Cambridge: Harvard
     University Press, 2009.
Daystar, Rosalie. "Dancing the Four Directions." In *Dancing on Earth: Special Issue
     of the Journal of Dance, Movement, and Spiritualities*, Vol. 4 (2) (2017).
de Bary, Wm. Theodore and Irene Bloom, eds. *Sources of Chinese Tradition*. New
     York: Columbia University Press, 1999.
de Bary, Wm. Theodore and John Chaffee, eds. *Neo-Confucian Education: The
     Formative Stage*. Berkeley: University of California Press, 1989.
de Bary, Wm. Theodore. "A Reappraisal of Neo-Confucianism." In *Studies in Chinese
     Thought: The American Anthropological Association*, Vol. 55 (5) (December 1953).
———. *Learning for One's Self: Essays on the Individual in Neo-Confucian
     Thought*. New York: Columbia University Press, 1991.
de la Torre, Miguel. *Santeria: The Beliefs and Rituals of a Growing Religion in
     America*. Grand Rapids: Eerdmans Publishing, 2004.
Deacon, Terrence. "Emergence: The Hole at the Wheel's Hub." In *The Re-
     emergence of Emergence: The Emergentist Hypothesis from Science to Religion*,
     edited by Philip Clayton and Paul Davies. Oxford: Oxford University Press, 2006.
———. *Incomplete Nature: How Mind Emerged from Matter*. New York: W.W.
     Norton, 2012.
———. *Symbolic Species: The Co-evolution of Language and the Brain*. New York:
     W.W. Norton, 1998.
Deane-Drummond, Celia and Agustin Fuentes, eds. *Theology and Evolutionary
     Anthropology: Dialogues in Wisdom, Humility, and Grace*. Abingdon: Routledge,
     2020.
Deleuze, Gilles and Felix Guattari. *What is Philosophy?* New York: Columbia
     University Press, 1994.
———. *A Thousand Plateaus: Capitalism and Schizophrenia*. Minneapolis: University
     of Minnesota Press, 1987.
Deloria, Vine, Jr. "Trouble in High Places: Erosion of American Indian Rights to
     Religious Freedom in the United States." In *The State of Native America:
     Genocide, Colonization, and Resistance*, edited by M. Annette Jaimes. Boston:
     South End Press, 1992.
———. *Red Earth, White Lies: Native Americans and the Myth of Scientific Fact*.
     New York: Scribner, 1995.
Dennett, Daniel C. *Kinds of Minds: Toward an Understanding of Consciousness*.
     New York: Basic Books, 1996.

Derrida, Jacques. "Hostipitality," In *Acts of Religion*, edited by Gil Anidjar. New York: Routledge, 2002.

———. *Specters of Marx: The State of the Debt, the Work of Mourning, and the New International*. Translated by Peggy Kamuf. New York: Routledge, 1994.

———. *The Animal That Therefore I Am*. Translated by David Wills, edited by Marie-Louise Mallet. New York: Fordham University Press, 2008.

Despret, Vinciane. *What Would Animals Say If We Asked the Right Questions?* Minneapolis: University of Minnesota Press, 2016.

Diab, Khaled. "The Art of Palestinian Resistance." In *HuffPost*. January 16, 2013. https://www.huffpost.com/entry/the-art-of-palestinian-resistance_b_2121029.

Dioszegi, Vilmos. *Tracing Shamans in Siberia*. New York: Humanities Press, 1968.

Dixon, John. *Images of Truth: Religion and the Art of Seeing*. Atlanta: Scholars Press, 1996.

Dochuk, Darren. *Anointed with Oil: How Christianity and Crude Made Modern America*. New York: Basic Books, 2019.

Dōgen Zenji. *Shōbōgenzō*. Tokyo: Iwanami Bunko, 1939.

Dolan, R. J. "Emotion, Cognition, and Behavior." *Science, Vol.* 298 (2002): 1191–4. DOI: 10.1126/science.1076358.

Dolphijn Rick and Iris van der Tuin. *New Materialism: Interviews & Cartographies*: Ann Arbor: Open Humanities Press, 2012.

Dore, Ronald, Tu Weiming, and Kim Kyong-Dong. "Modernization and Development." In *The Impact of Traditional Thought on Present Day Japan*, edited by Josef Kreiner. München: Iudicium-Verl, 1996.

Dowie, Mark. *Conservation Refugees: The Hundred-Year Conflict between Global Conservation and Native Peoples*. Cambridge: MIT Press, 2009.

Dussel, Enrique. *The Invention of the Americas: Eclipse of "the Other" and the Myth of Modernity*. Translated by Michael D. Barber. New York: Continuum, 1995.

———. "Europe, Modernity, and Eurocentrism." *Nepantla: Views from South*, Vol. 1(3) (2000): 465–478.

Eaton, Heather. "The Human Quest to Live in a Cosmos." In *Encountering Earth: Thinking Theologically with a More-Than-Human World*, edited by Trevor Bechtel, Matthew Eaton, and Timothy Harvie. Eugene: Wipf and Stock Publishers, 2018.

———. "An Ecological Imaginary: Evolution and Religion in an Ecological Era." In *Ecological Awareness: Exploring Religion, Ethics and Aesthetics*, edited by Sigurd Bergmann and Heather Eaton. Studies in Religion and the Environment, Volume 3/Studien zur Religion und Umwelt, Bd. Berlin: LIT Press, 2011.

———. "Global Visions and Common Ground: Biodemocracy, Postmodern Pressures, and The Earth Charter." *Zygon: Journal of Religion and Science*, Vol. 49 (4) (2014): 917–937.

———. "New Materialisms and Planetary Persistence, Purpose, and Politics." In *Earthly Things: Immanence, New Materialisms, and Planetary Thinking*, edited by Karen Bray, Heather Eaton, and Whitney Bauman. New York: Fordham University Press, 2023.

———. "The Challenges of Worldview Transformation: To Rethink and Refeel Our Origins and Destiny." In *Religion and Ecological Crisis: The Lyne White Thesis at Fifty*, edited by Todd LeVasseur and Anna Peterson. New York: Routledge, 2017.

———. "The Revolution of Evolution." *Worldviews: Environment, Culture, Religion*, Vol. 11 (1) (Spring 2007): 6–31.

———. ed. *The Intellectual Journey of Thomas Berry: Imagining the Earth Community*. Lanham: Lexington Press, 2014.

Ebrey, Patricia. *Chu Hsi's Family Rituals*. Princeton: Princeton University Press, 1991.

———. *Confucianism and Family Rituals in Imperial China*. Princeton: Princeton University Press, 1991.

Eckel, David Malcolm. "Is There a Buddhist Philosophy of Nature?" In *Buddhism and Ecology: The Interconnection of Dharma and Deeds*, edited by Mary Evelyn Tucker and Duncan Ryūken Williams. Cambridge: Harvard University Press, 1997.

Edwards, Jason. "The Materialism of Historical Materialism." In *New Materialisms: Ontology, Agency, Politics*, edited by Diana Coole and Samantha Frost. Durham: Duke University Press, 2010.

Elder-Vass, Dave. *The Causal Power of Social Structures: Emergence, Structure and Agency*. Cambridge: Cambridge University Press, 2010.

Eliade, Mircea. *Patterns in Comparative Religion*. Lincoln: University of Nebraska Press, 1996.

Eno, Robert. *The Confucian Creation of Heaven*. Albany: SUNY Press, 1990.

Epstein, Barbara. *Political Protest and Cultural Revolution: Nonviolent Direct Action in the 1970s and 1980s*. Berkeley: University of California Press, 1991.

Erni, Christian, ed. *The Concept of Indigenous Peoples in Asia: A Resource Book*. Copenhagen, Denmark, and Chiang Mai, Thailand: International Work Group for Indigenous Affairs and Asia Indigenous Peoples Pact Foundation, 2008.

Fackenheim, Emil L. *A Political Philosophy for the State of Israel: Fragments*. Jerusalem: Jerusalem Center for Public Affairs, 1988.

———. *God's Presence in History: Jewish Affirmations and Philosophical Reflections*. New York: New York University Press, 1970.

———. *To Mend the World: Foundations of Post-Holocaust Jewish Thought*. Indianapolis: Indiana University Press, 1994.

Farrell, Justin. *The Battle for Yellowstone*. Princeton: Princeton University Press, 2015.

Fletcher, Jeannine Hill. "We Are All Hybrids." In *Monopoly on Salvation?: A Feminist Approach to Religious Pluralism*. New York: Continuum, 2005.

Fogel, Alan. *The Psychophysiology of Self-Awareness: Rediscovering the Lost Art of Body Sense*. New York: W.W. Norton, 2009.

"Form Vs. Matter." In *Stanford Encyclopedia of Philosophy*. February 8, 2016. https://plato.stanford.edu/entries/form-matter/.

Fox, Michael W. "What Future for Man and Earth? Toward a Biospiritual Ethic." In *On the Fifth Day: Animal Rights and Human Ethics*, edited by Richard Knowles Morris and Michael W. Fox. New York: Acropolis Books, 1978.

Francis. *Laudato Si': On Care for Our Common Home.* Encyclical letter. Huntington: *Our Sunday Visitor,* 2015.

Freire, Paulo. "Cultural Action and Conscientization." *Harvard Educational Review,* Vol. 69 (4) (Winter 1998).

Friedman, Thomas. "Global Weirding Is Here," In *The New York Times,* February 17, 2010. https://www.nytimes.com/2010/02/17/opinion/17friedman.html.

Gabirol, Ibn. *The Font of Life (Fons Vitae). Mediaeval Philosophical Texts in Translation.* Milwaukee: Marquette University Press, 2014.

Gare, Arran. "Grand Narrative of the Age of Re-Embodiments: Beyond Modernism and Postmodernism, Cosmos and History." *The Journal of Natural and Social Philosophy,* Vol. 9 (1) (2013): 327–357.

Garuba, Harry. "Explorations in Animist Materialism: Notes on Reading/Writing African Literature, Culture, and Society." *Public Culture, Vol.* 15 (2) (2003): 261–85.

Gedicks, Al. *The New Resource Wars: Native and Environmental Struggles against Multinational Companies.* Boston, South End Press, 1993.

Ghosh, Amitav. "Petrofiction." *The New Republic* (1992): 29–34.

———. *The Great Derangement: Climate Change and the Unthinkable.* Chicago: University of Chicago Press, 2017.

Gilbert, Scott F. "When 'Personhood' Begins in the Embryo: Avoiding a Syllabus of Errors." *Birth Defects Research, Part C, Vol.* 84 (2) (2008): 164–73.

———. "Holobiont by Birth: Multilineage Individuals as the Concretion of Cooperative Processes" In *Arts of Living on a Damaged Planet: Ghosts and Monsters of the Anthropocene,* edited by Anna Tsing, Heather Swanson, Elaine Gan, and Nils Bubandt. Minneapolis: University of Minnesota Press, 2017.

Gilboff, Sander. *HG Bronn, Ernst Haeckel, and the Origins of German Darwinism.* Cambridge: MIT Press, 2008.

Gill, Bikrum. "Race, Nature, and Accumulation: A Decolonial World-Ecological Analysis of Indian Land Grabbing in the Gambella Province of Ethiopia." PhD Dissertation. Toronto: York University, 2016.

Goldenberg, Robert. "The Destruction of the Jerusalem Temple: Its Meaning and Its Consequences." In *The Cambridge History of Judaism,* 8 Volumes (4), edited by W.D. Davies and Louis Finkelstein. Cambridge: Cambridge University Press, 1984–2018.

Goodenough, Ursula. *The Sacred Depths of Nature.* New York: Oxford University Press, 2000.

Gottschall, Jonathan. *The Storytelling Animal: How Stories Make Us Human.* New York: Houghton Mifflin Harcourt Publishers, 2012.

Gowans, Christopher W. *Buddhist Moral Philosophy: An Introduction.* New York: Routledge, 2014.

Grainger, Brett Malcom. *Church in the Wild: Evangelicals in Antebellum America.* Cambridge: Harvard University Press, 2019.

Green, William Scott and Jed Silverstein. "Messiah." In *The Encyclopedia of Judaism,* 3 Volumes, edited by Alan J. Avery-Peck, Jacob Neusner, and William Scott Green. New York: Bloomsbury, 1999.

Gregg, Melissa and Gregory J. Seigworth, eds. *The Affect Theory Reader*. Durham: Duke University Press, 2010.

Grim, John and Mary Evelyn Tucker. *Ecology and Religion*. Washington, D.C.: Island Press, 2014.

Grim, John. "Indigenous Cosmovisions and a Humanist Perspective on Materialism." In *Earthly Things: Immanence, New Materialisms, and Planetary Thinking*, edited by Karen Bray, Heather Eaton, and Whitney Bauman. New York: Fordham University Press, 2023.

――――. ed. *Indigenous Traditions and Ecology: The Interbeing of Cosmology and Community*. Cambridge: Harvard Divinity School Center for the Study of World Religions, 2001.

――――. *Living Cosmology: Christian Responses to Journey of the Universe*. New York: Orbis Books, 2016.

――――. *The Shaman: Patterns of Religious Healing among the Ojibway Indians*. Norman: University of Oklahoma, 1983.

Grosz, Elizabeth. *The Incorporeal: Ontology, Ethics, and the Limits of Materialism*. New York: Columbia University Press, 2017.

Gruen, Lori and Robert C. Jones. "Veganism as an Aspiration." In *The Moral Complexities of Eating Meat*, edited by B. Bramble and B. Fischer. New York: Oxford University Press, 2015.

Guenther, Mathias. *Tricksters and Trancers: Bushman Religion and Society*. Bloomington: Indiana University Press, 1999.

Haberman, David L. *People Trees: Worship of Trees in Northern India*. New York: Oxford University Press, 2013.

Habito, Ruben. "Mountains and Rivers and the Great Earth." In *Buddhism and Ecology: The Interconnection of Dharma and Deeds*, edited by Mary Evelyn Tucker and Duncan Ryūken Williams. Cambridge: Harvard University Press, 1997.

Haeckel, Ernst. *Monism as Connecting Religion and Science: The Confession of Faith of a Man of Science*. London: Adam and Charles Black, 1895.

――――. *The Wonders of Life: A Popular Study of Biological Philosophy*. New York: Harper and Brothers, 1905.

Halberstam, Jack. *The Queer Art of Failure*. Durham: Duke University Press, 2011.

Halbmayer, Ernst, ed. "Debating Animism, Perspectivism and the Construction of Ontologies." *Indiana*, Vol. 29 (2012): 9–23.

Hallowell, A. Irving. "Ojibwa Ontology, Behavior, and World View." In *Primitive Views of the World*, edited by Stanley Diamond. New York: Columbia University Press, 1964.

――――. *The Ojibwa of Berens River, Manitoba*. Edited by Jennifer S.H. Brown. Fort Worth: Harcourt Brace, 1992.

Hanly, Elizabeth. "Symbol Truths." *South Florida* (April 1994).

Haraway, Donna. "Ecco Homo, Ain't (Ar'n't) I A Woman, and Inappropriate/d Others: The Human in a Post-Humanist Landscape." In *The Haraway Reader*. New York: Routledge, 2004.

———. *When Species Meet*. Minneapolis: University of Minnesota Press, 2008.

———. "Anthropocene, Capitalocene, Plantationocene, Chthulucene: Making Kin." *Environmental Humanities, Vol.* 6 (2015): 159–65.

———. *Staying with the Trouble: Making Kin in the Chthulucene*. Durham: Duke University Press, 2016,

Harding, Sandra. "Beyond Postcolonial Theory: Two Under Theorized Perspectives on Science and Technology." In *The Postcolonial Science and Technology Studies Reader*, edited by Sandra Harding. Durham: Duke University Press, 2011.

Harding, Stephan. *Animate Earth: Science, Intuition and Gaia*. White River Junction: Chelsea Green, 2006.

Harman, Graham. *Immaterialism: Objects and Social Theory*. Cambridge: Polity Press, 2016.

———. *Object-Oriented Ontology: A New Theory of Everything*. London: Pelican, 2018.

Harney, Stefano and Fred Moten. *The Undercommons: Fugitive Planning and Black Study*. Colchester, England: Minor Compositions, 2013.

Hartman, David. "Maimonides' Approach to Messianism and its Contemporary Implications." *Daat: A Journal of Jewish Philosophy & Kabbalah, Vol.* 2 (3) (1978): 5–33.

Hartman, Matthew R. "The Entangled Relations of Our Ecological Crisis: Religion, Capitalism's Logic, and New Forms of Planetary Thinking." In *Earthly Things: Immanence, New Materialisms, and Planetary Thinking*, edited by Karen Bray, Heather Eaton, and Whitney Bauman. New York: Fordham University Press, 2023.

Harvey, Graham, ed. *Indigenous Religions: A Companion*. London and New York: Cassell, 2000.

———, ed. *The Handbook of Contemporary Animism*. New York: Routledge, 2013.

———. "Performing Indigeneity and Performing Guesthood." In *Religious Categories and the Construction of the Indigenous*, edited by Christopher Hartney and Daniel J. Tower. Leiden: Brill, 2017.

———. *Animism: Respecting the Living World*. New York: Columbia University Press, 2016.

———. "We have always been animists . . ." In *Earthly Things: Immanence, New Materialisms, and Planetary Thinking*, edited by Karen Bray, Heather Eaton, and Whitney Bauman. New York: Fordham University Press, 2023.

Hauer, Mathew E., Jason M. Evans, and Deepak R. Mishra. "Millions Projected to Be at Risk from Sea-Level Rise in the Continental United States." *Nature Climate Change, Vol.* 6 (2016): 691–695. https://doi.org/10.1038/nclimate2961.

Hedges, Paul. "Multiple Religious Belonging after Religion: Theorising Strategic Religious Participation in a Shared Religious Landscape as a Chinese Mode." *Open Theology, Vol.* 3 (2017): 48–72.

Hegel, Georg Wilhelm Friedrich. *Phenomenology of Spirit*. Translated by A. V. Miller. Oxford: Oxford University Press, 1977.

Heidegger, Martin. *Being and Time*. Translated by John Macquarrie and Edward
    Robinson. New York: Harper & Row, 1962.
Heil, John. "Emergence and Panpsychism." In *The Routledge Handbook of
    Emergence*, edited by Sophie Gibb, Robin Findlay Hendry, and Tom Lancaster.
    London: Routledge, 2019.
Heine, Steven. "Introduction: Dōgen Studies on Both Sides of the Pacific." In *Dōgen:
    Textual and Historical Studies*, edited by Steven Heine. New York: Oxford
    University Press, 2012.
———. "What Is on the Other Side? Delusion and Realization in Dōgen's
    'Genjōkōan'." In *Dōgen: Textual and Historical Studies*, edited by Steven Heine.
    New York: Oxford University Press, 2012.
Henderson, John. *The Development and Decline of Chinese Cosmological Thought*.
    New York: Columbia University Press, 1984.
Hick, John. *God Has Many Names*. Philadelphia: The Westminster Press, 1982.
Hill, Julia Butterfly. *The Legacy of Luna: The Story of a Tree, a Woman, and the
    Struggle to Save the Redwoods*. San Francisco: Harper San Francisco, 2000.
Hirt-Manheimer, Aron. "On God, Indifference, and Hope: A Conversation with
    Elie Wiesel." *Reform Judaism* (2005). https://reformjudaism.org/jewish-life/arts
    - culture/literature/god-indifference-and-hope-conversation-elie-wiesel.
Hogan, Linda. "We Call It *Tradition*." In *The Handbook of Contemporary Animism*,
    edited by Graham Harvey. New York: Routledge, 2013.
Hogue, Michael. *American Immanence: Democracy for an Uncertain World*. New
    York: Columbia University Press, 2018.
———. *The Promises of Religious Naturalism*. Lanham: Rowman & Littlefield, 2010.
Horton, Robin. *Patterns of Thought in Africa and the West*. Boston: Cambridge
    University Press, 1997.
Houseman, Michael J. "Relationality." In *Theorizing Rituals, Volume I: Issues,
    Topics, Approaches, Concepts*, edited by Jens Kreinath, Jan Snoek, and Michael
    Stausberg. London: Brill, 2006.
Hrdy, Sarah. *Mothers and Others*. Cambridge: Harvard University Press, 2009.
Hucks, Tracey E. *Yoruba Traditions and African American Religious Nationalism*.
    Albuquerque: University of New Mexico Press, 2014.
Hunt, Carl. *Oyotunji Village: The Yoruba Movement in America*. Washington, D.C:
    University Press of America, 1979.
Idle No More Manifesto. idlenomore.ca/manifesto Indigenous Environmental
    Network, www.ien.org.
Ingold, Tim. *Perception of the Environment: Essays in Livelihood, Dwelling, and
    Skill*. New York: Routledge, 2000.
———. "Rethinking the Animate, Re-animating Thought." *Ethnos*, Vol. 71(1) (2011):
    9–20
———. "Earth, Sky, Wind, and Weather." *Journal of the Royal Anthropological
    Association*, Vol. 13 (1) (2007): S19–S38.

"Interactive Map." *B'Tselem: The Israeli Information Center for Human Rights in the Occupied Territories*. https://www.btselem.org/map.

Isasi-Diaz, Ada Maria. *Mujerista Theology: A Theology for the Twenty-First Century*. Maryknoll: Orbis Books, 1996.

Ivanhoe, Phillip. J. *Confucian Moral Self Cultivation*. Second Edition. Indianapolis: Hackett, 2000.

Ives, Christopher. "Mountains Preach the Dharma: Immanence in Mahāyāna Buddhism." In *Earthly Things: Immanence, New Materialisms, and Planetary Thinking*, edited by Karen Bray, Heather Eaton, and Whitney Bauman. New York: Fordham University Press, 2023.

Jain, S. Lochlann. "Living in Prognosis: Toward an Elegiac Politics." *Representations*, Vol. 98 (1) (2007): 77–92.

Jain, Sulelkh C. *An Ahimsa Crisis: You Decide*. Jaipur: Prakrit Bharati Academy, 2016.

James, William. *The Varieties of Religious Experience*. New York: Cromwell-Collier, 1961.

Jefferson-Tatum, Elana. "Africana Sacred Matters: Religious Materialities in Africa, the Caribbean, and the Americas." In *Earthly Things: Immanence, New Materialisms, and Planetary Thinking*, edited by Karen Bray, Heather Eaton, and Whitney Bauman. New York: Fordham University Press, 2023.

Jenkins, Willis, Mary Evelyn Tucker, and John Grim, eds. *Routledge Handbook of Religion and Ecology*. London and New York: Routledge, 2017.

Johnston, Basil. *The Manitous: The Spiritual World of the Ojibway*. New York, Harper Collins, 1995.

Jordan, Chris. "Midway: Message from the Gyre." *The New York Review of Books*, November 11, 2009. http://www.nybooks.com/daily/2009/11/11/chris-jordan/.

"Journey of the Universe." https://www.journeyoftheuniverse.org.

Kafer, Alison. *Feminist, Queer, Crip*. Bloomington: Indiana University Press, 2013.

Kaiser, David. *How the Hippies Saved Physics: Science, Counterculture, and the Quantum Revival*. New York: W.W. Norton, 2011.

Kant, Immanuel. *Critique of Judgment*, translated by Werner S. Pluhar. Indianapolis: Hackett, 1987.

Katz, Richard, Megan Biesele, and Verna St. Denis, *Healing Makes Our Hearts Happy: Spirituality and Cultural Transformations among the Kalahari Ju|'hoansi*. Rochester, VT: Inner Traditions, 1997.

Kearns, Laurel and Catherine Keller. "Introduction." In *EcoSpirit: Religions and Philosophies for the Earth*. New York: Fordham University Press, 2007.

Keeney, Bradford and Hillary Keeney, eds. *Way of the Bushmen as Told by the Elders: Spiritual Teachings and Practices of the Kalahari Ju/hoansi*. Rochester, VT: Inner Traditions, 2016.

Keller, Catherine and John J. Thatamanil. "Is This an Apocalypse? We Certainly Hope So—You Should Too." *Religion & Ethics* (2020). https://www.abc.net.au/religion/catherine-keller-and-john-thatamanil-why-we-hope-this-is-an-apo

/12151922?fbclid=IwAR0Yu6vmZqTVzdgwU4zHv83pB-A-9EFlpqrg15BtxuVr0Bu
8Ir4SAlsL7MU.

Keller, Catherine and Mary-Jane Rubenstein, eds. *Entangled Worlds: Religion, Science and New Materialisms*. New York: Fordham University Press, 2017.

Keller, Catherine. *A Political Theology of the Earth: Our Planetary Emergency and the Struggle for a New Public*. New York: Columbia University Press, 2018.

———. "Amorous Entanglements: The Matter of Christian Panentheism." In *Earthly Things: Immanence, New Materialisms, and Planetary Thinking*, edited by Karen Bray, Heather Eaton, and Whitney Bauman. New York: Fordham University Press, 2023.

———. *The Face of the Deep: A Theology of Becoming*. New York: Routledge, 2003.

———. *Cloud of the Impossible: Negative Theology and Planetary Entanglement*. New York: Columbia University Press, 2015.

———. *Intercarnations: Exercises in Theological Possibility*. New York: Fordham University Press, 2017.

Keller, Mary. "The Spirit of Climate Change." Conference paper presented at the meeting of the International Society for the Study of Religion, Nature, and Culture. Cork, Ireland: Cork University Campus, 2019.

Kendi, Ibram X. *Stamped from the Beginning: The Definitive History of Racist Ideas in America*. New York: Bold Type Books, 2017.

Kim, Hee-Jin. *Dōgen Kigen: Mystical Realist*. Tucson: University of Arizona Press, 1987.

Kimelman, Reuven. "Rabbinic Prayer in Late Antiquity." In *The Cambridge History of Judaism*, 8 Volumes, edited by W.D. Davies and Louis Finkelstein. Cambridge: Cambridge University Press, 1984–2018.

———. "The Rabbinic Theology of the Physical: Blessings, Body and Soul, Resurrection, and Covenant and Election." In *The Cambridge History of Judaism*, 8 Volumes, edited by W.D. Davies and Louis Finkelstein. Cambridge: Cambridge University Press, 1984–2018.

Kimmerer, Robin W. *Braiding Sweetgrass: Indigenous Wisdom, Scientific Knowledge, and the Teachings of Plants*. Minneapolis: Milkweed, 2013.

King, Thomas. *The Truth About Stories: A Native Narrative*. Minneapolis: University of Minnesota Press, 2003.

Klein, Naomi. *This Changes Everything: Capitalism vs. the Climate*. New York: Simon & Schuster, 2014.

Klein, Richard and Blake Edgar. *The Dawn of Human Culture: A Bold New Theory of What Sparked the "Big Bang" of Human Consciousness*. Hoboken: John Wiley & Sons, 2002.

Kohn, Eduardo. *How Forests Think: Toward an Anthropology Beyond the Human*. Berkeley: University of California Press, 2013.

Koosed, Jennifer L. and Stephen D. Moore. "Introduction: From Affect to Exegesis." In *Biblical Interpretation*, Vol. 22 (2014): 381–387.

Kopenawa, Davi Y. "Sonhos das origens." In *Povos indígenas no Brasil (1996–2000)*, edited by Carlos A. Ricardo. São Paulo: ISA, 2000.

Kopenawa, Davi Y. and Bruce Albert. *The Falling Sky: Words of a Yanomami Shaman*. Cambridge: Harvard University Press, 2013.

Kopf, Gereon. *Beyond Personal Identity: Dōgen, Nishida, and the Phenomenology of No-Self*. New York: Routledge, 2001.

Krech, Shepard, III. *The Ecological Indian: Myth and History*. New York: W.W. Norton, 2000.

Kreiner, Josef, ed. "Confucianism and Modernization in East Asia." In *The Impact of Traditional Thought on Present Day Japan*. München: IUDICIUM Verlag, 1996.

Kurath, Gertrude. *The Art of Tradition: Sacred Music, Dance, and Myth of Michigan's Anishinaabe, 1946–1955*. Edited by Michael D. McNally. East Lansing: University of Michigan Press, 2009.

Kurzweil, Ray. *The Singularity Is Near: When Humans Transcend Biology*. New York: Viking, 2005.

Kyong-Dong, Kim. "Confucianism and Modernization in East Asia." In *The Impact of Traditional Thought on Present-Day Japan*, edited by Josef Kreiner. München: Iudicium-Verl, 1996.

Lajee Center, http://www.lajee.org/.

LaMothe, Kimerer L. *A History of Theory and Method in the Study of Religion and Dance*. Leiden: Brill, 2018.

———. "Becoming a Bodily Self." In *Religious Experience and New Materialism*, edited by Joerg Rieger and Ed Waggoner. New York: Palgrave Macmillan, 2016.

———. "Can They Dance: Towards a Philosophy of Bodily Becoming." *Journal of Dance & Somatic Practices*, Vol. 4 (1) (2012): 93–107.

———. "Dancing Immanence: A Philosophy of Bodily Becoming." In *Earthly Things: Immanence, New Materialisms, and Planetary Thinking*, edited by Karen Bray, Heather Eaton, and Whitney Bauman. New York: Fordham University Press, 2023.

———. "Does Your God Dance? The Role of Rhythmic Bodily Movement in Friedrich Nietzsche's Revaluation of Values." In *Dance as Third Space: Interreligious, Intercultural, and Interdisciplinary Debates on Dance and Religion(s)*, edited by Heike Waltz. Gottingen, Germany: Vandenhoeck & Ruprecht, 2021.

———. *Nietzsche's Dancers: Isadora Duncan, Martha Graham, and the Revaluation of Christian Values*. New York: Palgrave Macmillan, 2006.

———. *Why We Dance: A Philosophy of Bodily Becoming*. New York: Columbia University Press, 2015.

Latour, Bruno. *Reassembling the Social: An Introduction to Actor-Network Theory*. Oxford: Oxford University Press, 2005.

———. "Fetish-Factish." *Material Religion*, Vol. 7 (1) (2015): 42–49.

———. *We Have Never Been Modern*. New York: Harvester Wheatsheaf, 1993.

————. "Whose Cosmos, Which Cosmopolitics? Comments on the Peace Terms of
   Ulrich Beck." *Common Knowledge*, Vol. 10 (3) (2004): 450–62.
————. "Thou Shall Not Freeze-Frame." In *Science, Religion, and the Human
   Experience*, edited by James Proctor. Oxford: Oxford University Press, 2005.
————. *An Inquiry into Modes of Existence: An Anthropology of the Moderns.*
   Cambridge: Harvard University Press, 2013.
LeBlanc, Charles. *Huai-nan tzu: Philosophical Studies in Early Han Thought.* Hong
   Kong: Hong Kong University Press, 1985.
Legat, Allice. *Walking the Land, Feeding the Fire: Knowledge and Stewardship
   Among the Tlicho Dene.* Tucson: University of Arizona Press, 2012.
Leighton, Taigen Dan. *Just This Is It: Dongshan and the Practice of Suchness.*
   Boston: Shambhala Publications, 2015.
LeMenager, Stephanie. *Living Oil: Petroleum Culture in the American Century.*
   New York: Oxford University Press, 2014.
Leopold, Aldo. *Sand County Almanac: And Sketches Here and There.* New York:
   Oxford University Press, 1987.
Leslie, Heather A., et al. "Discovery and Quantification of Plastic Particle Pollution
   in Human Blood." *Environment International.* https://doi.org/10.1016/j.
   envint.2022.107199.
Lévi-Strauss, Claude. *The Savage Mind.* Chicago: University of Chicago Press,
   1966.
Levinas, Emmanuel. *Difficult Freedom: Essays on Judaism.* Translated by Seán
   Hand. Baltimore: Johns Hopkins University Press, 1990.
————. *Existence and Existents.* Translated by A. Lingis. Pittsburgh: Duquesne
   University, 1978.
————. *Nine Talmudic Readings.* Translated by Annette Aronowicz. Bloomington:
   Indiana University Press, 1990.
Lévy-Bruhl, Lucien. *Primitive Mentality*, translated by Lilian A. Clare. New York:
   Macmillan, 1923.
Lewis-Williams, David. *A Cosmos in Stone: Interpreting Religion and Society
   through Rock Art.* Walnut Creek: Altamira Press, 2002.
————. *The Mind in the Cave: Consciousness and The Origins of Art.* London:
   Thames & Hudson, 2002.
Lidskog, Rolf and Claire Waterton. "Anthropocene—A Cautious Welcome from
   Environmental Sociology?" *Environmental Sociology*, Vol. 2, (4) (2016): 395–406.
   DOI: 10.1080/23251042.2016.1210841.
Llinas, Rodolfo. *I of the Vortex: From Neurons to Self.* Boston: MIT Press, 2001.
Lovejoy, Arthur O. "The Dialectic of Bruno and Spinoza." In *The Summum Bonum*,
   edited by Evander Bradley McGilvary. Berkeley: The University of California
   Press, 1904.
Lovelock, James. "Hands up for the Gaia Hypothesis." *Nature*, Vol. 344 (6262)
   (March 1990): 100–02. https://doi.org/10.1038/344100a0.
————. *Gaia: A New Look at Life on Earth.* Oxford: Oxford University Press, 1987.

Lynch, Thomas. *Apocalyptic Political Theology: Hegel, Taubes and Malabou*. London: Bloomsbury, 2019.

Maccagnolo, Enzo. "David of Dinant and the Beginnings of Aristotelianism in Paris." In *A History of Twelfth-Century Western Philosophy*, edited by Peter Dronke. New York: Cambridge University Press, 1992.

MacGaffey, Wyatt. "Complexity, Astonishment and Power: The Visual Vocabulary of Kongo Minkisi." *Journal of Southern African Studies*, Vol. 14 (2) (January 1988): 188–203.

———. "The Personhood of Ritual Objects: Kongo 'Minkisi.'" *Etnofoor, Vol.* 3 (1) (1990): 45–61.

———. *Kongo Political Culture: The Conceptual Challenge of the Particular*. Bloomington: Indiana University Press, 2000.

Machle, Edward. *Nature and Heaven in Xunzi: A Study of the Tian Lun*. Albany: SUNY Press, 1993.

MacKendrick, Karmen. "Remember—When?" In *Sexual (Dis)Orientations: Queer Temporalities, Affects, Theologies*, edited by Kent L. Brintnall, Joseph A. Marchal, and Stephen D. Moore. New York: Fordham University Press, 2017.

Magnus, Albertus. "Summa Theologiae Sive Scientia De Mirabili Scientia Dei." Edited by E. Borgnet. Paris: Vives, 1894. http://albertusmagnus.uwaterloo.ca/Downloading.html.

Malebranche, Nicolas. *The Search for Truth and Elucidations of the Search for Truth*. Translated by Thomas M. Lennon and Paul J. Olscamp. Cambridge: Cambridge University Press, 1997.

Margulis, Lynn and René Fester. *Symbiosis as a Source of Evolutionary Innovation*. Cambridge: MIT Press, 1991.

Marriott, McKim. "Hindu Transactions: Diversity Without Dualism." In *Transaction and Meaning: Directions in the Anthropology of Exchange and Symbolic Behavior*, edited by Bruce Kapferer. Philadelphia: ISHI, 1976.

Marrow Long, Carolyn. *Spiritual Merchants: Religion, Magic, and Commerce*. Knoxville: University of Tennessee Press, 2001.

Marx, Karl, and Frederick Engels. *Collected Works. Volume* 37. New York: International Publisher, 1975.

Marx, Karl. *Capital: A Critique of Political Economy. Volume I* (1867), translated by Ben Fowkes. New York: Random House, 1977.

Masuzawa, Tomoko. *The Invention of World Religions: Or, How European Universalism Was Preserved in the Language of Pluralism*. Chicago: University of Chicago Press, 2005.

Matthews, Maureen and Roger Roulette. "'Are All Stones Alive?': Anthropological and Anishinaabe Approaches to Personhood." In *Rethinking Personhood: Animism and Materiality*, edited by Miguel Astor-Aguilera and Graham Harvey. New York: Routledge, 2018.

Mazis, Glen. *Earth Bodies: Rediscovering our Planetary Senses*. Albany: SUNY Press, 2002.

McCarthy Brown, Karen. "Making Wanga: Reality Construction and the Magical Manipulation of Power." In *Religion and Healing in America*, edited by Linda L. Barnes and Susan S. Sered. New York: Oxford University Press, 2005.

McCleary, Timothy. *The Stars We Know: Crow Indian Astronomy and Lifeways.* Prospect Heights: Waveland Press, 1996.

McDaniel, Jay B. *Of God and Pelicans: A Theology of Reverence for Life.* Louisville: Westminster/John Knox Press, 1989.

McFague, Sallie. *Life Abundant: Rethinking Theology and Economy for a Planet in Peril.* Minneapolis: Fortress Press, 2001.

McIntyre, J. Lewis. *Giordano Bruno.* London: Macmillan, 1903.

McLaurin, John J. *Sketches in Crude Oil: Some Accidents and Incidents of the Petroleum Development in All Parts of the Globe.* Harrisburg: Published by the author, 1896.

McMahan, David L. *The Making of Buddhist Modernism.* New York: Oxford University Press, 2008.

McNeill, William. *Keeping Together in Time.* Cambridge: Harvard University Press, 1997.

McPherson, Dennis and J. Douglas Rabb. *Indian from the Inside: A Study in Ethno-Metaphysics.* Thunder Bay, Ontario: Lakehead University, Centre for Northern Studies, 1993.

Merchant, Carolyn. *The Death of Nature: Women, Ecology, and the Scientific Revolution.* New York: HarperCollins, 1980.

Merleau-Ponty, Maurice. *Phenomenology of Perception.* Translated by Colin Smith. London: Routledge, 2002.

Mickey, Sam, Mary Evelyn Tucker, and John Grim, eds. *Living Earth Community: Multiple Ways of Being and Knowing.* Cambridge: Open Book Publishers, 2020.

Mickey, Sam. "Solidarity with Nonhumans: Being Ecological with Object-Oriented Ontology." In *Earthly Things: Immanence, New Materialisms, and Planetary Thinking*, edited by Karen Bray, Heather Eaton, and Whitney Bauman. New York: Fordham University Press, 2023.

Mignolo, Walter. *The Darker Side of Western Modernity: Global Futures, Decolonial Options.* Durham: Duke University Press, 2011.

Mill, John Stuart. *The Logic of the Moral Sciences.* Chicago: Open Court, 1988.

Minister, Kevin. "Decolonizing Interreligious Studies." In *Interreligious Studies: Dispatches from an Emerging Field*, edited by Hans Gustafson. Waco: Baylor University Press, 2020.

———. "Interreligious Approaches to Sustainability Without a Future: Two New Materialist Proposals for Religion and Ecology." In *Earthly Things: Immanence, New Materialisms, and Planetary Thinking*, edited by Karen Bray, Heather Eaton, and Whitney Bauman. New York: Fordham University Press, 2023.

———. "Organizing Bodies." In *Religious Experience and New Materialism: Movement Matters*, edited by Joerg Rieger and Edward Waggoner. New York: Palgrave Macmillan, 2016.

Monier-Williams, Monier. *A Sanskrit-English Dictionary: Etymologically and Philologically Arranged with Special Reference to Cognate Indo-European Languages*. Oxford: Clarendon Press, 1899.

Moore, Jason W. "The Capitalocene, Part I: On the Nature and Origins of Our Ecological Crisis." *The Journal of Peasant Studies*, Vol. 44 (3) (2017): 594–630.

———. "The Capitalocene, Part II: Accumulation by Appropriation and the Centrality of Unpaid Work/Energy." *The Journal of Peasant Studies*, Vol. 45 (2) (2018): 237–279.

Morton, Timothy. "Buddhaphobia: Nothingness and the Fear of Things." In *Nothing: Three Inquiries in Buddhism*, edited by Marcus Boon, Eric Cazdyn, and Timothy Morton. Chicago: University of Chicago Press, 2015.

———. "Queer Ecology," in *PMLA*, Vol. 125 (2) (March 2010): 273–282.

———. *Being Ecological*. Cambridge: MIT Press, 2018

———. *Dark Ecology: Toward a Logic of Future Coexistence*. New York: Columbia University Press, 2016.

———. *Ecology without Nature: Rethinking Environmental Aesthetics*. Cambridge: Harvard University Press, 2007.

———. *Humankind: Solidarity with Non-Human People*. New York & London: Verso, 2017.

———. *Hyperobjects: Philosophy and Ecology after the End of the World*. Minneapolis: University of Minnesota Press, 2013.

———. *Realist Magic: Objects, Ontology, Causality*. Ann Arbor: Open Humanities Press, 2013.

———. *The Ecological Thought*. Cambridge: Harvard University Press, 2010.

Moyers, Bill. "Interview with James Cone." *Bill Moyers The Journal*. November 23, 2017. http://billmoyers.com/content/james-cone-on-the-cross-and-the-lynching-tree/.

Mugambi, Jesse. "Africa: African Heritage and Ecological Stewardship." In *Routledge Handbook of Religion and Ecology*. London and New York: Routledge, 2017.

Murphy, Joseph M. *Working the Spirit: Ceremonies of the African Diaspora*. Boston: Beacon Press, 1994.

Nadler, Steven. *Spinoza: A Life*. Cambridge: Cambridge University Press, 2001.

Nagarjuna. *The Fundamental Wisdom of the Middle Way*. Translated with commentary by Jay L. Garfield. Oxford: Oxford University Press, 1995.

Najita, Tetsuo. *Visions of Virtue in Tokugawa Japan: The Kaitokudo Merchant Academy of Osaka*. Chicago: University of Chicago Press, 1987.

Nash, Roderick. *The Wilderness and the American Mind*. New Haven: Yale University Press, 2014.

Naveh, Danny and Nurit Bird-David. "Animism, Conservation and Immediacy." In *The Handbook of Contemporary Animism*, edited by Graham Harvey. London: Routledge, 2013.

Needham, Joseph. *Science and Civilization in China. Volume 2, History of Scientific Thought*. Cambridge: Cambridge University Press, 1956.

Nelson, Melissa, ed. *Original Instructions: Indigenous Teachings for a Sustainable Future*. Rochester, VT: Bear and Company, 2008.

Nelson, Melissa. "North America: Native Ecologies and Cosmovisions Renew Treaties with the Earth and Fuel Indigenous Movements." In *Routledge Handbook of Religion and Ecology*. London and New York: Routledge, 2017.

Neusner, Jacob. "Emergent Rabbinic Judaism in a Time of Crisis." In *Early Rabbinic Judaism: Historical Studies in Religion, Literature and Art*. Leiden: Brill, 1975.

———. *Messiah in Context: Israel's History and Destiny in Formative Judaism*. Philadelphia: Fortress Press, 1984.

Neville, Robert. *Boston Confucianism*. Albany: SUNY Press, 2000.

———. *Hsun Tzu, Basic Writings*. Translated by Burton Watson. New York: Columbia University Press,1963.

Newcomb, Steven. *Pagans in the Promised Land: Decoding the Doctrine of Christian Discovery*. Golden: Fulcrum Publishing, 2008.

Nicholas of Cusa. "On Learned Ignorance." In *Nicholas of Cusa: Selected Writings*, translated by H. Lawrence Bond. New York: Paulist Press, 1997.

Nikiforuk, Andrew. *The Energy of Slaves: Oil and the New Servitude*. Vancouver, CA: Greystone Books, 2012.

Norberg, Johan. *Progress: Ten Reasons to Look Forward to the Future*. London: Oneworld, 2016.

Ogbonnaya, Okechukwu A. *On Communitarian Divinity: An African Interpretation of the Trinity*. New York: Paragon House, 1994.

Okumura, Shohaku. *Realizing Genjōkōan: A Key to Dōgen's Shōbōgenzō*. Boston: Wisdom Publications, 2010.

Otto, Rudolf. *The Idea of the Holy*. New York: Oxford University Press, 1958.

Outka, Paul. *Race and Nature: From Transcendence to the Harlem Renaissance*. New York: Palgrave Macmillan, 2008.

Palestine Institute for Biodiversity and Sustainability of Bethlehem University. https://www.palestinenature.org/.

Pike, Sarah M. *For the Wild: Ritual and Commitment in Radical Eco-Activism*. Oakland: University of California Press, 2017.

———. *New Age and Neopagan Religions in America*. New York: Columbia University Press, 2004.

———. "Rewilding Hearts and Habits in the Ancestral Skills Movement." *Religions*, Vol. 9 (2018): https://www.mdpi.com/2077-1444/9/10/300.

———. "Rewilding Religion for a Primeval Future." In *Earthly Things: Immanence, New Materialisms, and Planetary Thinking*, edited by Karen Bray, Heather Eaton, and Whitney Bauman. New York: Fordham University Press, 2023.

Pirkei Avot, Sefaria. https://www.sefaria.org/Pirkei_Avot.

Plumwood, Val. *Feminism and the Mastery of Nature*. New York: Routledge, 1993.

Polsky, Colin and Hallie Eakin. "Global Change Vulnerability Assessments: Definitions, Challenges, and Opportunities." In *The Oxford Handbook of Climate Change and Society*, edited by John S. Dryzek, Richard B. Norgaard, and David Schlosberg. New York: Oxford University Press, 2013.

Porter, Tom. *Kanatsiohareke: Traditional Mohawk Indians Return to Their Ancestral Homeland.* Greenfield Center: Bowman Books, 1998.

Posey, Darrell. *Indigenous Knowledge and Ethics: A Darrell Posey Reader.* Edited by Kristina Plenderleith. New York, Routledge, 2004.

Povinelli, Elizabeth A. *Geontologies: A Requiem to Late Liberalism.* Durham: Duke University Press, 2016.

Powers, Richard. *The Overstory.* New York: W.W. Norton, 2018

Premawardhana, Devaka. "The Unremarkable Hybrid: Aloysius Pieris and the Redundancy of Multiple Religious Belonging." *Journal of Ecumenical Studies,* Vol. 46 (1) (2011): 76–101.

Puar, Jasbir. *The Right to Maim: Debility, Capacity, Disability.* Durham: Duke University Press, 2017.

Queen, Sarah. *From Chronicle to Canon: The Hermeneutics of the Spring and Autumn According to Tung Chung-shu.* New York: Cambridge University Press, 1996.

Ramachandran, V.K. *The Tell-Tale Brain: A Neuroscientist's Quest for What Makes Us Human.* New York: W.W. Norton, 2011.

Raymo, Chet. *When God Is Gone, Everything Is Holy.* Notre Dame: Sorin Books, 2008.

Richards, Robert J. *The Tragic Sense of Life: Ernst Haeckel and the Struggle Over Evolutionary Thought.* Chicago: University of Chicago Press, 2008.

Ridington, Robin and Jillian Ridington. *When You Sing It Now, Just Like New: First Nations Poetics, Voices, and Representations.* Lincoln: University of Nebraska Press, 2006.

Rieger, Annika and Joerg Rieger. "Working with Environmental Economists." In *T&T Clark Handbook of Christian Theology and Climate Change,* edited by Ernst Conradie and Hilda Koster. London: Bloomsbury/T&T Clark, 2020.

Rieger, Joerg and Kwok Pui-Lan. *Occupy Religion: Theology of Multitude, Religion in the Modern World.* Lanham: Rowman & Littlefield, 2012.

Rieger, Joerg and Rosemarie Henkel Rieger. *Unified We Are a Force: How Faith and Labour Can Overcome America's Inequalities.* St. Louis: Chalice, 2016.

Rieger, Joerg and Edward Waggoner, eds. *Religious Experience and the New Materialism: Movement Matters.* New York: Palgrave Macmillan, 2015.

Rieger, Joerg. "Why Movements Matter Most: Rethinking the New Materialism for Religion and Theology." In *Religious Experience and New Materialism: Movement Matters, Radical Theologues,* edited by Joerg Rieger and Edward Waggoner. New York: Palgrave Macmillan, 2015.

———. "Christian Theology and Empire." In *Empire and the Christian Tradition: New Readings of Classical Theologians,* edited by Kwok Pui-lan, Don Compier, and Joerg Rieger. Minneapolis: Fortress Press, 2007.

———. "Which Materialism, Whose Planetary Thinking?" In *Earthly Things: Immanence, New Materialisms, and Planetary Thinking,* edited by Karen Bray, Heather Eaton, and Whitney Bauman. New York: Fordham University Press, 2023.

———. *Jesus vs. Caesar: For People Tired of Serving the Wrong God.* Nashville: Abingdon, 2018.

———. *No Rising Tide: Theology, Economics, and the Future.* Minneapolis: Fortress Press, 2009.

———. *Religion, Theology, and Class: Fresh Conversations After Long Silence. New Approaches to Religion and Power,* edited by Joerg Rieger. New York: Palgrave Macmillan, 2013.

———. *Theology in the Capitalocene: Ecology, Identity, Class, and Solidarity.* Minneapolis: Fortress Press, 2022.

Rigby, Kate. *Reclaiming Romanticism: Towards an Ecopoetics of Decolonization.* London: Bloomsbury, 2021.

Ritchie, Sarah Lane. "Integrated Physicality and the Absence of God: Spiritual Technologies in Theological Context." *Modern Theology,* Vol. 37 (2) (April 2021): 296–315. https://doi.org/10.1111/moth.12684.

Roberts, Michelle Voss. "Religious Belonging and the Multiple." *Journal of Feminist Studies in Religion,* Vol. 26 (1) (2010): 43–62.

Roberts, Tyler. *Encountering Religion: Responsibility and Criticism After Secularism.* New York: Columbia University Press, 2013.

Rogers-Vaughn, Bruce. *Caring for Souls in a Neoliberal Age: New Approaches to Religion and Power.* New York: Palgrave Macmillan, 2016.

Rogowska-Stangret, Monika. "Corpor(e)al Cartographies of New Materialism: Meeting the Elsewhere Halfway." *Minnesota Review,* Vol. 88 (2017).

Rolston Holmes, III. "Environmental Ethics and Religion/Science." In *The Oxford Handbook of Religion and Science,* edited by P. Clayton & Z. Simpson. New York: Oxford University Press, 2006.

———. "Science and Religion in the Face of the Environmental Crisis." In *The Oxford Handbook of Religion and Ecology,* edited by Roger S. Gottlieb. Oxford: Oxford University Press, 2006.

Rose, Deborah B. "Shimmer: When All You Love is Being Trashed." In *Arts of Living on a Damaged Planet: Ghosts and Monsters of the Anthropocene,* edited by Anna Tsing, Heather Swanson, Elaine Gan, and Nils Bubandt. Minneapolis: University of Minnesota Press, 2017.

Rosemont, Henry, ed. *Explorations of Early Chinese Cosmology.* Missoula: Scholars Press, 1984.

Rosenthal, Rudy. *Possession, Ecstasy, and Law in Ewe Voodoo.* Charlottesville: University of Virginia Press, 1998.

Rozman, Gilbert, ed. *The East Asian Region: Confucian Heritage and Its Modern Adaptation.* Princeton: Princeton University Press, 1991.

Rubenstein, Mary-Jane. "End without End: Cosmology and Infinity in Nicholas of Cusa." In *Desire, Faith, and the Darkness of God: Essays in Honor of Denys Turner,* edited by Eric Bugyis and David Newheiser. Notre Dame: University of Notre Dame Press, 2016.

———. *Pantheologies: Gods, Worlds, Monsters.* New York: Columbia University Press, 2018.

———. "The Animist, Almost Feminist, Quite Nearly Pantheist Old Materialism of Giordano Bruno." In *Earthly Things: Immanence, New Materialisms, and*

Planetary Thinking, edited by Karen Bray, Heather Eaton, and Whitney Bauman. New York: Fordham University Press, 2023.

————. "The Matter with Pantheism: On Shepherds and Hybrids and Goat-Gods and Monsters." In *Entangled Worlds: Religion, Science, and New Materialisms*, edited by Catherine Keller and Mary-Jane Rubenstein. New York: Fordham University Press, 2017.

————. *Worlds without End: The Many Lives of the Multiverse*. New York: Columbia University Press, 2014.

Rue, Loyal. *Religion Is Not about God: How Spiritual Traditions Nurture Our Biological Nature and What to Expect When They Fail*. New Brunswick: Rutgers University Press, 2006.

Sabin, Paul. "'A Dive Into Nature's Great Grab-bag': Nature, Gender, and Capitalism in the Early Pennsylvania Oil Industry." *Pennsylvania History*, Vol. 66 (1999): 472–505.

Sagan, Dorion and Lynn Margulis. "Gaian Views." In *Ecological Prospects: Scientific, Religious, and Aesthetic Perspectives*, edited by Christopher Key Chapple. Albany: SUNY Press, 1994.

Sagan, Lynn. "On the Origin of Mitosing Cells." *Journal of Theoretical Biology*, Vol. 14 (3) (1967): 255–74.

Sahlins, Marshall. *Stone Age Economics*. Chicago: Aldine-Atherton, 1972.

Sahtouris, Elisabet. *EarthDance: Living Systems in Evolution*. San Jose: iUniversity, 2000.

Schaefer, Donovan. *Religious Affects: Animality, Evolution, and Power*. Durham: Duke University Press, 2015.

Schilbrack, Kevin. "An Emergence Theory of Religion." *Journal of the America Academy of Religion*. Forthcoming.

————. "Emergence Theory and the New Materialisms." In *Earthly Things: Immanence, New Materialisms, and Planetary Thinking*, edited by Karen Bray, Heather Eaton, and Whitney Bauman. New York: Fordham University Press, 2023.

Schlesier. Karl. *The Wolves of Heaven: Cheyenne Shamanism, Ceremonies, and Prehistoric Origins*. Norman: University of Oklahoma Press, 1987.

Schmidt, Jeremy J., Peter G. Brown, and Christopher Orr. "Ethics in the Anthropocene: A Research Agenda." *The Anthropocene Review*, Vol. 3, (3) (2016): 188–200. https://doi.org/10.1177/2053019616662052.

Scholem, Gershom. *The Messianic Idea in Judaism and other Essays on Jewish Spirituality*. New York: Schocken Books, 1971.

Schor, Juliet B. and Craig Thompson, eds. *Sustainable Lifestyles and the Quest for Plenitude: Case Studies of the New Economy*. New Haven: Yale University Press, 2014.

Schwerin Rowe, Terra. *Of Modern Extraction: Experiments in Critical Petro-theology*. New York: Bloomsbury/T&T Clark, 2022.

————. "Oily Animations: On Protestantism and Petroleum." In *Earthly Things: Immanence, New Materialisms, and Planetary Thinking*, edited by Karen Bray,

Heather Eaton, and Whitney Bauman. New York: Fordham University Press, 2023.

Severson, Eric R. *Levinas's Philosophy of Time: Gift, Responsibility, Diachrony, Hope.* Pittsburgh: Duquesne University Press, 2013.

Seymour, Nicole. "The Queerness of Environmental Affect." In *Affective Ecocriticism: Emotion, Embodiment, Environment,* edited by Kyle Bladow and Jennifer Ladino. Lincoln: University of Nebraska Press, 2018.

Sherman, Jeremy. *Neither Ghost nor Machine: The Emergence and Nature of Selves.* Columbia: Columbia University Press, 2017.

Shu-hsien, Liu. "The Confucian Approach to the Problem of Transcendence and Immanence." *Philosophy East and West, Vol.* 22 (1) (1972): 45–52.

Shubin, Neil. *Your Inner Fish: A Journey into the 35 Million-Year History of the Human Body.* New York: Vintage Books, 2008.

Shults, F. LeRon. *Iconoclastic Theology: Gilles Deleuze and the Secretion of Atheism.* Edinburgh: Edinburgh University Press, 2014.

Siegel, Daniel. *The Developing Mind.* New York: W.W. Norton, 1997.

Slater, Candace. *Entangled Edens: Visions of the Amazon.* Berkley: University of California Press, 2003.

Smith, Jonathan Z. *Relating Religion: Essays in the Study of Religion.* Chicago: University of Chicago Press, 2004.

Smith, Kidder Jr., Peter K. Bol, Joseph A. Adler, and Don J. Wyatt, eds. *Sung Dynasty Uses of the I Ching.* Princeton: Princeton University Press, 1990.

Snyder, Gary. *Practice of the Wild: Essays by Gary Snyder.* San Francisco: North Point Press, 1990.

Spiegel, Marjorie. *The Dreaded Comparison: Human and Animal Slavery.* New York: Mirror Books, 1997.

Spivak, Gayatri. *Death of a Discipline.* New York: Columbia University Press, 2003.

Spratt, David and Ian Dunlop. "Existential Climate-Related Security Risk: A Scenario Approach." *Breakthrough,* National Centre for Climate Restoration, Melbourne, Australia (May 2019): 1–10. Accessed August 13, 2020.

Standing, Guy. *The Precariat: The New Dangerous Class.* London: Bloomsbury Publishing, 2011.

Stengers, Isabelle. "The Cosmopolitical Proposal." In *Making Things Public: Atmospheres of Democracy,* edited by Bruno Latour and Peter Weibel. Cambridge: MIT Press, 2005.

———. *Cosmopolitics I & II.* Minneapolis: University of Minnesota Press, 2010 and 2011.

———. "Reclaiming Animism," *e-flux,* Vol. 36. (2012). http://worker01.e-flux.com /pdf/article_8955850.pdf. accessed January 16, 2020.

Stenmark, Lisa and Whitney Bauman, eds. *Unsettling Science and Religion: Contributions and Questions from Queer Studies.* Lanham: Lexington Books, 2018.

Stone, Jerome and Donald Crosby. "Introduction." In *The Routledge Handbook of Religious Naturalism,* edited by Jerome Stone and Donald Crosby. New York: Routledge, 2018.

Stone, Jerome. *Religious Naturalism Today: The Rebirth of a Forgotten Alternative.* Albany: SUNY Press, 2008.

Strathern, Marilyn. *The Gender of the Gift: Problems with Women and Problems with Society in Melanesia.* Berkeley: University of California Press, 1988.

Sweet, James H. *Recreating Africa: Culture, Kinship, and Religion in the Afro-Portuguese World.* Chapel Hill: University of North Carolina Press, 2013.

Swimme, Brian and Mary Evelyn Tucker. *Journey of the Universe.* New Haven: Yale University Press, 2011.

Swimme, Brian and Thomas Berry. *The Universe Story: From the Primordial Flaring Forth to the Ecozoic Era—A Celebration of the Unfolding of the Cosmos.* San Francisco: HarperSanFrancisco, 1994.

Szeman, Imre and Dominic Boyer, *Energy Humanities: An Anthology.* Baltimore: Johns Hopkins University Press, 2017.

Szeman, Imre and Petroculture Research Group. "Introduction." In *After Oil,* Petrocultures Research Group. Edmonton, Alberta: Petrocultures Research Group, 2016.

Tanner, Katherine. *Christianity and the New Spirit of Capitalism.* New Haven: Yale University Press, 2019.

Taylor, Rodney, ed. *The Ways of Heaven.* Leiden: Brill, 1986.

———. *The Cultivation of Sagehood as a Religious Goal in Neo-Confucianism: A Study of Selected Writings of Kao P'an-lung (1562–1626).* Missoula: Scholars Press, 1978.

———. *The Religious Dimensions of Confucianism.* Albany: SUNY Press, 1989.

"Tent of Nations: People Building Bridges." Tent of Nations Educational and Environmental Farm. http://www.tentofnations.org/.

Thanissaro Bhikkhu. "Romancing the Buddha." *Tricycle,* Vol. 12, (2) (Winter 2002).

The Worldviews Group http://www.vub.ac.be/CLEA/dissemination/groups-archive/vzw_worldviews/.

Thich Nhat Hanh. *Being Peace.* Berkeley: Parallax Press, 1990.

Todd, Zoe. "An Indigenous Feminist's Take on the Ontological Turn: 'Ontology' Is Just Another Word for Colonialism." *Journal of Historical Sociology, Vol.* 29 (1) (2016): 4–22.

Tremlett, Paul, Liam Sutherland, and Graham Harvey, eds. *Edward Tylor, Religion and Culture.* London: Bloomsbury, 2017.

Trungpa, Chögyam. *Mindfulness in Action: Making Friends with Yourself through Meditation and Everyday Awareness.* Boston: Shambhala Publications, 2015.

———. *The Future Is Open: Good Karma, Bad Karma, and Beyond Karma.* Boulder: Shambhala Publications, 2018.

Tsing, Anna L. *Friction: An Ethnography of Global Connection.* Princeton: Princeton University Press, 2004.

———. *The Mushroom at the End of the World: On the Possibility of Life in Capitalist Ruins.* Princeton: Princeton University Press, 2015.

Tsing, Anna, Heather Swanson, Elaine Gan, and Nils Bubandt, eds. *Arts of Living on a Damaged Planet: Ghosts and Monsters of the Anthropocene*. Minneapolis: University of Minnesota Press, 2017.

Tucker, Mary Evelyn. "Confucianism as a Form of Immanental Naturalism." In *Earthly Things: Immanence, New Materialisms, and Planetary Thinking*, edited by Karen Bray, Heather Eaton, and Whitney Bauman. New York: Fordham University Press, 2023.

Turner, William. "David of Dinant." In *The Catholic Encyclopedia*. New York: Robert Appleton Company, 2017. http://www.newadvent.org/cathen/04645a.htm.

Tuveson, Ernest Lee. *Redeemer Nation: The Idea of America's Millennial Role*. Chicago: University of Chicago Press, 1968.

Tylor, Edward B. *Primitive Culture*, 2 Volumes. London: John Murray, 1871.

UNDRIP http://www.un.org/esa/socdev/unpfii/documents/DRIPS_en.pdf.

Van Horn, O'neil. "On the Matter of Hope: Weaving Threads of Jewish Wisdom for the Sake of the Planetary." In *Earthly Things: Immanence, New Materialisms, and Planetary Thinking*, edited by Karen Bray, Heather Eaton, and Whitney Bauman. New York: Fordham University Press, 2023.

van Huyssteen, Wentzel. *Alone in the World? Science and Theology on Human Uniqueness*. Grand Rapids: Eerdmans Publishing, 2004.

Vitebsky, Piers. *The Reindeer People: Living with Animals and Spirits in Siberia*. New York: Houghton Mifflin Company, 2006.

Viveiros de Castro, Eduardo B. "Cosmological Deixis and Amerindian Perspectivism." *Journal of the Royal Anthropological Institute*, Vol. 4 (1998): 469–88.

———. "Exchanging Perspectives: The Transformation of Objects into Subjects in Amerindian Ontologies." *Common Knowledge*, Vol. 10 (3) (2004): 463–84.

———. "The Crystal Forest: Notes on the Ontology of Amazonian spirits." *Inner Asia*, Vol. 9 (2007): 13–33.

Vizenor, Gerald. *Manifest Manners: Narratives on Postindian Survivance*. Lincoln: University of Nebraska, 1999.

von Stuckrad, Kocku. *A Cultural History of the Soul: Europe and North America from 1870 to the Present*. New York: Columbia University Press, 2022.

Wallace, Mark I. *When God Was a Bird: Christianity, Animism, and the Re-Enchantment of the World*. New York: Fordham University Press, 2019.

Walter, Chip. *Thumbs, Toes, and Tears: And Other Traits That Make Us Human*. New York: Walker and Co., 2006.

Watson, Burton. *Hsun Tzu: Basic Writings*. New York: Columbia Press, 1963.

Weaver, Jace, ed. *Defending Mother Earth: Native American Perspectives on Environmental Justice*. Maryknoll: Orbis Books, 1996.

Weeks, Kathi. *The Problem with Work: Feminism, Marxism, Antiwork Politics, and Postwork Imaginaries*. Durham: Duke University Press, 2011.

Weheliye, Alexander G. *Habeas Viscus: Racializing Assemblages, Biopolitics, and Black Feminist Theories of the Human*. Durham: Duke University Press, 2014.

Weiming, Tu and Mary Evelyn Tucker, eds. *Confucian Spirituality.* 2 Volumes. New York: Crossroad Publishing, 2003–04.

Weiming, Tu. "Confucian Traditions in East Asian Modernity." *Bulletin of the American Academy of Arts and Sciences,* Vol. 50 (2) (1996): 12–39. https://doi .org/10.2307/3824246.

———. *Centrality and Commonality: An Essay on Confucian Religiousness.* Albany: SUNY Press, 1989.

———. *Confucian Thought: Selfhood as Creative Transformation.* Albany: SUNY Press, 1989

Wheeler, Demian and David E. Conner, eds. *Conceiving an Alternative: Philosophical Resources for an Ecological Civilization.* Anoka: Process Century Press, 2019.

White, Carol Wayne. *Black Lives and Sacred Humanity: Toward an African American Religious Naturalism.* New York: Fordham Press, 2016.

———. "Planetary Thinking, Agency, and Relationality: Religious Naturalism's Plea." In *Earthly Things: Immanence, New Materialisms, and Planetary Thinking,* edited by Karen Bray, Heather Eaton, and Whitney Bauman. New York: Fordham University Press, 2023.

———. "Re-envisioning Hope: Anthropogenic Climate Change, Learned Ignorance, and Religious Naturalism." *Zygon: Journal of Religion and Science,* Vol. 52 (2) (June 2018): 579.

White, Kyle Powys. "Indigenous Science (Fiction) for the Anthropocene: Ancestral Dystopias and Fantasies of Climate Change Crises." *Environment and Planning: Nature and Space,* Vol. 1, (2018): 1–2, 224–242.

White, Lynn Jr. "The Historical Roots of our Ecologic Crisis." *Science, Vol.* 155 (3767) (March 1967): 1203–1207

Whitehead, Alfred North. *Science and the Modern World.* New York: The Free Press, 1953.

———. *Process and Reality: Corrected Edition.* Edited by David Ray Griffin and Donald W. Sherburne. New York: The Free Press, 1985.

"Who We Are: Alrowwad Cultural and Arts Society." In *Alrowwad Cultural and Arts Society.* https://www.alrowwad.org/en/?page_id=9423.

Wiesel, Elie. "Art and Culture after the Holocaust." In *Auschwitz: The Beginning of an Era?,* edited by Eva Fleishner. New York: Ktav Publishing, 1977.

———. *Night.* New York: Hill and Wang, 2006.

———. *The Gates of the Forest.* Translated by Frances Frenaye. New York: Schocken Books, 1966.

Wildman, Wesley. "Religious Naturalism: What It Can Be, and What It Need Not Be." *Philosophy, Theology, and the Sciences,* Vol. 1 (2014): 36–58.

Wilford, John Noble. "When Humans Became Human." Published in *The New York Times,* February 26, 2002.

Wilhelm, Hellmut. *Heaven Earth and Man in the Book of Changes.* Seattle: University of Washington Press, 1977.

Willerslev, Rane. *Soul Hunters: Hunting, Animism, and Personhood Among the Siberian Yukaghirs*. Berkeley: University of California Press, 2007.

Williams, Paul. *Buddhist Thought: A Complete Introduction to the Indian Tradition*. New York: Routledge, 2000.

Wilson, Elizabeth. *Gut Feminism*. Durham: Duke University Press, 2012.

Wilson, Sheena, Adam Carlson, and Imre Szeman, eds. *Petrocultures: Oil, Politics, Culture*. Montreal and Chicago: McGill-Queen's University Press, 2017.

Wilson, Thomas, ed. *Genealogy of the Way: The Construction and Uses of the Confucian Tradition in Late Imperial China*. Redwood City: Stanford University Press, 1995.

Wilson, William. "Hope." In *New Wilson's Old Testament Word Studies*. Grand Rapids: Kregel Publications, 1987.

Wiredu, Kwasi. *Cultural Universals and Particulars: An African Perspective*. Bloomington: Indiana University Press, 1997

Witherspoon, Gary. *Language and Art in the Navajo Universe*. Ann Arbor: The University of Michigan Press, 1977.

Wong, Kate. "Mammoth Kill." *Scientific American, Vol.* 284 (2) (2001): 22.

Working Groups of the Intergovernmental Panel on Climate Change. "Climate Change 2014 Synthesis Report Summary for Policy Makers." Accessed October 18, 2019. https://www.ipcc.ch/site/assets/uploads/2018/02/AR5_SYR_FINAL _SPM.pdf.

Yeats, William Butler. "The Second Coming." In *The Dial*. 1920.

Yusoff, Kathryn. *A Billion Black Anthropocenes or None*. Minneapolis: University of Minnesota Press, 2018.

Zeebe, R.E., A. Ridgewell, and J. C. Zachos. "Anthropogenic Carbon Release Rate Unprecedented during the Past 66 Million Years." *Nature Geoscience*, Vol. 9 (2016): 325–329. http:/doi:10.1038/NGEO2681.

Zuck, Rochelle Raineri. "The Wizard of Oil: Abraham James, the Harmonial Wells, and the Psychometric History of the Oil Industry." In *Oil Culture*, edited by Ross Barrett and Daniel Worden. Minneapolis: University of Minnesota Press, 2014.

Contributors

Whitney Bauman is professor of religious studies at Florida International University in Miami, FL. He is also co-founder and co-director of *Counterpoint: Navigating Knowledge*, a non-profit based in Berlin, Germany that holds public discussions over social and ecological issues related to globalization and climate change. His areas of research interest fall under the theme of "religion, science, and globalization." He is the recipient of a Fulbright Fellowship and a Humboldt Fellowship. His publications include *Religion and Ecology: Developing a Planetary Ethic* (Columbia University Press, 2014), and co-authored with Kevin O'Brien, *Environmental Ethics and Uncertainty: Tackling Wicked Problems* (Routledge, 2019). He is currently finishing a manuscript entitled, *A Critical Planetary Romanticism: Literary and Scientific Origins of New Materialism* (Columbia University Press, Forthcoming 2023).

Karen Bray is an associate professor of religion and philosophy at Wesleyan College. She received her MDiv from Harvard Divinity School in 2010 and her PhD in Theological and Philosophical Studies in Religion from Drew University in 2016. Her research areas include continental philosophy of religion; feminist, critical disability, critical race, queer, political, and decolonial theories and theologies; and secularism and the post-secular. She is particularly interested in exploring how secular institutions and cultures behave theologically. Bray's work has been published in journals such as *The American Journal of Theology and Philosophy*, *The Journal for Cultural and Religious Theory*, and *Palgrave Communications* and in volumes published by Fordham University Press and Palgrave Macmillan. Dr. Bray is also a candidate for ordination with the Unitarian Universalist Association. Her most recent book is: *Grave Attending: A Political Theology for the Unredeemed* (Fordham University Press, 2019).

Christopher Key Chapple is Doshi Professor of Indic and Comparative Theology and founding director of the Master of Arts in Yoga Studies at Loyola Marymount University in Los Angeles. A specialist in the religions of India, he has published more than twenty books, including the recent *Living Landscapes: Meditations on the Five Elements in Hindu, Buddhist, and Jain Yogas* (SUNY Press, 2020). He serves as advisor to multiple organizations including: the Forum on Religion and Ecology (Yale); the Ahimsa Center (Pomona); the Dharma Academy of North America (Berkeley); the Jain Studies Centre (SOAS, London); the South Asian Studies Association; and International School for Jain Studies (New Delhi). He teaches online through the Center for Religion and Spirituality (LMU) and YogaGlo. Recent book: http://www.sunypress.edu/p-6860-living -landscapes.aspx.

Philip Clayton holds the Ingraham Chair at Claremont School of Theology, where he directs the PhD program in comparative theologies and philosophies; he is also affiliated faculty at Claremont Graduate University. A graduate of Yale University, he has also taught at Williams College and The California State University, as well as holding guest professorships at the University of Munich, the University of Cambridge, and Harvard University. He has published two dozen books and some 350 articles. Philip is president of the Institute for Ecological Civilization (EcoCiv.org), which works internationally to support multi-sector innovations toward a sustainable society through collaborations between governments, businesses, policy experts, and NGOs. He is also president of the Institute for the Postmodern Development of China, which works with universities and government officials to promote the concept of ecological civilization through conferences, publications, educational projects, and ecovillages. He has previously served as dean, provost, and executive vice president of a small university. In 2018 he helped to organize the Justice track for the Parliament of the World's Religions.

Heather Eaton is full professor of conflict studies at St. Paul University in Ottawa. She teaches and writes on issues of peace and conflict studies, particularly as they relate to ecological issues, ecofeminism, and religion and ecology in general. She has written and edited many different books and articles including *Introducing Ecofeminist Theologies* (T&T Clark, 2005); with Lauren Levesque, (eds.) *Advancing Nonviolence and Social Transformation* (Equinox, 2016); and as editor, *The Intellectual Journey of Thomas Berry: Imagining the Earth Community* (Lexington Press, 2014).

John Grim is a senior lecturer and research scholar at Yale School of the Environment, Yale Divinity School, and Yale's Religious Studies Department. He co-directs the Forum on Religion and Ecology at Yale. He has published, *The Shaman: Patterns of Religious Healing Among the Ojibway Indians* (University of Oklahoma Press, 1983) and *"Shamans and Preachers, Color Symbolism and Commercial Evangelism" in American Indian Quarterly* (Nebraska, 1992). With Mary Evelyn Tucker, he edited the series, *"World Religions and Ecology,"* (Harvard University Press, 1997–2000). In that series he edited *Indigenous Traditions and Ecology: The Interbeing of Cosmology and Community* (Harvard University Press, 2001). With Mary Evelyn Tucker he edited: *Worldviews and Ecology* (Orbis Books, 1994, fifth printing 2000); *Religion and Ecology: Can the Climate Change?* (*Daedalus*, 2001); *The Christian Future and the Fate of Earth* (Orbis Books, 2009); and *Living Cosmology: Christian Responses to Journey of the Universe* (Orbis Books, 2016). They are executive producers of the Emmy Award-winning film, *Journey of the Universe*, broadcast on PBS (journeyoftheuniverse.org). They published *Ecology and Religion* (Island Press, 2014), and with Willis Jenkins, they edited *Handbook of Religion and Ecology* (Routledge, 2017). They published *Thomas Berry: A Biography* (Columbia University Press, 2019). John is former president of the American Teilhard Association (1987–2020).

Matthew Hartman is a doctoral candidate in ethics at the Graduate Theological Union in Berkeley, California. His research engages critical ethical questions at the intersection of religion, identity, and culture within the broader environmental humanities. Matthew's dissertation examines the cultural implications of climate change denial in environmental imaginaries, with particular attention to the role of religion and race in shaping identity construction and maintenance on the American right. He is lead managing editor of the *Berkeley Journal of Religion and Theology*, and along with Devin Zuber and Rita Sherma, co-editor of *Sustainable Societies, Religious Ethics, and Planetary Restoration: Visions for a Viable Future* (Springer Nature, Forthcoming 2023).

Graham Harvey is emeritus professor of religious studies at The Open University, UK. His research largely concerns "the new animism," especially in the rituals and protocols through which Indigenous and other communities engage with the larger-than-human world. His publications include *Food, Sex & Strangers: Understanding Religion as Everyday Life* (2013), and *Animism: Respecting the Living World* (second edition 2017). He is editor of the Routledge series "Vitality of Indigenous Religions" and the *Equinox* series "Religion and the Senses".

Christopher Ives teaches in the area of Asian Religions and in his scholarship he focuses on modern Zen ethics. In 2009, he published *Imperial-Way Zen* (University of Hawaii Press, 2009), a book on Buddhist social ethics in light of Zen nationalism, especially as treated by Buddhist ethicist Ichikawa Hakugen (1902–86). Currently he is engaged in research and writing on Zen approaches to nature and Buddhist environmental ethics. His other book publications include *Zen on the Trail: Hiking as Pilgrimage* (Wisdom, 2018); *Zen Awakening and Society* (University of Hawaii Press, 1992); a translation of philosopher Nishida Kitaro's *An Inquiry into the Good* (co-translated with Abe Masao, Yale University Press, 1990); a translation of Hisamatsu Shin'ichi's *Critical Sermons of the Zen Tradition* (co-translated with Tokiwa Gishin, University of Hawaii Press, 2002); *The Emptying God* (co-edited with John B. Cobb, Jr., Orbis Books, 1990); and *Divine Emptiness and Historical Fullness* (edited volume, Trinity Press International, 1995).

Elana Jefferson-Tatum, PhD, is an independent practitioner-scholar of Africana religions, a hypnobirthing coach, and a full-spectrum doula. She received her Masters in Theological Studies (M.T.S.) from Harvard Divinity School and her PhD from Emory University. She also spent several years teaching at Hobart and William Smith Colleges and Tufts University. In addition to her scholarship, she has centered her commitment to the embodied materiality of Africana traditions around supporting her clients, especially BIPOC and LGBTQIA+ birthing people, through the initiatory process of pregnancy, labor, and parenthood.

Catherine Keller is professor of constructive theology at the Theological School of Drew University. In her teaching, lecturing, and writing, she develops the relational potential of a theology of becoming. Her books reconfigure ancient symbols of divinity for the sake of a planetary conviviality—a life together, across vast webs of difference. Thriving in the interplay of ecological and gender politics, process cosmology, poststructuralist philosophy, and religious pluralism, her work is both deconstructive and constructive in strategy. She is the author and editor of many publications including, *Cloud of the Impossible* (Columbia University Press, 2014) and *Facing Apocalypse: Climate, Democracy, and Other Last Chances* (Orbis Books, 2021).

Kimerer LaMothe is a dancer, philosopher, and scholar of religion, with a PhD in Theology of the Modern West from Harvard University, who taught at Brown and Harvard before following a dream to a small farm in upstate New York. A pioneer in the field of religion and dance, Kimerer is the author of six books, including: *Why We Dance: A Philosophy of Bodily Becoming* (Columbia University Press, 2015); *Nietzsche's Dancers: Isadora Duncan, Martha Graham, and the Revaluation of Christian Values* (Palgrave Macmillan, 2011); *Between Dancing and Writing: The Practice of Religious Studies* (Fordham University Press, 2004); and *A History of Theory and Method in the Study of Religion and Dance* (Brill Research Perspectives, 2018). She edited a special issue "Dancing on Earth" for the *Journal of Dance, Movement, and Spiritualities* (2017), and has received fellowships for her work from the Radcliffe Institute of Advanced Study, the Harvard Center for the Study of World Religions, and the Lower Adirondack Regional Arts Council (twice). She regularly lectures, teaches, and consults on the subject of religion and dance, and writes a monthly blog called "What a Body Knows" for *Psychology Today*.

Sam Mickey, PhD, is an educator and writer working at the intersection of religious, philosophical, and scientific perspectives on human-Earth relations. He is an adjunct professor in the Theology and Religious Studies department at the University of San Francisco. He is the Reviews Editor for the journal *Worldviews: Global Religions, Culture, and Ecology*, and he is an author and editor of several publications, including *On the Verge of a Planetary Civilization: A Philosophy of Integral Ecology* (Rowman and Littlefield International, 2014), and *Coexistentialism and the Unbearable Intimacy of Ecological Emergency* (Lexington Books, 2016)

Kevin Minister is associate professor of religion and participating faculty in environmental studies at Shenandoah University. Minister's research examines how we cultivate resilient environments with attention to religious and cultural differences, including how interreligious cooperation fosters resilient community environments and how we create resilient classroom environments.

Sarah M. Pike is professor of Comparative Religion at California State University, Chico. She is the author of *Earthly Bodies, Magical Selves: Contemporary Pagans and the Search for Community* (University of California Press, 2001), *New Age and Neopagan Religions in America* (Columbia University Press, 2004), and *For the Wild: Ritual and Commitment in Radical Eco-Activism* (University of California Press, 2017) and co-editor of *Reassembling*

Democracy: Ritual as Cultural Resource (Bloomsbury, 2020) and *Ritual and Democracy: Protests, Publics and Performances* (Equinox, 2020). She has written numerous articles and book chapters on contemporary Paganism, ritual, the New Age movement, the ancestral skills movement, Burning Man, spiritual dance, California wildfires, environmental activism, climate strikes, and youth culture. Her current research focuses on ritual, spirituality, and ecology in several different contexts, including a project on ritualized relationships with landscapes after wildfires.

Joerg Rieger is distinguished professor of theology, the Cal Turner Chancellor's Chair of Wesleyan Studies, and the founding director of the Wendland-Cook Program in Religion and Justice at Vanderbilt University. Previously he was the Wendland-Cook endowed professor of constructive theology at Perkins School of Theology, Southern Methodist University. Rieger's work brings together the study of theology and of the movements for liberation and justice that mark our age. Author and editor of twenty-six books and more than 180 academic articles, a selection of his books includes: *Theology in the Capitalocene: Ecology, Identity, Class, and Solidarity* (Fortress, 2022); *Jesus vs. Caesar: For People Tired of Serving the Wrong God* (Abingdon, 2018); *No Religion but Social Religion: Liberating Wesleyan Theology* (Wesley's Foundery, 2018); *Unified We are a Force: How Faith and Labor Can Overcome America's Inequalities* (with Rosemarie Henkel-Rieger, Chalice, 2016); and *Occupy Religion: Theology of the Multitude* (with Kwok Pui-lan, Rowman & Littlefield, 2012).

Terra Schwerin Rowe is associate professor in the Philosophy and Religion Department at the University of North Texas. She received a PhD in Theological and Philosophical Studies from Drew University as well as two master's degrees in the Protestant tradition and theology. Her current research focuses on energy, extraction, and religion. She is a member of the Petrocultures Research Group and co-director of the AAR Seminar on Energy, Extraction, and Religion. Her most recent book, *Of Modern Extraction: Experiments in Critical Petro-theology* (Bloomsbury / T&T Clark) was released in 2022.

Mary-Jane Rubenstein is professor of religion and science in society at Wesleyan University, and is affiliated with the Feminist, Gender, and Sexuality Studies Program. She holds a BA from Williams College, an MPhil from Cambridge University, and a PhD from Columbia University. Her research unearths the philosophies and histories of religion and science, especially in relation to cosmology, ecology, and space travel. She is the author of *Pantheologies: Gods, Worlds, Monsters* (Columbia University Press, 2018); *Worlds without End: The Many Lives of the Multiverse* (Columbia University Press, 2014); and *Strange Wonder: The Closure of Metaphysics and the Opening of Awe* (Columbia University Press, 2009). She is also co-editor with Catherine Keller of *Entangled Worlds: Religion, Science, and New Materialisms* (Fordham University Press, 2017), and co-author with Thomas A. Carlson and Mark C. Taylor of *Image: Three Inquiries in Technology and Imagination* (University of Chicago Press, 2021). Her latest book is titled *Astrotopia: The Dangerous Religion of the Corporate Space Race (University of Chicago Press, 2022).*

Kevin Schilbrack teaches and writes about the philosophical study of religions. A graduate of the University of Chicago Divinity School, he is the author of *Philosophy and the Study of Religions: A Manifesto* (Blackwell, 2014) and is presently interested in the relevance of embodied cognition and social ontology for understanding what religion is and how it works. If you share these interests, feel free to contact him at schilbrackke@appstate.edu and/or follow him on academia.edu.

Mary Evelyn Tucker teaches at the School of the Environment and the Divinity School at Yale University. She is co-director with John Grim of the Yale Forum on Religion and Ecology. With Grim, she organized ten conferences on World Religions and Ecology at Harvard, and they were series editors for the ten resulting volumes. In that series, she co-edited *Confucianism and Ecology; Buddhism and Ecology;* and *Hinduism and Ecology.* With Grim, she co-authored, *Ecology and Religion* (Island Press, 2014) and co-edited Thomas Berry's books, including *Selected Writings* (Orbis Books, 2014). Tucker and Grim published *Thomas Berry: A Biography* (Columbia University Press, 2019). With Brian Thomas Swimme, Tucker and Grim created a multi-media project "Journey of the Universe" that includes a book (Yale University Press, 2011), an Emmy Award-winning film, a series of podcast conversations, and free online courses from Yale/Coursera. Tucker was a member of the Earth Charter Drafting Committee and the International Earth Charter Council.

O'neil Van Horn is Assistant Professor of Theology at Xavier University. He is the author of *On the Ground: Terrestrial Theopoetics and Planetary Politics* (Fordham University Press, 2023), which explores questions of hope and pluralistic politics in the Anthropocene. He holds a PhD in Philosophical and Theological Studies from Drew University, and his research explores the intersections between constructive theological ethics and environmental justice. He has published articles and book chapters on issues such as bioregionalism and mysticism, Hip Hop and biblical interpretation, and environmental philosophy.

Carol Wayne White is Presidential Professor of Philosophy of Religion at Bucknell University. Her books include *Poststructuralism, Feminism, and Religion: Triangulating Positions* (Humanities, 2002); *The Legacy of Anne Conway (1631-70): Reverberations from a Mystical Naturalism* (SUNY Press, 2009); and *Black Lives and Sacred Humanity: Toward an African American Religious Naturalism* (Fordham University Press, 2016), which won a Choice Award for Outstanding Academic Titles. White has published many essays on the creative intersections of critical theory and religion, process philosophy, science and religion, and religious naturalism; her work has also appeared in *Zygon: The Journal of Religion and Science, The American Journal of Theology and Philosophy, Philosophia Africana,* and *Religion & Public Life.* White is currently finishing a book manuscript exploring a trajectory of modernist racial discourse that intimately conjoined White supremacy and speciesism in promoting views of Black animality and doing research for a new book project that explores the insights of religious naturalism expressed in contemporary North American nature poets and writers.

Index

Printed and bound by CPI Group (UK) Ltd, Croydon, CR0 4YY

27/10/2024

14580327-0005